Nature-Inspired Networking

Phan Cong Vinh

Nature-Inspired Networking: Theory and Applications

CRC Press
Taylor & Francis Group
Boca Raton London New York

CRC Press is an imprint of the
Taylor & Francis Group, an **informa** business

CRC Press
Taylor & Francis Group
6000 Broken Sound Parkway NW, Suite 300
Boca Raton, FL 33487-2742

© 2018 by Taylor & Francis Group, LLC
CRC Press is an imprint of Taylor & Francis Group, an Informa business

No claim to original U.S. Government works

Printed on acid-free paper

International Standard Book Number-13: 978-1-4987-6150-5 (Hardback)

Library of Congress Cataloging-in-Publication Data

Names: Cong-Vinh, Phan, editor.
Title: Nature-inspired networking : theory and applications / edited by Phan Cong Vinh.
Description: Boca Raton : Taylor & Francis, CRC Press, 2018. | Includes bibliographical references and index.
Identifiers: LCCN 2017044928 | ISBN 9781498761505 (hb : alk. paper)
Subjects: LCSH: Ad-hoc networks (Computer networks) | Neural networks (Computer science) | Biomimicry.
Classification: LCC TK5105.77 .N38 2018 | DDC 004.6–dc23

Visit the Taylor & Francis Web site at
http://www.taylorandfrancis.com

and the CRC Press Web site at
http://www.crcpress.com

Contents

List of Figures

List of Tables

Foreword

Nature-inspired networking (NiN)—which is inspired from nature such as biological, social, and physical phenomena—is a new networking paradigm in which inspiration from nature has emerged as a new characteristic of networking that makes communication activities smarter. Furthermore, for adaptive communication NiN has widely attracted attention from researchers in the networking field due to its unique advantages and has become mainstream in the networking society. Unfortunately, there are very few books specifically focused on theory and applications for NiN. But there has been a strong research need worldwide to strengthen information exchange in the area of NiN. This edited book is a salient document for disseminating research results related to theory and applications for NiN.

Nature-Inspired Networking: Theory and Applications contains original, peer-reviewed chapters reporting on new developments of interest to the NiN/computing communities. All chapters contain remarkable topics of NiN, from theory to applications for NiN based on rigorous interdisciplinary approaches in which theoretical contributions have been formally stated and justified, and practical applications which are based on their firm, formal basis. In other words, the major technical contents of the book include the following: autonomic systems, bio-inspired vehicular social networks, nature-inspired algorithms for wireless multi-hop ad hoc network optimization problems in disaster response scenarios, distributed system-based convergent information architecture, collective intelligence in networking, quailty of service (QoS) route search for mobile ad hoc networks using genetic algorithm, thermal network, and bio-inspired network on chip.

This book has come into being from a fine collection of chapters emphasizing the multidisciplinary character of investigations from the point of view of not only the NiN field involved but also the algebraic methods. This book, which is specifically dedicated to reporting on recent progress made in the field of autonomic systems, will usefully serve as a technical guide and reference material for engineers, scientists, practitioners, and researchers by providing them with state-of-the-art research results and future opportunities and trends. To the best of my knowledge, this is one of the first books that presents achievements and findings of NiN research covering the full spectrum of formalizing autonomic systems. These make the book unique and, in more than one respect, a truly valuable source of information that may be considered a landmark in the progress of NiN. Congratulations to those who have contributed to the highest technical quality of the book!

Nguyen Thanh Tung
Vietnam National University, Hanoi

Preface

A new networking paradigm, which is seen as a cutting-edge approach to networks, is currently a priority research area: nature-inspired networking (NiN), which is inspired from nature such as biological, social, and physical phenomena.

The book is a place for highly original ideas about how the nature is going to shape networking systems of the future. Hence, it focuses on theory and applications, which encompass rigorous approaches and cutting-edge solutions that take inspiration from nature for the development of novel problem solving techniques. To this end, we will take advantage of formal engineering methods and establish in this book formal and practical aspects of NiN to achieve foundations and practice of NiN.

The book is a reference for readers who already have a basic understanding of networking and are now ready to know how to use rigorous approaches to develop networking that is inspired by nature. Hence, the book includes both theoretical contributions and reports on applications. To keep a reasonable trade-off between theoretical and practical issues, chapters were carefully selected to, on one hand, cover a broad spectrum of formal and practical aspects and, on the other hand, achieve as much as possible in a self-contained book.

Formal and practical aspects are presented in a straightforward fashion by discussing in detail the necessary components and briefly touching on the more advanced components. Therefore, theory and applications demonstrating how to use the formal engineering methods for NiN will be described using sound judgment and reasonable justifications.

This book, with chapters contributed by prominent researchers from academia and industry, will serve as a technical guide and reference material for engineers, scientists, practitioners, and researchers by providing them with state-of-the-art research findings and future opportunities and trends. These contributions include state-of-the-art architectures, protocols, technologies, and applications in NiN. In particular, the book covers existing and emerging research issues in NiN. The book has nine chapters addressing various topics from theory to applications of NiN based on rigorous interdisciplinary approaches.

Chapter 1 by P. C. Vinh considers self-* as a foundation for autonomic computing. The notion of autonomic systems (ASs) and self-* serves as a basis on which to build the intuition about algebraic aspects of ASs in general. In this chapter the author specifies ASs and self-* and then moves on to consider some universal constructions such as products, coproducts, curried self-* actions, finite limits, colimits of ASs, and monoids of self-* actions. All of this material is taken as an investigation of the algebraic aspects of ASs.

Chapter 2 by F. Chiti, E. Dei, and R. Fantacci reviews the most relevant bio-inspired routing and clustering approaches with a special focus on vehicular scenarios. The underlying concept of mobile group management is adopted to select the more suited schemes, among which are the ant or bee colonies organization and the bacteria life cycle. Finally, the practical issues related to protocol design are investigated by relying on bacterial quorum sensing mechanism, which seems to be a promising approach to design decentralized and flexible joint clustering and routing schemes, while achieving a full context awareness.

Chapter 3 by J. Sánchez-García, J. M. García-Campos, D. G. Reina, S. L. Toral, and F. Barrero indicates that nowadays nature-inspired algorithms can be applied to many fields and provide optimal or pseudo-optimal solutions to many different optimization problems. Disaster scenarios, which present a highly dynamic nature, are one of these areas of application that can benefit from nature-inspired algorithms. Specifically, mobile ad hoc networks which are envisioned to be deployed in disaster scenarios present several aspects that need to be addressed with flexibility, accuracy, and efficiency. Nature-inspired algorithms present the required characteristics for efficiently dealing with several optimization problems related to multi-hop ad hoc networks in disaster scenarios. The authors propose the chapter for gathering and organizing the information about current works in this field.

Chapter 4 by A. M. Alberti, M. A. F. Casaroli, R. da Rosa Righi, and D. Singh focuses on exploring the synergies, dependencies, and common requirements between the future Internet and distributed systems, discussing the relationships between both research areas. It provides a detailed description of NovaGenesis and newer challenges in distributed systems (evolutionary and revolutionary). Furthermore, it addresses communication model, message passing, naming, identification, addressing, and locating issues, as well as approaches for modeling distributed systems and communication protocols under the concepts of NovaGenesis. With this scope, the chapter will naturally touch on other research areas, including cloud-based storage and networking, network management, virtualization, Internet of things systems, autonomic networking, context-awareness, evolutionary networking, and even hardware solutions. The contributions permeate not only networking protocols and architectures, but also distributed systems and software engineering.

Chapter 5 by Ratneshwer and V. Kushwaha indicates that while collective intelligence is considered as a powerful approach to enhance decision capabilities, very few attempts have been made to utilize its potential in the computer network domain. This chapter is intended to provide the general overview of collective intelligence with a discussion on how it can be effectively applied in different areas of computer networks. More studies are required to inculcate the idea of collective intelligence in various areas of computer networks. It is therefore very promising to identify opportunities for integration, and inculcation of collective intelligence methodologies and technologies into networking. This would require judiciously considering the various activities in a computer network for this purpose. Such an effort would finally aim at remodeling of computer networks in context of collective intelligence.

Chapter 6 by J. Abdullah presents QoS routing algorithm for mobile ad hoc networks (MANET) with specialized encoding, initialization, and route search using genetic algorithm (GA). The objective is to find the best QoS route in order to optimize the design of MANET routing protocols. This NP-hard problem is often highly constrained such that random initialization and standard genetic operators usually generate infeasible networks. Another complication is that the fitness function involves calculating all node reliability of the routes, a calculation that is computationally expensive. Therefore, it is imperative that the search balances the need to thoroughly explore the boundary between feasible and infeasible networks, along with calculating fitness on only the most promising candidate routes. The algorithm results are compared to the result of standard MANET protocols without the GA search algorithm.

Chapter 7 by S. Sarwari and S. Rao considers a thermal network, which is a network where the flow functionality of a node depends upon its temperature. This model is inspired by several types of real-life networks, and generalizes some conventional network models wherein nodes have fixed capacities, and the problem is to maximize the flow through the network. In a thermal network, the temperature of a node increases as traffic moves through it, and nodes may also cool spontaneously over time, or by employing cooling packets. The problems of maximizing the flow from a source to a sink are analyzed for both these cases, for a holistic view with respect to the single-source-single-sink dynamic flow problem in a thermal network. The chapter also studies certain properties that a thermal network exhibits, and gives closed-form solutions for the maximum flow that can be achieved through such a network.

Chapter 8 by M. A. J. Sethi, F. A. Hussin, and N. H. Hamid gets inspiration from biological brain fault tolerant techniques to make Network on Chip (NoC) communication reliable as the biological brain is highly robust and fault tolerant. The brain tries to work properly, even if some neurons, synapses, or some other part of the brain is damaged. Synaptogenesis and sprouting are two biological brain techniques used in this chapter to to implement the self-adapt and self-heal concepts in NoC.

Finally, Chapter 9 by M. A. J. Sethi, F. A. Hussin, and N.H. Hamid introduces two broad categories of QoS parameters in NoC: guaranteed throughput connection service (GT) and best-effort connection service (BE). The bio-inspired algorithm is implemented using these connection setups. The GT connections are implemented using time division multiplexing (TDM). TDM connections help divide the bandwidth of interconnect among multiple connections using slots. The BE connections are implemented using packet switching. In packet switching, routers decide at which direction the packet should be sent based on the situation of neighbor's routers, interconnects, and routing table. In this chapter, the bio-inspired NoC algorithm using GT connection is discussed.

This book has the following remarkable features:

■ Provides a comprehensive reference on NiN

■ Presents state-of-the-art techniques in NiN

■ Includes illustrative figures facilitating easy reading

■ Discusses emerging trends and open research problems in NiN

We owe our deepest gratitude to Professor Nguyen Manh Hung (Rector and Chairman of Nguyen Tat Thanh University, Vietnam) for his useful support, notably to all the authors for their valuable contributions to this book and their great efforts. All of them are extremely professional and cooperative. We wish to express our thanks to CRC Press, especially Richard O'Hanley, for their support and guidance during the preparation of this book. A special thank you also goes to our families and friends for their constant encouragement, patience, and understanding throughout this project.

The book serves as a comprehensive and essential reference on NiN and is intended as a textbook for senior undergraduate- and graduate-level courses. It can also be used as a supplementary textbook for undergraduate courses. The book is a useful resource for students and researchers to learn NiN. In addition, it will be valuable to professionals from both academia and industry and generally has instant appeal to people who would like to contribute to NiN technologies.

We highly welcome and greatly appreciate your feedback and hope you enjoy reading the book.

Phan Cong Vinh
Ho Chi Minh City, Vietnam

Editor

Phan Cong Vinh received a PhD in computer science from London South Bank University (LSBU) in the United Kingdom. He finished his PhD dissertation with the title of *Formal Aspects of Dynamic Reconfigurability in Reconfigurable Computing Systems* at LSBU where he was affiliated with the Center for Applied Formal Methods (CAFM) at the Institute for Computing Research (ICR). At present, he is an Associate Professor of Nguyen Tat Thanh University (NTTU) to take on the responsibility of a senior research scientist. He has been author or co-author of many refereed contributions published in prestigious journals, conference proceedings, or edited books. He is editor of two books titled, *Autonomic Networking-on-Chip: Bio-Inspired Specification, Development and Verification* (CRC Press, 2012) and *Formal and Practical Aspects of Autonomic Computing and Networking: Specification, Development and Verification* (IGI Global, 2011). He has served on many conference program committees and has been general or technical (co)chair and (co)organizer of several international conferences such as ICCASA and ICTCC series. His research interests center on all aspects of formal methods in computing, context-awareness, nature of computation and communication, and applied categorical structures in computer science.

List of Contributors

J. Abdullah
Universiti Tun Hussein Onn Malaysia
Johor, Malaysia

Antonio Marcos Alberti
Instituto Nacional de Telecomunicações
(INATEL)
Santa Rita do Sapucaí, Brazil

Federico Barrero
University of Seville
Seville, Spain

Marco Aurelio Favoreto Casaroli
Instituto Nacional de Telecomunicações
(INATEL)
Santa Rita do Sapucaí, Brazil

Francesco Chiti
University of Florence
Florence, Italy

Enrico Dei
University of Florence
Florence, Italy

Romano Fantacci
University of Florence
Florence, Italy

José Manuel García-Campo
University of Seville
Seville, Spain

Daniel Gutiérrez Reina
University of Seville
Seville, Spain

Nor Hisham Hamid
Universiti Teknologi
PETRONAS
Perak, Malaysia

Fawnizu Azmadi Hussin
Universiti Teknologi
PETRONAS
Perak, Malaysia

Vandana Kushwaha
Banaras Hindu University
Varanasi, India

Ratneshwer
Jawaharlal Nehru University
New Delhi, India

Rodrigo da Rosa Righi
Universidade do Vale do Rio dos
Sinos
São Leopoldo, Brazil

Jesús Sánchez-García
University of Seville
Seville, Spain

Muhammad Athar Javed Sethi
University of Engineering and
  Technology, Peshawar
Peshawar, Pakistan

Dhananjay Singh
Hankuk University of Foreign Studies
Seoul, South Korea

Sergio Luis Toral
University of Seville
Seville, Spain

Phan Cong Vinh
Nguyen Tat Thanh University
Ho Chi Minh City, Vietnam

Chapter 1

Algebraic Aspects of Autonomic Systems

Phan Cong Vinh

Nguyen Tat Thanh University

CONTENTS

Self-* is widely considered as a foundation for autonomic computing. The notion of autonomic systems (ASs) and self-* serves as a basis on which to build our intuition about algebraic aspects of ASs in general. In this chapter we will define ASs and self-* and then move on to consider some universal constructions such as products, coproducts, curried self-* actions, finite limits, colimits of ASs, and monoids of self-* actions. All of this material is taken as an investigation of the algebraic aspects of ASs.

1.1 Introduction

Autonomic computing (AC) imitates and simulates the natural intelligence possessed by the human autonomic nervous system using generic computers. This indicates that the nature of software in AC is the simulation and embodiment of human behaviors, and the extension of human capability, reachability, persistency, memory, and information processing speed. AC was first proposed by IBM in 2001 where it was defined as

> Autonomic computing is an approach to self-managed computing systems with a minimum of human interference. The term derives from the body's autonomic nervous system, which controls key functions without conscious awareness or involvement [23].

AC in our recent investigations [44–49,54,55] is generally described as self-*. Formally, let self-* be the set of self-_'s. Each self-_ that is an element in self-* is called a *self-* action*. That is,

$$\text{self-*} = \{\text{self-}\_ \mid \text{self-}\_ \text{ is a self-* action}\} \tag{1.1}$$

We see that self-CHOP is composed of four self-* actions of self-configuration, self-healing, self-optimization, and self-protection. Hence, self-CHOP is a subset of self-*. That is, self-CHOP = {self-configuration, self-healing, self-optimization, self-protection} ⊂ self-*. Every self-* action must satisfy some certain criteria, so-called *self-* properties*.

In its AC manifesto, IBM proposed eight actions setting forth an AS; they are *self-awareness, self-configuration, self-optimization, self-maintenance, self-protection (security and integrity), self-adaptation, self-resource allocation* and *open-standard-based* [23]. In other words, consciousness (self-awareness) and non-imperative (goal-driven) behaviors are the main features of ASs.

In this chapter we will define ASs and self-* and then move on to consider some universal constructions such as products, coproducts, curried self-* actions, finite limits, colimits of ASs and monoids of self-* actions. All of this material is taken as an investigation on algebraic aspects of ASs [46,50,53].

1.2   Outline

In the chapter, we attempt to make the presentation as self-contained as possible, although familiarity with the notions of self-* and ASs is assumed. Acquaintance with the associated notion of algebraic language is useful for recognizing the results, but is not strictly necessary in most examples.

The rest of this chapter is organized as follows: Section 1.3 introduces work related to autonomic computing. Section 1.4 presents the notion of ASs. In Section 1.5, self-* actions in ASs are specified. In Section 1.6, products and coproducts of ASs are considered. In Section 1.7 we investigate curried self-* actions of ASs. In Section 1.8, we consider some finte limits such as pullbacks of ASs, spans on ASs, and equalizers of self-* actions. In Section 1.9, we consider some finte colimits such as pushouts of ASs and coequalizers of self-* actions. Monoids of self-* actions are exploited in Section 1.10. Finally, a short summary is given in Section 1.11.

1.3   Related Work

The topic of AC has seen a number of developments through various research investigations following the IBM initiative, such as AC paradigm [4,19,32,36, 40]; different approaches and infrastructures [1,2,37,39,59] enabling autonomic behaviors [42,43,51,52]; core enabling systems, technologies, and services [9,10, 15,16,22,28] to support the realization of self-* properties in autonomic systems and applications; specific realizations of self-* properties in autonomic systems and applications [3,8,12,21,25,30,34]; architectures and modeling strategies of autonomic networks [11,18,29]; middleware and service infrastructure as facilitators of autonomic communications [6,13,20]; approaches in [14,17,35] to equipping current networks with autonomic functionality for migrating these types of networks to autonomic networks.

Moreover, AC also has been intensely studied by various areas of engineering including artificial intelligence, control systems, and human orientated systems [24,31,57,58]. Autonomic computing has been set as an important requirement for systems devised to work in new generation global networked and distributed environments like wireless networks, P2P networks, Web systems, multi-agent systems, grids, and so on [5,7,26,33,60]. Such systems pose new challenges for the development and application of autonomic computing techniques, due to their special characteristics including: *nondeterminism, context-awareness and goal- and inference-driven adaptability* [44,45,56,57].

Finally, the choice of the underlying formalization requires a close look at models for AC. Hence, our interest centers on, formal approach to AC taking advantage of category theory [44]. In fact, categories were first described by Samuel Eilenberg and Saunders Mac Lane in 1945 [27], but have since grown substantially to become a branch of modern mathematics. Category theory spreads its influence over the development of both mathematics and theoretical computer science [38]. The categorical structures themselves are still the subject of active research, including work to increase their range of practical applicability.

1.4 Autonomic Systems (ASs)

We can think of an AS as a collection of states $x \in AS$, each of which is recognizable as being in AS and such that for each pair of named states $x, y \in AS$ we can tell whether or not $x = y$. The symbol $\oslash$ denotes the AS with no states.

If AS_1 and AS_2 are ASs, we say that AS_1 is a subsystem of AS_2, and write $AS_1 \subseteq AS_2$, if every state of AS_1 is a state of AS_2. Checking the definition, we see that for any system AS, we have subsystems $\oslash \subseteq AS$ and $AS \subseteq AS$.

We can use system-builder notation to denote subsystems. For example, the autonomic system can be written $\{x \in AS \mid x \text{ is a state of AS}\}$.

The symbol $\exists$ means "there exists." So we can write the autonomic system as $\{x \in AS \mid \exists y \text{ is a final state such that } self\text{-}*action(x) = y\}$

The symbol $\exists!$ means "there exists a unique." So the statement "$\exists!x \in AS$ is an initial state" means that there is one and only one state to start from; that is, the state of the autonomic system before any self-* action is processed.

Finally, the symbol $\forall$ means "for all." So the statement "$\forall x \in AS \; \exists y \in AS$ such that $self\text{-}* \, action(x) = y$" means that for every state of autonomic system there is the next one.

In this chapter, we use the $\overset{def}{=}$ notation "$AS_1 \overset{def}{=} AS_2$" to mean to "define AS_1 to be AS_2". That is, a $\overset{def}{=}$ declaration is not denoting a fact of nature (like $1 + 2 = 3$), but our formal notation. It just so happens that the notation above, such as Self-CHOP $\overset{def}{=}$ {self-configuration, self-healing, self-optimization, self-protection}, is a widely-held choice.

1.5 Self-* Actions in Autonomic Systems

If AS and AS' are sets of autonomic system states, then a self-* action $self\text{-}*action$ from AS to AS', denoted $self\text{-}*action: AS \rightarrow AS'$, is a mapping that sends each state $x \in AS$ to a state of AS', denoted $self\text{-}*action(x) \in AS'$. We call AS the domain of $self\text{-}*action$ and we call AS' the codomain of $self\text{-}*action$.

Note that the symbol AS', read "AS-prime," has nothing to do with calculus or derivatives. It is simply a notation that we use to name a symbol that is suggested as being somehow like AS. This suggestion of consanguinity between AS and AS' is meant only as an aid for human cognition, and not as part of the mathematics.

For every state $x \in AS$, there is exactly one arrow emanating from x, but for a state $y \in AS'$, there can be several arrows pointing to y, or there can be no arrows pointing to y.

Suppose that $AS' \subseteq AS$ is a subsystem. Then we can consider the self-* action $AS' \rightarrow AS$ given by sending every state of AS' to "itself" as a state of AS. For example if $AS = \{a, b, c, d, e, f\}$ and $AS' = \{b, d, e\}$ then $AS' \subseteq AS$, and we turn that into the self-* action $AS' \rightarrow AS$ given by $b \mapsto b, d \mapsto d, e \mapsto e$. This kind of arrow, $\mapsto$, is read aloud as "maps to." A self-* action *self-*action*: $AS \rightarrow AS'$ means a rule for assigning to each state $x \in AS$ a state *self-*action*$(x) \in AS'$. We say that "x maps to *self-*action*(x)" and write $x \mapsto$ *self-*action*(x).

As a matter of notation, we can sometimes say something like the following: Let *self-*action*: $AS' \subseteq AS$ be a subsystem. Here we are making clear that AS' is a subsystem of AS, but that *self-*action* is the name of the associated self-* action.

Given a self-* action *self-*action*: $AS \rightarrow AS'$, the states of AS' that have at least one arrow pointing to them are said to be in the image of *self-*action*; that is we have

$$\text{im}(\textit{self-*action}) \overset{def}{=} \{y \in AS' \mid \exists x \in AS \text{ such that } \textit{self-*action}(x) = y\} \quad (1.2)$$

Given *self-*action*: $AS \rightarrow AS'$ and *self-*action'* : $AS' \rightarrow AS''$, where the codomain of *self-*action* is the same set of autonomic system states as the domain of *self-*action'* (namely AS'), we say that *self-*action* and *self-*action'* are composable

$$AS \xrightarrow{\textit{self-*action}} AS' \xrightarrow{\textit{self-*action'}} AS''$$

The composition of *self-*action* and *self-*action'* is denoted by *self-*action'* $\circ$ *self-*action*: $AS \rightarrow AS''$.

We write $\text{Hom}_{\mathbf{AS}}(AS, AS')$ to denote the set of *self-*actions* $AS \rightarrow AS'$. Two self-* actions *self-*action*, *self-*action'* : $AS \rightarrow AS'$ are equal if and only if for every state $x \in AS$ we have *self-*action*$(x) =$ *self-*action'*(x).

We define the identity *self-*action* on AS, denoted $id_{AS} : AS \rightarrow AS$, to be the self-* action such that for all $x \in AS$ we have $id_{AS}(x) = x$.

A *self-*action*: $AS \rightarrow AS'$ is called an *isomorphism*, denoted *self-*action*: $AS \overset{\cong}{\rightarrow} AS'$, if there exists a self-* action *self-*action'* : $AS' \rightarrow AS$ such that *self-*action'* $\circ$ *self-*action*$= id_{AS}$ and *self-*action* $\circ$ *self-*action'* $= id_{AS'}$. We also say that *self-*action* is *invertible* and we say that *self-*action'* is the *inverse* of *self-*action*. If there exists an isomorphism $AS \overset{\cong}{\rightarrow} AS'$ we say that AS and AS' are isomorphic autonomic systems and may write $AS \cong AS'$.

Proposition 1.1 *The following facts hold about isomorphism.*

1. *Any autonomic system AS is isomorphic to itself (i.e., there exists an isomorphism $AS \overset{\cong}{\rightarrow} AS$).*

2. *For any autonomic systems AS and AS', if AS is isomorphic to AS' then AS' is isomorphic to AS.*

3. *For any autonomic systems AS, AS', and AS'', if AS is isomorphic to AS' and AS' is isomorphic to AS'' then AS is isomorphic to AS''.*

Proof 1.1

1. The identity self-* action $id_{AS} : AS \to AS$ is invertible; its inverse is id_{AS} because $id_{AS} \circ id_{AS} = id_{AS}$.

2. If *self-*action*: $AS \to AS'$ is invertible with inverse *self-*action'* : $AS' \to AS$ then *self-*action'* is an isomorphism with inverse *self-*action*.

3. If *self-*action*: $AS \to AS'$ and $\widehat{self\text{-}*action} : AS' \to AS''$ are each invertible with inverses *self-*action'* : $AS' \to AS$ and $\widehat{self\text{-}*action}{}' : AS'' \to AS'$ then the following calculations show that $\widehat{self\text{-}*action} \circ self\text{-}*action$ is invertible with inverse *self-*action'* $\circ \widehat{self\text{-}*action}{}'$:

$$(\widehat{self\text{-}*action} \circ self\text{-}*action) \circ (self\text{-}*action' \circ \widehat{self\text{-}*action}{}') =$$
$$\widehat{self\text{-}*action} \circ (self\text{-}*action \circ self\text{-}*action') \circ \widehat{self\text{-}*action}{}' =$$
$$\widehat{self\text{-}*action} \circ id_{AS'} \circ \widehat{self\text{-}*action}{}' = \widehat{self\text{-}*action} \circ \widehat{self\text{-}*action}{}' = id_{AS''}$$

and

$$(self\text{-}*action' \circ \widehat{self\text{-}*action}{}') \circ (\widehat{self\text{-}*action} \circ self\text{-}*action) =$$
$$self\text{-}*action' \circ (\widehat{self\text{-}*action}{}' \circ \widehat{self\text{-}*action}) \circ self\text{-}*action =$$
$$self\text{-}*action' \circ id_{AS'} \circ self\text{-}*action = self\text{-}*action' \circ self\text{-}*action = id_{AS}$$

For any natural number $n \in \mathbb{N}$, define a set $\underline{n} = \{1, 2, \ldots, n\}$. So, in particular, $\underline{0} = \oslash$. A function $f : \underline{n} \to AS$ can be written as a sequence $f = (f(1), f(2), \ldots, f(n))$. We say that AS has cardinality n, denoted $|AS| = n$, if there exists an isomorphism $AS \cong \underline{n}$. If there exists some $n \in \mathbb{N}$ such that AS has cardinality n, then we say that AS is finite. Otherwise, we say that AS is infinite and write $|AS| \geqslant \infty$.

Proposition 1.2 *Suppose that AS and AS' are finite. If there is an isomorphism of autonomic systems $f : AS \to AS'$ then the two autonomic systems have the same cardinality, $|AS| = |AS'|$.*

Proof 1.2 Suppose that $f : AS \to AS'$ is an isomorphism. If there exists natural numbers $m, n \in \mathbb{N}$ and isomorphisms $\alpha : \underline{m} \xrightarrow{\cong} AS$ and $\beta : \underline{n} \xrightarrow{\cong} AS'$ then

$$\underline{m} \xrightarrow{\alpha} AS \xrightarrow{f} AS' \xrightarrow{\beta^{-1}} \underline{n}$$

is an isomorphism. We can prove by induction that the sets $\underline{m}$ and $\underline{n}$ are isomorphic if and only if $m = n$.

Consider the following diagram:

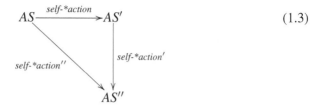

$$(1.3)$$

We say this is a diagram of ASs if each of AS, AS', AS'' is an AS and each of *self-*action*, *self-*action'*, *self-*action''* is a self-* action. We say this diagram commutes if *self-*action'* $\circ$ *self-*action* = *self-*action''*. In this case we refer to it as a commutative triangle of ASs. Diagram 1.3 is considered to be the same diagram as each of the following:

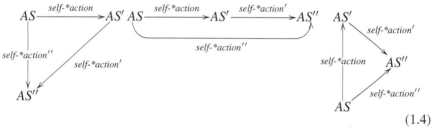

$$(1.4)$$

Consider the following picture:

$$\begin{array}{ccc}
AS & \xrightarrow{\text{self-}*action} & AS' \\
\downarrow \text{self-}*action'' & & \downarrow \text{self-}*action' \\
AS'' & \xrightarrow{\text{self-}*action'''} & AS'''
\end{array} \qquad (1.5)$$

We say this is a diagram of ASs if each of AS, AS', AS'', AS''' is an AS and each of *self-*action*, *self-*action'*, *self-*action''*, *self-*action'''* is a self-* action. We say this diagram commutes if *self-*action'* $\circ$ *self-*action* = *self-*action'''* $\circ$ *self-*action''*. In this case we refer to it as a commutative square of ASs.

1.6 Products and Coproducts of Autonomic Systems

Let AS and AS' be ASs. The product of AS and AS', denoted $AS \times AS'$, is defined as the AS of ordered pairs (x, y) where states of $x \in AS$ and $y \in AS'$. Symbolically,

$AS \times AS' = \{(x,y)|x \in AS, y \in AS'\}$. There are two natural projection actions of self-* to be $self\text{-}*action_1 : AS \times AS' \to AS$ and $self\text{-}*action_2 : AS \times AS' \to AS'$

$$
\begin{array}{c}
AS \times AS' \\
\swarrow \quad\quad \searrow \\
self\text{-}*action_1 \quad\quad self\text{-}*action_2 \\
AS \quad\quad\quad\quad\quad AS'
\end{array}
\tag{1.6}
$$

For illustration, suppose that $\{a,b,c\}$ are states in AS and $\{d,e\}$ are states in AS' and the states are happening in such ASs. Thus, AS and AS', which are running concurrently, can be specified by $AS|AS' \overset{def}{=} \{(a|d),(a|e),(b|d),(b|e), (c|d),(c|e)\}$. Note that the symbol "|" is used to denote concurrency of states existing at the same time. We define self-* actions as $disable(d,e)$ and $disable(a,b,c)$ to be able to drop out relevant states.

$$
\begin{array}{c}
\{(a|d),(a|e),(b|d),(b|e),(c|d),(c|e)\} \\
\swarrow \quad\quad\quad\quad\quad\quad\quad\quad \searrow \\
disable(d,e) \quad\quad\quad\quad\quad disable(a,b,c) \\
\{a,b,c\} \quad\quad\quad\quad\quad\quad\quad\quad\quad \{d,e\}
\end{array}
\tag{1.7}
$$

It is possible to take the product of more than two ASs as well. For example, if AS_1, AS_2, and AS_3 are ASs then $AS_1|AS_2|AS_3$ is the system of triples,

$$
AS_1|AS_2|AS_3 \overset{def}{=} \{(a|b|c)|a \in AS_1, b \in AS_2, c \in AS_3\}
$$

Proposition 1.3 *Let AS and AS' be autonomic systems. For any autonomic system AS'' and actions self-*action$_3$: AS'' → AS and self-*action$_4$: AS'' → AS', there exists a unique action AS'' → AS × AS' such that the following diagram commutes*

$$
\begin{array}{c}
AS \times AS' \\
self\text{-}*action_1 \nearrow \quad \uparrow \quad \nwarrow self\text{-}*action_2 \\
AS \quad\quad \exists! \quad\quad AS' \\
\forall \, self\text{-}*action_3 \nwarrow \quad | \quad \nearrow \forall \, self\text{-}*action_4 \\
AS''
\end{array}
\tag{1.8}
$$

We might write the unique action as

$$
\langle self\text{-}*action_3, self\text{-}*action_4 \rangle : AS'' \to AS \times AS'
$$

Proof 1.3 Suppose given $self\text{-}*action_3$ and $self\text{-}*action_4$ as above. To provide an action $z : AS'' \to AS \times AS'$ is equivalent to providing a state $z(a) \in AS \times AS'$ for each $a \in AS''$. We need such an action for which $self\text{-}*action_1 \circ z = self\text{-}*action_3$ and $self\text{-}*action_2 \circ z = self\text{-}*action_4$. A state of $AS \times AS'$ is an ordered pair (x,y), and we can use $z(a) = (x, y)$ if and only if $x = \textit{self-*action}_1(x, y) = \textit{self-*action}_3(a)$ and $y = \textit{self-*action}_2(x, y) = \textit{self-*action}_4(a)$. So it is necessary and sufficient to define

$$\langle \textit{self-*action}_3, \textit{self-*action}_4 \rangle \stackrel{def}{=} (\textit{self-*action}_3(a), \textit{self-*action}_4(a))$$

for all $a \in AS''$.

Given autonomic systems AS, AS', and AS'', and actions $\textit{self-*action}_3$: $AS'' \to AS$ and $\textit{self-*action}_4 : AS'' \to AS'$, there is a unique action $AS'' \to AS \times AS'$ that commutes with $\textit{self-*action}_3$ and $\textit{self-*action}_4$. We call it the *induced action* $AS'' \to AS \times AS'$, meaning the one that arises in light of $\textit{self-*action}_3$ and $\textit{self-*action}_4$.

For example, as mentioned above autonomic systems $AS = \{a, b, c\}$, $AS' = \{d, e\}$ and $AS|AS' \stackrel{def}{=} \{(a|d), (a|e), (b|d), (b|e), (c|d), (c|e)\}$. For an autonomic system $AS'' = \varnothing$, which stops running, we define self-* actions as $enable(d, e)$ and $enable(a, b, c)$ to be able to add further relevant states. Then there exists a unique action

$$enable((a|d), (a|e), (b|d), (b|e), (c|d), (c|e))$$

such that the following diagram commutes

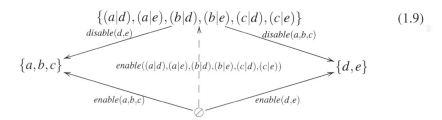

$$(1.9)$$

Let AS and AS' be autonomic systems. The coproduct of AS and AS', denoted $AS \sqcup AS'$, is defined as the "disjoint union" of AS and AS' (i.e., the autonomic system for which a state is either a state of AS or a state of AS'). If something is a state of both AS and AS' then we include both copies, and distinguish between them, in $AS \sqcup AS'$. There are two natural inclusion actions $\textit{self-*action}_1 : AS \to AS \sqcup AS'$ and $\textit{self-*action}_2 : AS' \to AS \sqcup AS'$.

$$(1.10)$$

$$AS \quad\quad\quad AS'$$
$$\textit{self-*action}_1 \quad \textit{self-*action}_2$$
$$AS \sqcup AS'$$

For illustration, suppose that $\{a,b,c\}$ are states in autonomic system AS and $\{d,e\}$ in AS'. Thus, $AS \sqcup AS'$, which is a disjointed union, can be specified by $AS \sqcup AS' \overset{def}{=} \{a,b,c,d,e,\}$. We define self-* actions as $ensable(d,e)$ and $enable(a,b,c)$ to be able to add further relevant states.

$$\{a,b,c\} \qquad\qquad\qquad \{d,e\} \tag{1.11}$$
$$\searrow \;{\scriptstyle enable(d,e)} \qquad {\scriptstyle enable(a,b,c)}\; \swarrow$$
$$\{a,b,c,d,e\}$$

Proposition 1.4 *Let AS and AS' be autonomic systems. For any autonomic system AS'' and actions self-*action$_3$: AS → AS'' and self-*action$_4$: AS' → AS'', there exists a unique action AS ⊔ AS' → AS'' such that the following diagram commutes*

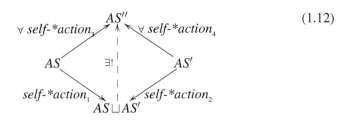

$$(1.12)$$

We might write the unique action as

$$[self\text{-}*action_3, self\text{-}*action_4] : AS \sqcup AS' \to AS''$$

Proof 1.4   Suppose given *self-*action*$_3$, *self-*action*$_4$ as above. To provide an action $z : AS \sqcup AS' \to AS''$ is equivalent to providing a state *self-*action*$_3(m) \in AS''$ for each $m \in AS \sqcup AS'$. We need such an action such that $z \circ self\text{-}*action_1 = self\text{-}*action_3$ and $z \circ self\text{-}*action_2 = self\text{-}*action_4$. But each state $m \in AS \sqcup AS'$ is either of the form *self-*action*$_1 x$ or *self-*action*$_2 y$, and cannot be of both forms. So we assign

$$[self\text{-}*action_3, self\text{-}*action_4](m) = \begin{cases} self\text{-}*action_3(x) & \text{if } m = self\text{-}*action_1 x \\ self\text{-}*action_4(y) & \text{if } m = self\text{-}*action_2 y \end{cases}$$
$$(1.13)$$

This assignment is necessary and sufficient to make all relevant diagrams commute.

For example, as mentioned above autonomic systems $AS = \{a,b,c\}$, $AS' = \{d,e\}$ and $AS \sqcup AS' \overset{def}{=} \{a,b,c,d,e\}$. For an autonomic system $AS'' = \oslash$, which stops running, we define self-* actions as $disable(d,e)$ and $disable(a,b,c)$ to drop out relevant states. Then there exists a unique action $disable(a,b,c,d,e)$ such that the following diagram commutes

(1.14)

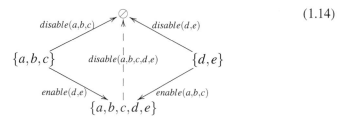

We denote the coproduct of two autonomic systems AS and AS' by the notation $AS + AS'$ rather than $AS \sqcup AS'$. It is a reasonable notation in general, and one that is often used.

There are some universal properties so that the following isomorphisms exist for any autonomic systems AS, AS', and AS''

$$AS + \underline{0} \cong AS$$
$$AS + AS' \cong AS' + AS$$
$$(AS + AS') + AS'' \cong AS + (AS' + AS'')$$
$$AS \times \underline{0} \cong \underline{0}$$
$$AS \times \underline{1} \cong AS$$
$$AS \times AS' \cong AS' \times AS$$
$$(AS \times AS') \times AS'' \cong AS \times (AS' \times AS'')$$
$$AS \times (AS' + AS'') \cong (AS \times AS') + (AS \times AS'')$$
$$AS^{\underline{0}} \cong \underline{1}$$
$$AS^{\underline{1}} \cong AS$$
$$\underline{0}^{AS} \cong \underline{0}$$
$$\underline{1}^{AS} \cong \underline{1}$$
$$AS^{AS' + AS''} \cong AS^{AS'} \times AS^{AS''}$$
$$(AS^{AS'})^{AS''} \cong AS^{AS' \times AS''}$$

If $n \in \mathbb{N}$ is a natural number and $\underline{n} = \{1, 2, \ldots, n\}$, then AS^n is an abbreviation for $\prod_n AS$ and AS^n is an abbreviation for $AS^{\underline{n}}$. Thus, we have $AS^{\underline{n}} \cong \prod_n AS$

In the case of $\underline{0}^{\underline{0}}$, we get conflicting answers, because for any autonomic system AS, including $AS = \oslash = \underline{0}$, we have claimed both that $AS^{\underline{0}} \cong \underline{1}$ and that $\underline{0}^{AS} \cong \underline{0}$. Based on the definitions of 0, 1 and $AS^{AS'}$ given in Section 1.5, the correct answer for $\underline{0}^{\underline{0}}$ is $\underline{0}^{\underline{0}} \cong \underline{1}$. The universal properties, which are considered in this section, are in some sense about isomorphisms. It says that understanding isomorphisms of autonomic systems reduces to understanding natural numbers.

1.7 Currying Self-* Actions

Currying is the idea that when a self-* takes action on many ASs, we can let the self-* take action on one system at a time or all systems at once. For example, consider self-* that takes action on AS and AS' and returns AS''. This is a self-* action *self-*action*: $AS \times AS' \to AS''$. This self-* takes action on two ASs at once, but it is convenient to curry the second AS. Currying transforms *self-*action* into a self-* action

$$curry(self\text{-}*action) : AS \to \mathrm{Hom}_{\mathbf{AS}}(AS',AS'')$$

This is a good way to represent the same information in another fashion. For any AS', we can represent the self-* that takes action on AS and returns AS''. This is a self-* action

$$curry(self\text{-}*action)' : AS' \to \mathrm{Hom}_{\mathbf{AS}}(AS,AS'')$$

Note that sometimes we denote the set of self-* actions from AS to AS' by

$$AS'^{AS} \overset{def}{=} \mathrm{Hom}_{\mathbf{AS}}(AS,AS')$$

If AS and AS' are both finite (so one or both are empty), then $|AS'^{AS}| = |AS'|^{|AS|}$. For any AS and AS', there is an isomorphism

$$\mathrm{Hom}_{\mathbf{AS}}(AS \times AS', AS'') \cong \mathrm{Hom}_{\mathbf{AS}}(AS, AS''^{AS'})$$

Let $AS = \{a,b\}$, $AS' = \{c,d\}$ and $AS'' = \{1,0\}$. Suppose that we have the following self-* action *self-*action*: $AS \times AS' \to AS''$

$$self\text{-}*action : \begin{pmatrix} (a,c) \mapsto 1 \\ (a,d) \mapsto 0 \\ (b,c) \mapsto 0 \\ (b,d) \mapsto 1 \end{pmatrix}$$

Currying transforms *self-*action* mentioned above into another self-* action with the same semantics

$$curry(self\text{-}*action) : \begin{pmatrix} a \mapsto \begin{pmatrix} c \mapsto 1 \\ d \mapsto 0 \end{pmatrix} \\ \\ b \mapsto \begin{pmatrix} c \mapsto 0 \\ d \mapsto 1 \end{pmatrix} \end{pmatrix}$$

1.8   Finite Limits in Autonomic Systems

In this section, we consider limits of variously shaped diagrams of ASs.

1.8.1 Pullbacks of autonomic systems

Consider the diagram of ASs and self-*actions below.

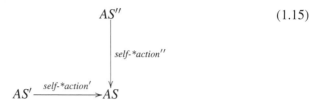

$$\text{(1.15)}$$

Its fiber product is the AS

$$AS' \times_{AS} AS'' \overset{def}{=} \{(x,w,y)\,|\,self\text{-}*action'(x) = w = self\text{-}*action''(y)\}$$

There are obvious projections $self\text{-}*action_1 : AS' \times_{AS} AS'' \to AS'$ and $self\text{-}*action_2 :$ $AS' \times_{AS} AS'' \to AS''$. Note that if $AS''' = AS' \times_{AS} AS''$ then the following diagram commutes

$$
\begin{array}{ccc}
AS''' & \xrightarrow{\;self\text{-}*action_2\;} & AS'' \\
{\scriptstyle self\text{-}*action_1}\big\downarrow & \lrcorner & \big\downarrow{\scriptstyle self\text{-}*action''} \\
AS' & \xrightarrow{\;self\text{-}*action'\;} & AS
\end{array}
\qquad (1.16)
$$

Given the setup of diagram 1.16 we come to the pullback of AS' and AS'' over AS to be any AS''' for which we have an isomorphism $AS''' \overset{\cong}{\to} AS' \times_{AS} AS''$. The corner symbol "⌟" in diagram 1.16 indicates that AS''' is the pullback.

Some may prefer to denote this fiber product by $self\text{-}*action' \times_{AS}$ $self\text{-}*action''$ rather than $AS' \times_{AS} AS''$. The former is the mathematically better notation, but human-readability is often enhanced by the latter, which is also more common in the literature. We use whichever is more convenient.

Consider the diagram of ASs and self-actions as in Equation 1.17.

$$
\begin{array}{ccc}
 & & AS'' \\
 & & \big\downarrow{\scriptstyle self\text{-}*action_4} \\
AS' & \xrightarrow{\;self\text{-}*action_3\;} & AS
\end{array}
\qquad (1.17)
$$

For any AS''' and commutative solid arrow diagram as in Equation 1.18 $self\text{-}*action_1 : AS''' \to AS'$ and $self\text{-}*action_2 : AS''' \to AS''$ such that $self\text{-}*action_3 \circ$ $self\text{-}*action_1 = self\text{-}*action_4 \circ self\text{-}*action_2$ there exists a unique arrow

$$< self\text{-}*action_1, self\text{-}*action_1 >_{AS} : AS''' \to AS' \times_{AS} AS''$$

making everything commute,

$$self\text{-}*action_1 = self\text{-}*action' \circ <self\text{-}*action_1, self\text{-}*action_1>_{AS}$$

and

$$self\text{-}*action_2 = self\text{-}*action'' \circ <self\text{-}*action_1, self\text{-}*action_1>_{AS}$$

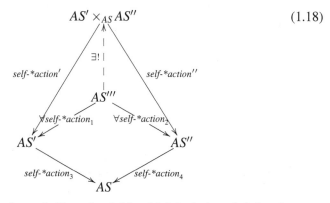 (1.18)

Consider the diagram drawn in Equation 1.19, which includes a left-hand square, a right-hand square, and a big rectangle

$$AS'_1 \xrightarrow{self\text{-}*action_1} AS'_2 \xrightarrow{self\text{-}*action_2} AS'_3 \qquad (1.19)$$

$$\left. \begin{array}{ccc} \downarrow self\text{-}*action_3 & \lrcorner & \downarrow self\text{-}*action_4 & \lrcorner & \downarrow self\text{-}*action_5 \\ AS_1 \xrightarrow{self\text{-}*action_6} AS_2 \xrightarrow{self\text{-}*action_7} AS_3 \end{array} \right.$$

If $AS'_2 \cong AS_2 \times_{AS_3} AS'_3$ then the right-hand square is a pullback. The right-hand square has a corner symbol indicating that $AS'_2 \cong AS_2 \times_{AS_3} AS'_3$ is a pullback. But the corner symbol on the left might be indicating that the left-hand square is a pullback, or the big rectangle is a pullback. Thus, If $AS'_2 \cong AS_2 \times_{AS_3} AS'_3$ then the left-hand square is a pullback if and only if the big rectangle also is.

Consider the diagram drawn in Equation 1.20

$$AS'_2 \xrightarrow{self\text{-}*action_2} AS'_3 \qquad (1.20)$$

$$\left. \begin{array}{ccc} \downarrow self\text{-}*action_4 & \lrcorner & \downarrow self\text{-}*action_5 \\ AS_1 \xrightarrow{self\text{-}*action_6} AS_2 \xrightarrow{self\text{-}*action_7} AS_3 \end{array} \right.$$

where $AS_2' \cong AS_2 \times_{AS_3} AS_3'$ is a pullback. Then there is an isomorphism

$$AS_1 \times_{AS_2} AS_2' \cong AS_1 \times_{AS_3} AS_3'$$

In other words,

$$AS_1 \times_{AS_2} (AS_2 \times_{AS_3} AS_3') \cong AS_1 \times_{AS_3} AS_3'$$

1.8.2 Spans on autonomic systems

Consider AS_1 and AS_2; a span on AS_1 and AS_2 is an AS together with self-* actions $self\text{-}*action_1 : AS \to AS_1$ and $self\text{-}*action_2 : AS \to AS_2$.

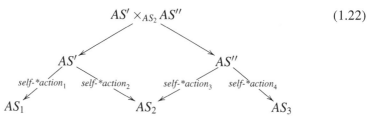

$$(1.21)$$

Let AS_1, AS_2, and AS_3 be autonomic systems, and let

$$AS_1 \overset{self\text{-}*action_1}{\leftarrow} AS' \overset{self\text{-}*action_2}{\to} AS_2$$

and

$$AS_2 \overset{self\text{-}*action_3}{\leftarrow} AS'' \overset{self\text{-}*action_4}{\to} AS_3$$

be spans. Their composite span is given by the fiber product $AS' \times_{AS_2} AS''$ as in the diagram 1.22:

$$AS' \times_{AS_2} AS'' \qquad (1.22)$$

If there is a span as $AS_1 \leftarrow AS \to AS_2$ then by the universal property of products [49], we have a unique map $AS \overset{\exists!}{\to} AS_1 \times AS_2$.

If there are two spans as $AS_1 \leftarrow AS' \to AS_2$ and $AS_1 \leftarrow AS'' \to AS_2$, we can take the disjoint union $AS' \sqcup AS''$ and by the universal property of coproducts, we have a unique span $AS_1 \leftarrow AS' \sqcup AS'' \to AS_2$ making the diagram 1.23 commute.

$$(1.23)$$

Given a span $AS_1 \xleftarrow{self\text{-}*action_1} AS \xrightarrow{self\text{-}*action_2} AS_2$, we can draw a bipartite graph with each state of AS_1 drawn as a dot on the left, each state of AS_2 drawn as a dot on the right, and each state a in AS drawn as an arrow connecting vertex $self\text{-}*action_1(a)$ on the left to vertex $self\text{-}*action_2(a)$ on the right.

1.8.3 Equalizers of self-* actions

Consider two parallel self-* actions

$$AS_1 \underset{self\text{-}*action_2}{\overset{self\text{-}*action_1}{\rightrightarrows}} AS_2$$

The equalizer of $self\text{-}*action_1$ and $self\text{-}*action_2$ is the commutative diagram in Equation 1.24,

$$Eq(self\text{-}*action_1, self\text{-}*action_2) \xrightarrow{\;p\;} AS_1 \underset{self\text{-}*action_2}{\overset{self\text{-}*action_1}{\rightrightarrows}} AS_2 \qquad (1.24)$$

where we define

$$Eq(self\text{-}*action_1, self\text{-}*action_2) \overset{def}{=} \{a \in AS_1 \mid self\text{-}*action_1(a) = self\text{-}*action_2(a)\}$$

and where p is the canonical inclusion.

1.9 Finite Colimits in Autonomic Systems

We consider several types of finite colimits to obtain some intuition about them, without formally defining them yet.

1.9.1 Pushouts of autonomic systems

Consider the diagram 1.25 of ASs and self-* actions below:

$$
\begin{array}{ccc}
AS & \xrightarrow{self\text{-}*action_2} & AS_2 \\
\big\downarrow{\scriptstyle self\text{-}*action_1} & & \\
AS_1 & &
\end{array}
\qquad (1.25)
$$

Its fiber sum, denoted $AS_1 \sqcup_{AS} AS_2$, is defined as the quotient of $AS_1 \sqcup AS \sqcup AS_2$ by the equivalence relation $\sim$ generated by $a \sim self\text{-}*action_1(a)$ and $a \sim self\text{-}*action_2(a)$ for all states a in AS. In other words,

$$AS_1 \sqcup_{AS} AS_2 \overset{def}{=} (AS_1 \sqcup AS \sqcup AS_2)/\sim$$

where $\forall a \in AS, a \sim self\text{-}*action_1(a)$ and $a \sim self\text{-}*action_2(a)$

There are obvious inclusions *self-*action₃* : $AS_1 \to AS_1 \sqcup_{AS} AS_2$ and *self-*action₄* : $AS_2 \to AS_1 \sqcup_{AS} AS_2$. Note that if $AS_3 = AS_1 \sqcup_{AS} AS_2$ then diagram 1.26 commutes.

$$
\begin{array}{ccc}
AS & \xrightarrow{\textit{self-*action}_2} & AS_2 \\
\textit{self-*action}_1 \downarrow & \ulcorner & \downarrow \textit{self-*action}_4 \\
AS_1 & \xrightarrow[\textit{self-*action}_3]{} & AS_3
\end{array}
\qquad (1.26)
$$

Given the setup of diagram 1.26, we define the pushout of AS_1 and AS_2 over AS to be any autonomic system AS_3 for which we have an isomorphism $AS_3 \xrightarrow{\cong} AS_1 \sqcup_{AS} AS_2$. The corner symbol "$\ulcorner$" in diagram 1.26 indicates that AS_3 is the pushout.

For diagram 1.25, any autonomic system AS_3 and commutative solid arrow diagram in 1.27, self-* actions *self-*action₃* : $AS_1 \to AS_3$ and *self-*action₄* : $AS_2 \to AS_3$ such that *self-*action₃* ∘ *self-*action₁* = *self-*action₄* ∘ *self-*action₂*, there exists a unique arrow

$$\ll \textit{self-*action}_3, \textit{self-*action}_4 \gg : AS_1 \sqcup_{AS} AS_2 \to AS_3$$

making everything commute. In other words,

$$\textit{self-*action}_3 = \ll \textit{self-*action}_3, \textit{self-*action}_4 \gg \circ \textit{self-*action}'$$

and

$$\textit{self-*action}_4 = \ll \textit{self-*action}_3, \textit{self-*action}_4 \gg \circ \textit{self-*action}''$$

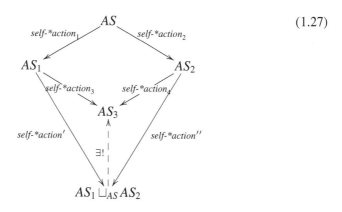

$$(1.27)$$

1.9.2 Coequalizers of self- actions*

Consider two parallel self-* actions

$$AS_1 \xrightarrow[\text{self-*action}_2]{\text{self-*action}_1} AS_2$$

The coequalizer of *self-*action*$_1$ and *self-*action*$_2$ is the commutative diagram in 1.28,

$$AS_1 \xrightarrow[\text{self-*action}_2]{\text{self-*action}_1} AS_2 \xrightarrow{q} Coeq(self\text{-}*action_1, self\text{-}*action_2) \qquad (1.28)$$

where we define the coequalizer of *self-*action*$_1$ and *self-*action*$_2$ as the quotient of AS_2 by the equivalence relation generated by

$$\{(self\text{-}*action_1(a), self\text{-}*action_2(a)) | a \in AS_1\} \subseteq AS_2 \times AS_2$$

In other words,

$$Coeq(self\text{-}*action_1, self\text{-}*action_2) \overset{def}{=} AS_2 / self\text{-}*action_1(a) \sim self\text{-}*action_2(a)$$

1.10 Monoids of Self-* Actions

A common way to interpret self-* in ASs is to say that self-* actions are running on ASs. Triggers of self-* actions from self-* can be performed concurrently to transform one AS state into another. A first rule for self-* actions is this: the performance of a sequence of several self-* actions is itself the performance of a self-* action—a more complex self-* action, but a self-* action nonetheless. Algebraic objects called monoids are tasked with encoding the self-* action's perspective in all this (i.e., what the self-* action can do, and what happens when different self-* actions are done in succession). A monoid can be construed as a set of self-* actions, together with a formula that encodes how a sequence of self-* actions is itself considered a self-* action. In this section we concentrate on monoids.

A monoid of self-* is a sequence $(SELF\text{-}*, skip, |)$, where $SELF\text{-}*$ is a set of self-* actions, $skip \in SELF\text{-}*$ is an action, and $|: SELF\text{-}* \times SELF\text{-}* \rightarrow SELF\text{-}*$ is a concurrence, such that the following conditions hold for all $se\text{-}m, se\text{-}n, se\text{-}p \in SELF\text{-}*$:

■ $se\text{-}m \mid skip = se\text{-}m,$

■ $skip \mid se\text{-}m = se\text{-}m,$ and

■ $(se\text{-}m \mid se\text{-}n) \mid se\text{-}p = se\text{-}m \mid (se\text{-}n \mid se\text{-}p)$

The way they are written here is called *infix notation*. We refer to *skip* as the identity action and to | as the concurrence formula for the monoid. We call the first two rules *identity laws* and the third rule the *associativity law* for monoids.

Alternatively, the rules of identity and associativity can be stated

■ $| (se\text{-}m, skip) = se\text{-}m,$

■ $| (skip, se\text{-}m) = se\text{-}m,$ and

■ $| (| (se\text{-}m, se\text{-}n), se\text{-}p) =| (se\text{-}m, | (se\text{-}n, se\text{-}p))$

The way they are written above is called *prefix notation*. Note that we often use infix notation without mentioning it. That is, given a concurrence $|: SELF\text{-}* \times SELF\text{-}* \to SELF\text{-}*$, we may write $se\text{-}a \mid se\text{-}b$ rather than $| (se\text{-}a, se\text{-}b)$.

There is a monoid with only one action, $(\{skip\}, skip, |)$ where $|: \{skip\} \times \{skip\} \to \{skip\}$ is the unique concurrence. We call this monoid the *trivial monoid*, and sometimes denote it as $\underline{1}$.

In monoid $(SELF\text{-}*, skip, |)$, given actions $se\text{-}m_1, se\text{-}m_2, se\text{-}m_3, se\text{-}m_4$ there are five different ways to parenthesize the concurrence $se\text{-}m_1 \mid se\text{-}m_2 \mid se\text{-}m_3 \mid se\text{-}m_4$, and the associativity law for monoids will show them all to be the same. We have

$$
\begin{aligned}
((se\text{-}m_1 \mid se\text{-}m_2) \mid se\text{-}m_3) \mid se\text{-}m_4 &= (se\text{-}m_1 \mid se\text{-}m_2) \mid (se\text{-}m_3 \mid se\text{-}m_4) \\
&= (se\text{-}m_1 \mid (se\text{-}m_2 \mid se\text{-}m_3)) \mid se\text{-}m_4 \\
&= se\text{-}m_1 \mid (se\text{-}m_2 \mid (se\text{-}m_3 \mid se\text{-}m_4)) \\
&= se\text{-}m_1 \mid ((se\text{-}m_2 \mid se\text{-}m_3) \mid se\text{-}m_4)
\end{aligned}
$$

In fact, the concurrence of any list of monoid self-* actions is the same, regardless of parenthesization. Therefore, we can unambiguously write $se\text{-}m_1 \mid se\text{-}m_2 \mid se\text{-}m_3 \mid se\text{-}m_4$ rather than any given parenthesization of it. This is known as the *coherence theorem* and can be found in [41].

1.10.1   Free monoids of self-* actions

Let $SELF\text{-}*$ be a set of self-* actions. A list in $SELF\text{-}*$ is a pair (n, f) where $n \in \mathbb{N}$ is a natural number (called the length of the list) and $f : \underline{n} \to SELF\text{-}*$ is a function, where $\underline{n} = \{1, 2, \ldots, n\}$. We may denote such a list by

$$(n, f) = [f(1), f(2), \ldots, f(n)]$$

The empty list is the unique list in which $n = 0$; we may denote it by []. Given a self-* action $se\text{-}x \in SELF\text{-}*$ the singleton list on $se\text{-}x$ is the list $[se\text{-}x]$. Given a list $L = (n, f)$ and a number $i \in \mathbb{N}$ with $i \leqslant n$, the ith entry of L is the self-* action $f(i) \in SELF\text{-}*$.

Given two lists $L = (n, f)$ and $L' = (n', f')$, define the concatenation of L and L', denoted $L \ddagger L'$, to be the list $(n + n', f \ddagger f')$, where $f \ddagger f' : \underline{n + n'} \to SELF\text{-}*$ is given on $i \leqslant n + n'$ by

$$(f \ddagger f')(i) = \begin{cases} f(i) & \text{if } i \leqslant n \\ f'(i - n) & \text{if } i \geqslant n + 1 \end{cases} \tag{1.29}$$

Let $SELF\text{-}* = \{se\text{-}a, se\text{-}b, se\text{-}c, \ldots, se\text{-}z\}$. The following are self-* actions of $List(SELF\text{-}*)$:

$$[se\text{-}a, se\text{-}b], [se\text{-}p], [se\text{-}p, se\text{-}a, se\text{-}a], \ldots$$

The concatenation of $[se\text{-}a, se\text{-}b]$ and $[se\text{-}p, se\text{-}a, se\text{-}a]$ is $[se\text{-}a, se\text{-}b, se\text{-}p, se\text{-}a, se\text{-}a]$. The concatenation of any list with [] is just itself.

A free monoid generated by $SELF\text{-}*$ is the sequence $(List(SELF\text{-}*), [], \ddagger)$, where $List(SELF\text{-}*)$ is the set of lists of self-* actions in $SELF\text{-}*$, where $[] \in List(SELF\text{-}*)$ is the empty list, and where $\ddagger$ is the operation of list concatenation. We refer to $SELF\text{-}*$ as the set of generators for the free monoid $(List(SELF\text{-}*), [], \ddagger)$.

A free monoid generated by $\varnothing$ is the sequence $(List(\varnothing), [], \ddagger)$ where $List(\varnothing)$ consists only of the empty list. It is the trivial free monoid.

In the section below, we will define the monoid $(List(SELF\text{-}*), [], \ddagger)$ by specifying some generators and some relations. Lists of generators provide us all the possible ways to write self-* actions of $(List(SELF\text{-}*), [], \ddagger)$. The relations allow us to have two ways of writing the same self-* actions.

1.10.2   Presented monoids of self-* actions

Let $SELF\text{-}*$ be a finite set of self-* actions, let $n \in \mathbb{N}$ be the number of relations we declare, and for each $1 \leqslant i \leqslant n$, let m_i and m_i' be self-* actions of $List(SELF\text{-}*)$. The *monoid presented by generators SELF-* and relations* $\{(m_i, m_i')$ where $1 \leqslant i \leqslant n\}$ is the monoid $(List(SELF\text{-}*)/\sim, [], \ddagger)$ defined fully when $\sim$ denotes the equivalence relation on $(List(SELF\text{-}*)$ generated by $\{xm_i y \sim xm_i' y$ where $x, y \in List(SELF\text{-}*), 1 \leqslant i \leqslant n\}$.

Every free monoid $(List(SELF\text{-}*), [], \ddagger)$ is a presented monoid, because we can just take the set of relations to be empty.

Let $SELF\text{-}* = \{se\text{-}a, se\text{-}b, se\text{-}c, se\text{-}d\}$. The idea of presented monoids is that you notice that the list of self-* actions $[se\text{-}a, se\text{-}a, se\text{-}c]$ always gives the same result as the list of self-* actions $[se\text{-}d, se\text{-}d]$. You also notice that the list of self-* actions $[se\text{-}c, se\text{-}a, se\text{-}c, se\text{-}a]$ is the same thing as doing nothing. In this case, we have $m_1 = [se\text{-}a, se\text{-}a, se\text{-}c]$, $m_1' = [se\text{-}d, se\text{-}d]$, and $m_2 = [se\text{-}c, se\text{-}a, se\text{-}c, se\text{-}a]$, $m_2' = []$ and relations $\{(m_1, m_1'), (m_2, m_2')\}$. What this really means is that we are equating m_1 with m_1' and m_2 with m_2', which for convenience we will write out $[se\text{-}a, se\text{-}a, se\text{-}c] = [se\text{-}d, se\text{-}d]$ and $[se\text{-}c, se\text{-}a, se\text{-}c, se\text{-}a] = []$. To see how this plays out, we give an example of a calculation in $List(SELF\text{-}*)/\sim$. Namely,

$$[se\text{-}b, \underline{se\text{-}d, se\text{-}d}, se\text{-}a, se\text{-}c, se\text{-}a, se\text{-}a, se\text{-}c, se\text{-}d] = [se\text{-}b, se\text{-}a, \underline{se\text{-}a, se\text{-}c}, se\text{-}a, se\text{-}c,$$
$$se\text{-}a, se\text{-}a, se\text{-}c, se\text{-}d]$$
$$= [se\text{-}b, se\text{-}a, \underline{se\text{-}a, se\text{-}a}, se\text{-}c, se\text{-}d]$$
$$= [se\text{-}b, se\text{-}a, se\text{-}d, se\text{-}d, se\text{-}d]$$

1.10.3   Cyclic monoids of self-* actions

A monoid is called cyclic if it has a presentation involving only one generator. Let *se-a* be a self-* action; we look at some cyclic monoids generated by {*se-a*}.

With no relations the monoid will be the free monoid on one generator and will have underlying set $\{[], [se\text{-}a], [se\text{-}a, se\text{-}a], [se\text{-}a, se\text{-}a, se\text{-}a], \ldots\}$, with identity list $[]$ and concatenation such that $[se\text{-}a, se\text{-}a] \ddagger [se\text{-}a] = [se\text{-}a, se\text{-}a, se\text{-}a, se\text{-}a] = se\text{-}a^4$. Note that $se\text{-}a^4$ is shorthand for $[se\text{-}a, se\text{-}a, se\text{-}a, se\text{-}a]$.

With the relation $se\text{-}a \sim []$ we will get the trivial monoid, a monoid having only one action. Consider the cyclic monoid with generator *se-a* and relation $se\text{-}a^7 = se\text{-}a^4$. This monoid has seven actions

$$\{[] = se\text{-}a^0, se\text{-}a = se\text{-}a^1, se\text{-}a^2, se\text{-}a^3, se\text{-}a^4, se\text{-}a^5, se\text{-}a^6\}$$

and we know that

$$se\text{-}a^6 \ddagger se\text{-}a^5 = se\text{-}a^7 \ddagger se\text{-}a^4 = se\text{-}a^4 \ddagger se\text{-}a^4 = se\text{-}a^7 \ddagger se\text{-}a = se\text{-}a^5$$

We can depict this monoid as follows

$$skip \longrightarrow se\text{-}a \longrightarrow se\text{-}a^2 \longrightarrow se\text{-}a^3 \longrightarrow se\text{-}a^4 \longrightarrow se\text{-}a^5 \longrightarrow se\text{-}a^6$$

Let *AS* be a set of autonomic system states. A self-* action of monoid $(SELF\text{-}*, skip, |)$ on *AS*, or simply a self-* action of *SELF-* on *AS* or a *SELF-* action on *AS*, is a function

$$\circlearrowleft: SELF\text{-}* \times AS \to AS$$

such that the following conditions hold for all $se\text{-}m, se\text{-}n \in SELF\text{-}*$ and all $s \in AS$:

■ $skip \circlearrowleft s = s$

■ $se\text{-}n \circlearrowleft (se\text{-}n \circlearrowleft s) = (se\text{-}n \mid se\text{-}n) \circlearrowleft s$

Alternately, we can rewrite $\circlearrowleft$ as $\alpha : SELF\text{-}* \times AS \to AS$ and restate the above conditions as

■ $\alpha(skip, s) = s$

■ $\alpha(se\text{-}n, \alpha(se\text{-}n, s)) = \alpha(se\text{-}n \mid se\text{-}n, s)$

The following proposition expresses the notion of autonomic system in terms of free monoids and their actions on finite sets.

Proposition 1.5 *Let SELF-* and AS be finite non-empty sets of self-* actions and autonomic system states, respectively. Giving a function $\alpha : SELF\text{-}* \times AS \to AS$ is equivalent to giving an action of the free monoid $List(SELF\text{-}*)$ on AS.*

Proof 1.5 We know that function $\delta : List(SELF\text{-}*) \times AS \to AS$ constitutes an action of the monoid $List(SELF\text{-}*)$ on the set AS if and only if for all $s \in AS$ we have $\delta([],s) = s$, and for any two actions $m, m' \in List(SELF\text{-}*)$ we have $\delta(m, \delta(m',s)) = \delta(m \ddagger m', s)$. Let

$$A = \{\delta : List(SELF\text{-}*) \times AS \to AS \parallel \delta \text{ constitutes an action}\}$$

We need to prove that there is an isomorphism of sets

$$\phi : A \xrightarrow{\cong} \mathrm{Hom}_{\mathbf{Set}}(SELF\text{-}* \times AS, AS)$$

Given an element $\delta : List(SELF\text{-}*) \times AS \to AS$ in A, define $\phi(\delta)$ on an element $(se\text{-}a, s) \in SELF\text{-}* \times AS$ by $\phi(\delta)(se\text{-}a, s) \stackrel{def}{=} \delta([se\text{-}a], s)$, where $[se\text{-}a]$ is the one-element list.

We now define $\psi : \mathrm{Hom}_{\mathbf{Set}}(SELF\text{-}* \times AS, AS) \to A$. Given an element $f \in \mathrm{Hom}_{\mathbf{Set}}(SELF\text{-}* \times AS, AS)$ define $\psi(f) : List(SELF\text{-}*) \times AS \to AS$ on a pair $(L, s) \in List(SELF\text{-}*) \times AS$, where $L = [\delta_1, \ldots, \delta_n]$ as follows. By induction, if $n = 0$, put $\psi(f)(L, s) = s$; if $n \geqslant 1$, let $L' = [\delta_1, \ldots, \delta_{n-1}]$ and put $\psi(f)(L, s) = \psi(f)(L', f(\delta_n, s))$.

We can easily check that $\psi(f)$ satisfies the two rules of action above, making it an action of $List(SELF\text{-}*)$ on AS. It is also easy to check that ϕ and ψ are mutually inverse, completing the proof.

It follows that *an autonomic system is an action of a free monoid on a finite set.*

Let $\mathcal{M} : (SELF\text{-}*, skip, |)$ and $\mathcal{M}' : (SELF\text{-}*', skip', |')$ be monoids. A monoid homomorphism f from $\mathcal{M}$ to $\mathcal{M}'$, denoted $f : \mathcal{M} \to \mathcal{M}'$, is a function $f : SELF\text{-}* \to SELF\text{-}*'$ satisfying two conditions:

■ $f(skip) = skip'$

■ $f(se\text{-}a \mid se\text{-}b) = f(se\text{-}a) \mid' f(se\text{-}b)$, for all $se\text{-}a, se\text{-}b \in SELF\text{-}*$

The set of monoid homomorphisms from $\mathcal{M}$ to $\mathcal{M}'$ is denoted $\mathrm{Hom}_{\mathbf{Mon}}(\mathcal{M}, \mathcal{M}')$.

Let $SELF\text{-}* = \{se\text{-}a, se\text{-}c, se\text{-}g, se\text{-}u\}$ and let $SELF\text{-}*' = SELF\text{-}*^3$, the set of triplets in $SELF\text{-}*$. Let $\mathcal{M} = List(SELF\text{-}*)$ be the free monoid on $SELF\text{-}*$ and let $\mathcal{M}' = List(SELF\text{-}*')$ denote the free monoid on $SELF\text{-}*'$. There is a monoid homomorphism $F : \mathcal{M}' \to \mathcal{M}$ given by sending $m = (se\text{-}a, se\text{-}b, se\text{-}c)$ to the list $[se\text{-}a, se\text{-}b, se\text{-}c]$.

Given any monoids $List(SELF\text{-}*)$ there is a unique monoid homomorphism from $List(SELF\text{-}*)$ to the trivial monoid $\underline{1}$. There is also a unique homomorphism $\underline{1} \to List(SELF\text{-}*)$. These facts together have an upshot: between any two monoids $List(SELF\text{-}*)$ and $List(SELF\text{-}*')$ we can always construct a homomorphism

$$List(SELF\text{-}*) \to \underline{1} \to List(SELF\text{-}*')$$

which we call the trivial homomorphism $List(SELF\text{-}*) \to List(SELF\text{-}*')$. A morphism $List(SELF\text{-}*) \to List(SELF\text{-}*')$ that is not trivial is called a nontrivial homomorphism.

Proposition 1.6 *Let* $F(SELF\text{-}*) : (List(SELF\text{-}*), [], \ddagger)$ *be the free monoid on* $SELF\text{-}*$, *and let* $\mathcal{M} : (M, skip, |)$ *be any monoid. There is a natural bijection*

$$Hom_{Mon}(F(SELF\text{-}*), \mathcal{M}) \xrightarrow{\;\cong\;} Hom_{Set}(SELF\text{-}*, M)$$

Proof 1.6 We provide a function $\phi : Hom_{\mathbf{Mon}}(F(SELF\text{-}*), \mathcal{M}) \longrightarrow Hom_{\mathbf{Set}}$ $(SELF\text{-}*, M)$ and a function $\psi : Hom_{\mathbf{Set}}(SELF\text{-}*, M) \longrightarrow Hom_{\mathbf{Mon}}(F(SELF\text{-}*), \mathcal{M})$ and show that they are mutually inverse. Let us first construct ϕ. Given a monoid homomorphism $f : F(SELF\text{-}*) \longrightarrow \mathcal{M}$, we need to provide $\phi(f) : SELF\text{-}* \longrightarrow M$. Given any $se\text{-}g \in SELF\text{-}*$ we define $\phi(f)(se\text{-}g) \overset{def}{=} f[se\text{-}g]$.

Now let us construct ψ. Given $p : SELF\text{-}* \longrightarrow M$, we need to provide $\psi(p) :$ $List(SELF\text{-}*) \longrightarrow \mathcal{M}$ such that $\psi(p)$ is a monoid homomorphism. For a list $L = [se\text{-}g_1, \ldots, se\text{-}g_n] \in List(SELF\text{-}*)$, define $\psi(p)(L) : p(se\text{-}g_1) | \ldots | p(se\text{-}g_n) \in M$. In particular, $\psi(p)([]) = skip$. It is not hard to see that this is a monoid homomorphism. It is also easy to see that $\phi; \psi(p) = p$ for all $p \in Hom_{\mathbf{Set}}(SELF\text{-}*, M)$. We show that $\psi; \phi(f) = f$ for all $f \in Hom_{\mathbf{Mon}}(F(SELF\text{-}*), \mathcal{M})$. Choose $L = [se\text{-}g_1, \ldots, se\text{-}g_n] \in List(SELF\text{-}*)$. Then

$$\psi(\phi f)(L) = (\phi f)(se\text{-}g_1) | \ldots | (\phi f)(se\text{-}g_n)$$

$$= f[se\text{-}g_1] | \ldots | f[se\text{-}g_n]$$

$$= f([se\text{-}g_1, \ldots, se\text{-}g_n])$$

$$= f(L)$$

1.11 Conclusions

The chapter is a reference material for readers who already have a basic understanding of self-* and ASs and are now ready to consider their algebraic aspects such as products, coproducts, curried self-* actions, finite limits, colimits of ASs, and monoids of self-* actions using algebraic language. Algebraic specification is presented in a straightforward fashion by discussing in detail the necessary components and briefly touching on the more advanced components.

Bibliography

[1] S. Abdelwahed and N. Kandasamy. A control-based approach to autonomic performance management in computing systems. In M. Parashar and S. Hariri, editors, *Autonomic Computing: Concepts, Infrastructure and Applications*, 1st edition, pp. 149–168. CRC Press, Boca Raton, FL, 2006.

[2] R. Anthony, A. Butler, and M. Ibrahim. Exploiting emergence in autonomic systems. In M. Parashar and S. Hariri, editors, *Autonomic Computing: Concepts, Infrastructure and Applications*, 1st edition, pp. 121–148. CRC Press, Boca Raton, FL, 2006.

[3] V. Bhat, M. Parashar, and N. Kandasamy. Autonomic data streaming for high-performance scientific applications. In M. Parashar and S. Hariri, editors, *Autonomic Computing: Concepts, Infrastructure and Applications*, 1st edition, pp. 413–434. CRC Press, Boca Raton, FL, 2006.

[4] D.W. Bustard and R. Sterritt. A requirements engineering perspective on autonomic systems development. In M. Parashar and S. Hariri, editors, *Autonomic Computing: Concepts, Infrastructure and Applications*, 1st edition, pp. 19–34. CRC Press, Boca Raton, 2006.

[5] W. Butera. Text display and graphics control on a paintable computer. In G.D.M. Serugendo, J.P.M. Flatin, and M. Jelasity, editors, *Proceedings of 1st International Conference on Self-Adaptive and Self-Organizing Systems (SASO'07)*, pp. 45–54, IEEE Computer Society Press, Boston, MA, 9–11 July 2007.

[6] M. Calisti, R. Ghizzioli, and D. Greenwood. Autonomic service access management for next generation converged networks. In M. Calisti, S.V.D. Meer, and J. Strassner, editors, *Advanced Autonomic Networking and Communication*, 1st edition, pp. 101–126. Whitestein Series in Software Agent Technologies and Autonomic Computing. Springer-Verlag, Heidelberg, 2008.

[7] M. Calisti, S.V.D. Meer, and J. Strassner, editors. *Advanced Autonomic Networking and Communication*. Whitestein Series in Software Agent Technologies and Autonomic Computing, 190 pp Springer-Verlag, Heidelberg, 2008.

[8] A. Chakravarti, G. Baumgartner, and M. Lauria. Self-organizing scheduling on the organic grid. In M. Parashar and S. Hariri, editors, *Autonomic Computing: Concepts, Infrastructure and Applications*, 1st edition, pp. 389–412. CRC Press, Boca Raton, FL, 2006.

[9] D.M. Chess, J.E. Hanson, J.O. Kephart, I. Whalley, and S.R. White. Dynamic collaboration in autonomic computing. In M. Parashar and

S. Hariri, editors, *Autonomic Computing: Concepts, Infrastructure and Applications*, 1st edition, pp. 253–274. CRC Press, Boca Raton, FL, 2006.

[10] L. Durham, M. Milenkovic, P. Cayton, and M. Yousif. Platform support for autonomic computing: A research vehicle. In M. Parashar and S. Hariri, editors, *Autonomic Computing: Concepts, Infrastructure and Applications*, 1st edition, pp. 329–350. CRC Press, Boca Raton, FL, 2006.

[11] C. Fahy et al. Modelling behaviour and distribution for the management of next generation networks. In M. Calisti, S.V.D. Meer, and J. Strassner, editors, *Advanced Autonomic Networking and Communication*, 1st edition, pp. 43–62. Whitestein Series in Software Agent Technologies and Autonomic Computing. Springer-Verlag, Heidelberg, 2008.

[12] G. Jiang et al. Trace analysis for fault detection in application servers. In M. Parashar and S. Hariri, editors, *Autonomic Computing: Concepts, Infrastructure and Applications*, 1st edition, pp. 471–492. CRC Press, Boca Raton, FL, 2006.

[13] G. Nguengang et al. Autonomic resource regulation in IP military networks: A situatedness based knowledge plane. In M. Calisti, S.V.D. Meer, and J. Strassner, editors, *Advanced Autonomic Networking and Communication*, 1st edition, pp. 81–100. Whitestein Series in Software Agent Technologies and Autonomic Computing. Springer-Verlag, Heidelberg, 2008.

[14] J. Chen et al. Game theoretic framework for autonomic spectrum management in heterogeneous wireless networks. In M. Calisti, S.V.D. Meer, and J. Strassner, editors, *Advanced Autonomic Networking and Communication*, 1st edition, pp. 169–190. Whitestein Series in Software Agent Technologies and Autonomic Computing. Springer-Verlag, Heidelberg, 2008.

[15] K. Schwan et al. AutoFlow: Autonomic information flows for critical information systems. In M. Parashar and S. Hariri, editors, *Autonomic Computing: Concepts, Infrastructure and Applications*, 1st edition, pp. 275–304. CRC Press, Boca Raton, FL, 2006.

[16] R. Adams et al. Scalable management—Technologies for management of large-scale, distributed systems. In M. Parashar and S. Hariri, editors, *Autonomic Computing: Concepts, Infrastructure and Applications*, 1st edition, pp. 305–328. CRC Press, Boca Raton, FL, 2006.

[17] R.R. Amoud et al. An autonomic MPLS DiffServ-TE domain. In M. Calisti, S.V.D. Meer, and J. Strassner, editors, *Advanced Autonomic Networking and Communication*, 1st edition, pp. 149–168. Whitestein Series in Software Agent Technologies and Autonomic Computing. Springer-Verlag, Heidelberg, 2008.

[18] S.V.D. Meer et al. Technology neutral principles and concepts for autonomic networking. In M. Calisti, S.V.D. Meer, and J. Strassner, editors, *Advanced Autonomic Networking and Communication*, 1st edition, pp. 1–25. Whitestein Series in Software Agent Technologies and Autonomic Computing. Springer-Verlag, Heidelberg, 2008.

[19] A. Ganek. Overview of autonomic computing: Origins, evolution, direction. In M. Parashar and S. Hariri, editors, *Autonomic Computing: Concepts, Infrastructure and Applications*, 1st edition, pp. 3–18. CRC Press, Boca Raton, FL, 2006.

[20] D. Greenwood and R. Ghizzioli. Autonomic communication with RASCAL hybrid connectivity management. In M. Calisti, S.V.D. Meer, and J. Strassner, editors, *Advanced Autonomic Networking and Communication*, 1st edition, pp. 63–80. Whitestein Series in Software Agent Technologies and Autonomic Computing. Springer-Verlag, Heidelberg, 2008.

[21] R. Griffith, G. Valetto, and G. Kaiser. Effecting runtime reconfiguration in managed execution environments. In M. Parashar and S. Hariri, editors, *Autonomic Computing: Concepts, Infrastructure and Applications*, 1st edition, pp. 369–388. CRC Press, Boca Raton, FL, 2006.

[22] T. Heinis, C. Pautasso, and G. Alonso. A self-configuring service composition engine. In M. Parashar and S. Hariri, editors, *Autonomic Computing: Concepts, Infrastructure and Applications*, 1st edition, pp. 237–252. CRC Press, Boca Raton, FL, 2006.

[23] IBM. Autonomic computing manifesto. http://www.research. ibm.com/autonomic/, 2001.

[24] X. Jin and J. Liu. From individual based modeling to autonomy oriented computation. In M. Nickles, M. Rovatsos, and G. Weiss, editors, *Agents and Computational Autonomy: Potential, Risks, and Solutions*, volume 2969 of *Lecture Notes in Computer Science*, pp. 151–169. Springer, Berlin, 2004.

[25] B. Khargharia and S. Hariri. Autonomic power and performance management of internet data. In M. Parashar and S. Hariri, editors, *Autonomic Computing: Concepts, Infrastructure and Applications*, 1st edition, pp. 435–470. CRC Press, Boca Raton, FL, 2006.

[26] S. Ko, I. Gupta, and Y. Jo. Novel mathematics-inspired algorithms for self-adaptive peer-to-peer computing. In G.D.M. Serugendo, J.P.M. Flatin, and M. Jelasity, editors, *Proceedings of 1st International Conference on Self-Adaptive and Self-Organizing Systems (SASO/07)*, pp. 3–12, IEEE Computer Society Press, Boston, MA, 9–11 July 2007.

[27] F.W. Lawvere and S.H. Schanuel. *Conceptual Mathematics: A First Introduction to Categories*, 1st edition. Cambridge University Press, Heidelberg, 1997.

[28] H. Liu and M. Parashar. A programming system for autonomic self-managing applications. In M. Parashar and S. Hariri, editors, *Autonomic Computing: Concepts, Infrastructure and Applications*, 1st edition, pp. 211–236. CRC Press, Boca Raton, FL, 2006.

[29] J.A.L López, J.M.G. Munoz, and J.M. Padial. A Telco Approach to Autonomic Infrastructure Management. In M. Calisti, S.V.D. Meer, and J. Strassner, editors, *Advanced Autonomic Networking and Communication*, 1st edition, pp. 27–42. Whitestein Series in Software Agent Technologies and Autonomic Computing. Springer-Verlag, Heidelberg, 2008.

[30] D.A. Menasće and M.N. Bennani. Dynamic server allocation for autonomic service centers in the presence of failures. In M. Parashar and S. Hariri, editors, *Autonomic Computing: Concepts, Infrastructure and Applications*, 1st edition, pp. 353–368. CRC Press, Boca Raton, FL, 2006.

[31] O. Pacheco. Autonomy in an organizational context. In M. Nickles, M. Rovatsos, and G. Weiss, editors, *Agents and Computational Autonomy: Potential, Risks, and Solutions*, volume 2969 of Lecture Notes in Computer Science, pp. 195–208. Springer, Berlin, 2004.

[32] M. Parashar. Autonomic grid computing: Concepts, requirements, and infrastructure. In M. Parashar and S. Hariri, editors, *Autonomic Computing: Concepts, Infrastructure and Applications*, 1st edition, pp. 49–70. CRC Press, Boca Raton, FL, 2006.

[33] M. Parashar and S. Hariri, editors. *Autonomic Computing: Concepts, Infrastructure and Applications*, 1st edition, 568 pp. CRC Press, Boca Raton, FL, 2006.

[34] G. Qu and S. Hariri. Anomaly-based self protection against network attacks. In M. Parashar and S. Hariri, editors, *Autonomic Computing: Concepts, Infrastructure and Applications*, 1st edition, pp. 493–522. CRC Press, Boca Raton, FL, 2006.

[35] M.A. Razzaque, S. Dobson, and P. Nixon. Cross-layer optimisations for autonomic networks. In M. Calisti, S.V.D. Meer, and J. Strassner, editors, *Advanced Autonomic Networking and Communication*, 1st edition, pp. 127–148. Whitestein Series in Software Agent Technologies and Autonomic Computing. Springer-Verlag, Heidelberg, 2008.

[36] R.V. Renesse and K.P. Birman. Autonomic computing: A system-wide perspective. In M. Parashar and S. Hariri, editors, *Autonomic Computing: Concepts, Infrastructure and Applications*, 1st edition, pp. 35–48. CRC Press, Boca Raton, FL, 2006.

[37] S.M. Sadjadi and P.K. McKinley. Transparent autonomization in composite systems. In M. Parashar and S. Hariri, editors, *Autonomic Computing: Concepts, Infrastructure and Applications*, 1st edition, pp. 169–188. CRC Press, Boca Raton, FL, 2006.

[38] D. I. Spivak. *Category Theory for the Sciences, 1st edition*. The MIT Press, 2014.

[39] P. Steenkiste and A.C. Huang. Recipe-based service configuration and adaptation. In M. Parashar and S. Hariri, editors, *Autonomic Computing: Concepts, Infrastructure and Applications*, 1st edition, pp. 189–208. CRC Press, Boca Raton, FL, 2006.

[40] J.W. Sweitzer and C. Draper. Architecture overview for autonomic computing. In M. Parashar and S. Hariri, editors, *Autonomic Computing: Concepts, Infrastructure and Applications*, 1st edition, pp. 71–98. CRC Press, Boca Raton, FL, 2006.

[41] J. van Oosten. *Basic Category Theory*. Department of Mathematics, Utrecht University, The Netherlands, 2002.

[42] P.C. Vinh. Formal aspects of dynamic reconfigurability in reconfigurable computing systems. PhD thesis, London South Bank University, London, UK, 4 May 2006.

[43] P.C. Vinh. Homomorphism between AOMRC and Hoare model of deterministic reconfiguration processes in reconfigurable computing systems. *Scientific Annals of Computer Science*, (XVII):113–145, 2007.

[44] P.C. Vinh. Formal aspects of self-* in autonomic networked computing systems. In M.K. Denko, L.T. Yang, and Y. Zhang, editors, *Autonomic Computing and Networking*, pp. 381–410. Springer, Heidelberg, 2009.

[45] P.C. Vinh. Toward formalized autonomic networking. *Mobile Networks and Applications*, 19(5):598–607, 2014. DOI:10.1007/s11036-014-0521-z.

[46] P.C. Vinh. Algebraically autonomic computing. *Mobile Networks and Applications*, 21(1):3–9, 2016. DOI:10.1007/s11036-015-0615-2.

[47] P.C. Vinh. Concurrency of self-* in autonomic systems. *Future Generation Computer Systems*, 56:140–152, 2016. DOI:10.1016/j.future.2015.04.017.

[48] P.C. Vinh. Finite limits and colimits in autonomic systems. In P.C. Vinh and V. Alagar, editors, *The 4th International Conference on Context-Aware Systems and Applications (ICCASA 2015)*, volume 165 of *Lecture Notes of the Institute for Computer Sciences, Social Informatics and Telecommunications Engineering (LNICST)*, pp. 10–20. Springer, Heidelberg, 2016.

[49] P.C. Vinh. Products and coproducts of autonomic systems. In P.C. Vinh and V. Alagar, editors, *The 4th International Conference on Context-Aware Systems and Applications (ICCASA 2015)*, volume 165 of *Lecture Notes of the Institute for Computer Sciences, Social Informatics and Telecommunications Engineering (LNICST)*, pp. 1–9. Springer, Heidelberg, 2016.

[50] P.C. Vinh. Some universal constructions of autonomic systems. *Mobile Networks and Applications*, 21(1):89–97, 2016. DOI:10.1007/s11036-015-0672-6.

[51] P.C. Vinh and J.P. Bowen. A formal approach to aspect-oriented modular reconfigurable computing. In *Proceedings of 1st IEEE and IFIP International Symposium on Theoretical Aspects of Software Engineering (TASE)*, pp. 369–378, IEEE Computer Society Press, Shanghai, China, 6–8 June 2007.

[52] P.C. Vinh and J.P. Bowen. Formalization of data flow computing and a coinductive approach to verifying flowware synthesis. *LNCS Transactions on Computational Science*, 1(4750):1–36, 2008.

[53] P.C. Vinh and N.M. Truong. Products, coproducts and universal properties of autonomic systems. *EAI Endorsed Transactions on Context-aware Systems and Applications*, 16(7): e5, 2016.

[54] P.C. Vinh and N.T. Tung. Coalgebraic aspects of context-awareness. *Mobile Networks and Applications*, 18(3):391–397, 2013. DOI:10.1007/s11036-012-0404-0.

[55] P.C. Vinh and E. Vassev. Nature-inspired computation and communication: A formal approach. *Future Generation Computer Systems*, 56:121–123, 2016. DOI:10.1016/j.future.2015.10.011.

[56] P. C. Vinh. Self-adaptation in collective adaptive systems. *Mobile Networks and Applications*, 19(5):626–633, 2014. DOI:10.1007/s11036-014-0529-4.

[57] Y. Wang. Toward theoretical foundations of autonomic computing. *The International Journal of Cognitive Informatics and Natural Intelligence (IJCiNi)*, 1(3):1–16, 2007.

[58] M. Witkowski and K. Stathis. A dialectic architecture for computational autonomy. In M. Nickles, M. Rovatsos, and G. Weiss, editors, *Agents and*

Computational Autonomy: Potential, Risks, and Solutions, volume 2969 of *Lecture Notes in Computer Science*, pp. 261–273. Springer, Berlin, 2004.

[59] T.D. Wolf and T. Holvoet. A taxonomy for self-* properties in decentralized autonomic computing. In M. Parashar and S. Hariri, editors, *Autonomic Computing: Concepts, Infrastructure and Applications*, 1st edition, pp. 101–120. CRC Press, Boca Raton, FL, 2006.

[60] B. Yang and J. Liu. An autonomy oriented computing (AOC) approach to distributed network community mining. In G.D.M. Serugendo, J.P.M. Flatin, and M. Jelasity, editors, *Proceedings of 1st International Conference on Self-Adaptive and Self-Organizing Systems (SASO'07)*, pp. 151–160, IEEE Computer Society Press, Boston, MA, 9–11 July 2007.

Chapter 2

Bio-Communities Communications Paradigms for Vehicular Social Networks

Francesco Chiti, Enrico Dei, and Romano Fantacci

University of Florence

CONTENTS

2.1 Introduction

Vehicular ad hoc networks (VANETs) have been envisioned by governmental organisations and vehicle manufacturers as a disruptive paradigm capable of providing increased convenience to drivers, with a wide variety of applications ranging from traffic safety and efficiency, up to socially inspired critical situations management. In addition, VANETs could represent a distributed and collaborative platform, providing personalised and context-sensitive services, which rely on accessible local information of other vehicles and drivers. The recent outcomes of standardisation activities — in particular IEEE 802.11p, IEEE 1609 in the United States, ETSI ITS (Intelligent Transportation System) and ISO CALM (Continuous Air-interface Long and Medium range) in the European Union — fostered this vision, allowing both vehicle-to-vehicle (V2V) communication and vehicle-to-infrastructure (V2I) communication. When compared to conventional ad hoc networks, VANETs exhibit specific features mainly in terms of mobility, which results in frequent and unpredictable network topology changes. In addition, a typical VANET wireless link is unreliable due to both deep fading and co-channel interference. This aspect requires the specific design of data communication, aggregation, and dissemination and routing protocols for vehicular applications. However, the development of suitable protocols, which effectively support information sharing, is still an open issue. Recent research on large-scale self-organised distributed networks, as well as in social communities, suggests an alternative perspective for routing protocol design. It relies on the evidence of scale-free properties for typical VANET deployments, which implies a *small-world* network structure. The analysis of the so called network clustering coefficient quantifies the clustering properties of a specific community of devices and motivates the introduction of clustering protocols. Generally speaking, a joint clustering and routing scheme dynamically organizes nodes into groups, referred to as *clusters*, where each cluster is coordinated by one of the vehicles, referred to as the cluster head (CH) and the rest of vehicles are denoted as the ordinary nodes (ONs). To address the aforementioned topic, this chapter reviews the most relevant bio-inspired routing and clustering approaches with a special focus on vehicular scenarios. The underlying concept of mobile group management is adopted to select the more suited schemes, among which are the ant or bee colonies organisation and the bacteria life cycle. The latter is a particularly promising approach, since it involves the autonomic (re)configuration of the system, including the transition from stand alone to clustered states, which could be accomplished with machine-to-machine (M2M) communications. In addition, the interaction with more complex systems such as plants or animals leads to context-aware dynamic protocols. Finally, the practical issues related to protocol design are investigated by relying on a bacterial quorum sensing mechanism, which seems to be a promising approach to designing decentralized and flexible joint clustering and routing schemes, while achieving a full context awareness.

2.2   Overview on Mobile Ad Hoc Networking

Mobile ad hoc networks (MANETs) are usually defined as wireless mobile networks formed *spontaneously* depending on some *opportunity* [5,14,19]. Communication in such a decentralised network typically involves temporary hop *relays*, which act as on demand routers without the need of any fixed *a priori* infrastructure. This innovative network paradigm is very flexible and suitable for applications such as temporary information sharing in dynamic and critical contexts. However, the main challenge in MANETs is to find a path between communicating nodes, since *multi-hop* routing, random movement of mobile nodes and other features unique to MANETs lead to enormous overhead for route discovery and maintenance. Furthermore, this problem is worsened by the resource constraints in energy, computational capacities and bandwidth. As a consequence, research interest in MANETs has been growing in the past few years, with the design of MANET routing protocols receiving significant attention. One of the reasons for this is that routing in MANETs is a particularly challenging task due to the fact that the network topology continuously changes, and paths which were initially efficient can quickly become inefficient or even infeasible. Moreover, control of information flow in the network is very restricted. This is because (1) the bandwidth of the wireless medium is very limited, and (2) the medium is shared, since the nodes are usually *half-duplex*. Many approaches have been proposed to solve the routing challenge in MANETs [1,29,34,43]; however they are usually sufficient for conventional applications such as file transfer or download. For instance, to support applications with a stringent requirement for delay, such as jitter and packet losses, support for Quality-of-Service (QoS) is needed in addition to basic routing functionality.

Traditionally, routing protocols in MANETs are classified into three categories; *proactive*, *reactive* and *hybrid* [16,18,23]. Proactive routing protocols often need to exchange control packets among mobile nodes and continuously update their routing tables, where each node maintains the state of the network in real time. This causes high overhead and congestion of the network, which requires a lot of memory. The advantage of proactive protocols is that nodes have correct and updated information; hence, when a path is required, it can be found directly in the memory and links can be established quickly. Reactive routing protocols only seek a route to the destination when it is needed. The advantage of these protocols is that the routing tables located in the memory are not continuously updated. However, their disadvantage is that they cannot establish connections in real time. The aim of these protocols is to save time in the route discovery process, since the reactive protocol is designed to reduce the latency, which is critical in this kind of protocol. It also aims to avoid the maintenance of routes to prevent long delay.

Nevertheless, it is important to redesign the above mentioned protocols to make them adaptive, robust and self-healing. Moreover, due to the lack of central control or infrastructure in the network, they should work in a localised way. To this end, *biology-inspired* techniques, such as the swarm intelligence principle, have been proven to be very adaptable in other problem domains. They have been widely applied to the MANET routing problem as they are a good fit for the problem [24,37]. As a matter of fact, the general characteristics of these biological systems, which include their capability for self-organisation, self-healing and local decision making, make them suitable for routing in MANETs. Nature's self-organising systems, such as insect societies, termite hills, bee colonies, bird flocks and fish schools, provide precisely these features and, hence, have been a source of inspiration for the design of many routing algorithms for MANETs. One well known example of biological swarm social behaviour is ant colony foraging. Many ant species have a trail-laying, trail-following behaviour when foraging: individual ants deposit a chemical substance called *pheromone* as they move from a food source to their nest, and foragers follow such pheromone trails. Subsequently, more ants are attracted by these pheromone trails and in turn reinforce them even more. As a result of this *catalytic* effect, the optimal solution rapidly emerges. In this food searching process, a phenomenon called *stigmergy* plays a key role in developing and manipulating local information. It describes the indirect communication of individuals through modifying the environment. From the self-organisation theory point of view, the behaviour of the social ant can be modelled based on four elements: positive feedback, negative feedback, randomness and multiple interactions. This model of social ants using self-organisation theories provides powerful tools to transfer knowledge about the social insects to the design of intelligent decentralised problem-solving systems. Specifically there exist several ant colony optimisation (ACO) based routing approaches, such as AntNet, hybrid ACO (AntHocNet), ACO based routing algorithm (ARA), imProved ant colony optimisation routing algorithm for mobile ad hoc NETworks (PACONET), ACO based on demand distance vector (Ant-AODV) and ACO based dynamic source routing (Ant-DSR), which are all based on forward ants (FANT) and backward ants (BANT) principles.

2.3   The Special Case of Vehicular Ad Hoc Networks

A VANET is considered as a particular kind of a MANET, consisting of a set of mobile nodes (Vehicles) and fixed nodes known as Road-Side Units

(RSUs) [15,36,42]. A VANET provides digital data communication among vehicles through V2V communication and, between vehicles and RSUs through V2I communication. Due to usual constraints in terms of directions and speeds, VANET vehicles move according to an organized mobility pattern with some differences between highways, urban or rural areas.

It is widely accepted that ITS safety applications are the most sensitive services in VANETs because of the significant impact they can have on human lives. These particular applications rely on the *aggregation* of VANET information using safety messages, which are sent by each vehicle in the vehicular ecosystem, carrying information such as the state of the other vehicles.

Specifically, there are two main transmission modes in VANET routing. The first one, called unicast, sends data packets to one destination. The second mode, known as *multicast*, aims to transmit data packets to multiple destinations at once. In addition to this, each vehicle can process the gathered information according to the Fog Computing (FC) paradigm to quickly reach a higher context awareness [2,17,21,35]. To ensure the timely reception of these kinds of safety messages, fast, reliable and efficient dissemination of messages should be guaranteed. In addition, multimedia information can also play a role in traffic management and road safety. As a consequence, the development of efficient routing algorithms supporting adequate QoS in terms of a minimum end-to-end delay, routing overheads and minimum dropping rate is requested.

Over the last years, several routing protocols that can be broadly classified into two categories have been investigated: *topology*-based routing protocols and *geography*-based ones [20,26,28,38,40,41]. Topology-based routing protocols determine routes based on topological links between nodes along the source-destination path with good accuracy, absence of loops and, consequently, a low dropping rate, at the expense of potential instability with increasing end-to-end delays and routing overheads. In contrast, geography-based protocols route data packets without any knowledge of the underlying network topology; they present a high robustness and a low number of dropped messages, while the latency is usually remarkable.

However, the previous solutions are often inadequate for large-scale VANETs because of their high complexity and low performance [10]. Recently, a new area, called *bio-inspired* computing (which looks at biology as a source of inspiration), has emerged. It mimics the laws and dynamics of natural species and applies their solutions to the theoretical and experimental problems for various applications [3,4,25]. In fact, bio-inspired approaches for routing in large-scale VANETs are capable of providing several advantages [11,25]:

■ *Efficiency* due to the similarity between the manner of finding VANET routes and natural behavior of species to discover food source paths for *ants* or *bees*;

■ Low computational *complexity*;

- *Adaptability* and *self-organization* which in perspective can cope with frequent topology changes;

- *Robustness* to hardware/software faults and potentially *security* against threats.

According to Ref. [3], bio-inspired VANET routing appraoches can be classified into three main categories:

- *Evolutionary* algorithms, consisting of computational techniques inspired by natural evolution such as inheritance, mutation, selection and crossover, which aim to find the optimal route. The Intersection-based Geographical Routing Protocol (IGRP) and the automatic configuration of the Optimized Link State Routing (OLSR) protocol applied to VANETs both belong to this class.

- *Swarm intelligence* algorithms are indeed inspired by swarm intelligence such as ants, bees or birds which originate three sub-classes: (1) ACO applied in designing the Mobility-aware Ant colony optimization Routing DYMO (MAR-DYMO), (2) Particle Swarm Optimization (PSO) leading to the parallel Particle Swarm Optimization (pPSO) improving the QoS parameters of the Ad hoc On-demand Distance Vector routing (AODV) protocol, and (3) Bees Colony Optimization algorithms (BCO) to which belongs the Hybrid Bee swarm Routing (HyBR).

- Various biologically inspired algorithms, including human social behavior, which motivated the design of the Fuzzy-Assisted Social-based routing (FAST) protocol. Moreover, the natural immune system was considered as an idea to conceive a routing protocol called Datataxis.

In conclusion, it can be noticed that all the above mentioned bio-inspired approaches have been applied to different application scenarios, always pointing out the potential to address the challenges associated with classic VANET routing such as vehicular network scalability, self-organized control, complexity of messages exchange patterns and robustness. However, they are not fully capable of focusing on some specific issues of VANETs, especially those arising in the so called *smart city* scenario, which will be in the next section.

2.4   Small Worlds Bio-Networking

2.4.1   Vehicular social networks

Vehicular Social Networks (VSNs) [7,27] are an emerging type of network that allows the exchange of information among drivers and passengers, as well as other vehicles. The exploitation of social relationships can improve and extend the added value provided by applications of vehicular networks (e.g. navigation safety applications, navigation efficiency, entertainment, participatory and urban sensing and emergency [12]). Indeed, leveraging the social dimension in vehicular networking can bring several advantages:

- Ease the dissemination of information by exploiting users' common interests and preferences, thus avoiding delivery of information to uninterested parties;

- Ease the on-time and continuous production of up-to-date and capillary information on the road conditions and traffic status by turning users and vehicles into producers of information;

- Enhance the exchange of information, media content and recommendations for entertainment and tourism;

- Support new mechanisms for reputation building and establishment of trust among nodes in VANETs [27]; for instance, the relationship degree in online social networks could contribute to define the degree of trust with an unknown node entering the network.

With respect to online and mobile social networks, VSNs pose several challenges:

- The topology of a VSN changes very often and it is formed by peers that move at high speed;

- Contact duration of vehicles on the move is short, in the order of tens of seconds [27];

- The technological infrastructure is made by heterogeneous components, such as On-Board Units (OBU), vehicle sensors and actuators, mobile phones, RSU and remote services.

2.4.1.1 Context-awareness

The nature of VSNs is inherently dynamic and dependent on context changes. In particular, *context-awareness* is usually defined as the capability of a system to use context for providing relevant information and/or services to the user, depending on the user's task. Context-awareness plays a relevant role in VSNs for several reasons:

- Context information can be used to dynamically detect and build communities. Relations in VSNs can be built by taking into account highly variable information such as common geographical location, final destination and/or part of the route; but also rarely changing information, such as habits (e.g., daily routes to work), friendship, family relationships, common interests and trust;

■ Members of VSNs can exchange context updates, such as change of speed, occurrence of congestion events or accidents and Point of interest (PoI) recommendations;

■ Provided services can adapt to current context (e.g., navigation services can suggest a new route, alerts are propagated only to nodes that can be affected by the alerted event).

It is expected that the evolution of VSNs will intersect with the evolution of smart phones and Wireless Sensor Networks (WSN) [12], thus extending the range of context sources and the type of context information that can be gathered in the vehicular environment, opening up the opportunity for novel context-aware services. Ad-hoc networking mechanisms provided by wireless and cellular technologies are key enablers for VSNs since they enable the establishment of opportunistic networks relying on the cooperation of dynamic groups of peers (i.e., smart phones, vehicle OBUs, smart objects along the road infrastructure). This context information can be used to trigger appropriate adaptation at different scopes (in-vehicle, vehicle-to-vehicle, vehicle-to-remote services) in order to enhance the quality of travel for drivers and passengers and assure their safety.

2.4.1.2 Service overlay and application scenarios

Wireless communications, indeed, may enable the exchange of mobile information among vehicles in a *participatory* way to enhance open information sharing and knowledge exchange processes. The appropriate use of context-aware adaptation paradigms in the operation of application, network and communication services is considered a key enabler not only for VSN but also, on a wider scope, for 5G technology [32]. Indeed, 5G infrastructure is envisioned as a *neural bearer* where *everything* (applications, processing resources, network, data) can be provided as a service and most of the intelligence, including context-aware adaptation logic, is going to be distributed at the edge (i.e., in the aggregation and access segments up to the end user premises) [32]. As a consequence, it is possible to envisage an overlay of heterogeneous service capabilities pervasively distributed, interconnected and deeply integrated through the 5G network infrastructure that can be dynamically composed to be adapted to context changes. Recently, investigation activities have been started on the possible use of 4G Long Term Evolution (LTE) standard to ensure connectivity between vehicles, roadside infrastructure and the people inside and around the so called *connected car* [6,13,31]. The definition of a generic interface, namely V2X, between a vehicle and a generic device has been addressed, generalizing the existing communication interfaces as V2V (between vehicles), V2P (between a vehicle and a device carried by a generic individual), and V2I (between a vehicle and RSUs).

The main goal in VSN communications protocols design is represented by the prompt achievement of a *global* context awareness; that is, a group of vehicles are expected to jointly cooperate to gather and refine information together while effectively distributing it. Cooperation among vehicles can enable a wide range of application services for the benefit of several categories of end users: drivers of private vehicles, drivers of public or collective transport vehicles, transportation authorities and public administrations. Regarding the aim of deriving an overall comprehensive communications scenario, three different networking use cases can be considered, starting with (1) critical information diffusion performed by a single vehicle towards a potential group of neighbors, up to (2) data exchange among convoys when a communication opportunity arises, and (3) data aggregation and fusion performed by a vehicle acting as gateway to the Internet, as depicted in Figure 2.1. As soon as communications become less sporadic and context aware, an ad hoc networking scheme is required: in particular we indicated (1) a basic flooding scheme or more advanced solutions relying on (2) disruption-tolerant networking (DTN) or classical MANET.

To this purpose, the analogy with the natural communications scheme could be extremely useful. Specifically, recent research on large scale self-organised systems, as well as in social communities, pointed out the so called *scale-free* properties for typical VANETs, which implies a *small-world* network structure [8]. In particular, the so called *network clustering coefficient*, which briefly represents the probability that two nodes are mutually connected, quantifies the

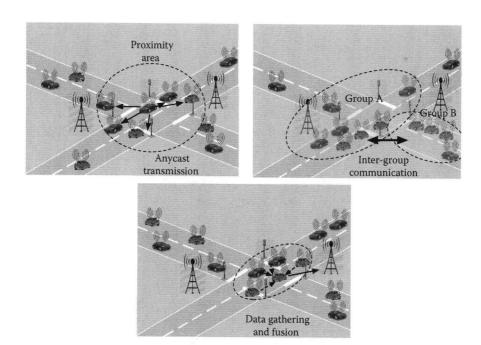

Figure 2.1 Potential use cases from a data exchanging and networking perspective.

clustering properties of a specific community of devices. We refer here to the way bacteria communicate and cooperate to form colonies and more complex systems as spores, while indirectly achieving a global payoff. This leads to a joint clustering and routing codesign [9], in order to successfully adopt the *small world* paradigm, which has been proven to be effective in describing animal, human of even computer interactions [39]. In all the above introduced cases, it is invariantly a high level of aggregation and a low level of separation with a relatively small number of links connecting two generic elements.

2.4.2   *Bacterial colony life cycle*

Bacteria have a tendency to attach to surfaces and self-organise into *micro-colonies*, which represent the first step toward the formation of *biofilms*, which can function as a structural scaffold and as a protective barrier to antimicrobials; in fact, biofilm communities exhibit enhanced antibiotic tolerance and biofilm infections are notoriously difficult to treat [33]. Upon particular stress conditions, mainly due to the scarcity of nutrition, previously isolated bacteria increase their spatial density and, as soon this parameter is above a certain threshold value, a colony is established, with a subsequent modification of basic bacteria behaviour that is no more *myopic* and merely selfish, but starts to be indirectly cooperating towards a common utility function. In particular, they adopt a specific communication code (*messages*), as recently discovered, with a complexity comparable to neurones in the brain [30]. Once all the resources of the colony are exhausted the bacteria aggregation is terminated and they become again free (or aggregated in more complex scaffold). The detection mechanism originating this transition is called *quorum sensing* (QS) [22].

The main phases involved in the bacteria life cycle are shown in Figure 2.2, including (1) attachment, (2) growth and (3) detachment. In addition to this basic

Figure 2.2 Bacterial life cycle main phases.

behaviour, several bacteria are able to form a high level and complex community (usually called spores), which has an additional set of specialised and differentiated functions, and can even interact with advanced organisms as plants.

2.4.3 Bacterial networking for social VANETs

In light of the previous consideration, the information exchange occurring in a typical SVN could be consequently modelled. In particular, each car could be modelled as a finite state machine, whose states and transitions are depicted in Figure 2.3.

The analogy between bacterial and VANETs life cycles is based on the fact that vehicles (depending on the traffic conditions) are usually aggregated in groups and, because of this, have the chance to cooperate and achieve a deeper insight into the surrounding context and to adapt their driving style accordingly. In particular, this kind of *opportunistic* and *sporadic* networking depends upon the vehicles' spatial density and the amount of information provided and processed by each one. Once a context variation is implicitly detected via the increased density, a cluster is formed; further, the different clusters could be

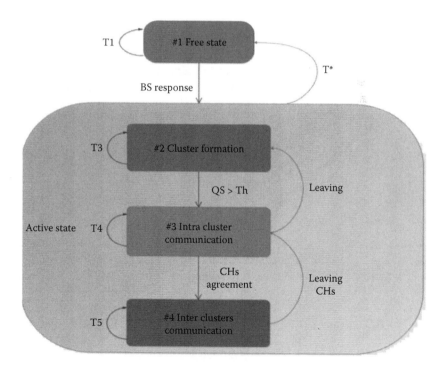

Figure 2.3 Vehicle life cycle main phases representation via finite state machine.

aggregated into more complex topologies with a hierarchical role specialisation (e.g., cluster heads, relay, gateway, proxy and ordinary nodes).

In more detail, the phases involved in a generic vehicle life cycle are the following:

1. *Free state*: This represents the initial default condition of a vehicle entering the flow of traffic. It has no link established with neighbour vehicles, but it can be assumed to be connected and associated to a cellular Base Station (BS), which, considering the terminal mobility pattern and profile, can assign a geo-based IPv6 address and a local density threshold value (*Th*) to possibly perform the QS. The timeout value T^* is then set every time the connection with the BS is lost.

2. *Active state*: This is a macro-state entered when BS successfully associates the vehicle. It includes the following operation (sub-states):

3. *Cluster formation*: Within this phase the cluster is formed via network discover messages disseminated by each vehicle. Once all the requested information is collected, and the QS outcome is positive, (i.e., QX > Th) the cluster is originated and the next stage is entered.

4. *Intra cluster communication*: Upon the cluster establishment, each vehicle can share its context view with other ones and jointly process it to allow for the reconciliation of the partial view and achieve a full awareness. In addition to this basic function, the vehicles joining and leaving are managed, which accomplishes the formation of a fully dynamic network.

5. *Inter clusters communication*: This phase typically represents the opportunity to communicate among different clusters to fuse different micro contexts into wider view; it usually requires the presence of Cluster Heads (CHs) and additional messages.

2.4.4 Quorum sensing mechanism

As previously outlined, the main aspect that needs to be addressed to successfully design a bacterial inspired general scheme is the underlying QS mechanism, whose completion drives the transition towards the Active State. To mimic the bacterial behaviour, the ith node, via a proper message passing, collects several pieces of information regarding the neighbourhood and stores it in the list QS_i, whose generic element $QS_{i,j}$ is evaluated as:

$$QS_{i,j} \doteq \varepsilon_i \cdot \dim\{\mathcal{N}_i \cap \mathcal{N}_j\}, \ 1 \leq j \leq v_i \qquad (2.1)$$

where $\dim(\mathcal{S})$ is the number of elements of the set $\mathcal{S}$, v_i is the number of the ith node neighbors collected during the network discovery phase, $0 \leq \varepsilon_i \leq 1$ is a normalized weighting factor representing the relevance of the ith node to the context awareness achievement (e.g., a vehicle could be private or public), $\mathcal{N}_i$ and $\mathcal{N}_j$ represent the neighbors list of the ith and jth node, respectively. As a consequence, the QS procedure performed by the ith node is the following:

$$QS_{i,j} \gtrless Th \qquad (2.2)$$

where the threshold value Th is a priori provided by the BS depending on the *local* traffic pattern characteristics, including vehicles speed, density and topographic map. As soon as $QS_{i,j} > Th$ the ith and the jth nodes originate a new cluster or join an existing one.

The choice of the optimal Th value depends upon the objective function required to be optimised; usually it is effectively represented by the end-to-end message delivering *delay* δ, as it is the latency experienced by a couple of nodes willing to share their contextual partial knowledge. It can be noticed that δ is affected by the following factors: (1) the number of unconnected vehicles, (2) the numbers of clusters within and end-to-end intra-cluster paths, and (3) the hop length of an intra-cluster path. To this purpose, the simulated δ values as a function of Th are pointed out in Figure 2.4 in the light curve, together with an analytical model (dark curve), pointing out a similar trend that allows us to

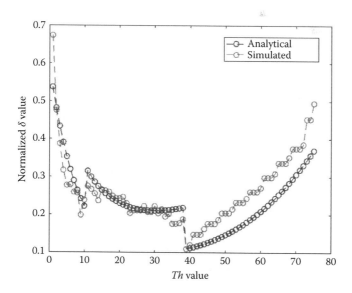

Figure 2.4 **End-to-end delivering** *latency* δ **as a function of** *Th* **for both an analytical model and a simulated scenario.**

express the optimal value $Th^* \doteq \arg\min \delta$ as it follows:

$$Th^* \approx \alpha \left(\frac{n\,l\,w}{v} - k \right) \tag{2.3}$$

where $0 \leq \alpha \leq 1$ represents the normalised congestion level of the area under analysis, n is the number of lanes, l and w represent the length and width, respectively of each lane, v is the vehicle length (including the safety distance), and k is a correction factor usually dependent on the spatial variance of the above paramenters, which makes this threshold adapatable to different scenarios occurring in a typical smart city environment.

Symbol Description

α	To solve the generator maintenance scheduling, in the past, several mathematical techniques have been applied.
σ^2	These include integer programming, integer linear programming, dynamic programming, branch and bound etc.
$\sum$	Several heuristic search algorithms have also been developed. In recent years expert systems,
abc	fuzzy approaches, simulated annealing and genetic algorithms have also been tested.
$\theta\sqrt{abc}$	This paper presents a survey of the literature
ζ	over the past fifteen years in the generator
∂	maintenance scheduling. The objective is to
sdf	present a clear picture of the available recent literature
ewq	of the problem, the constraints and the other aspects of
bvcn	the generator maintenance schedule.

Bibliography

[1] L. Abusalah, A. Khokhar, and M. Guizani. A survey of secure mobile ad hoc routing protocols. *IEEE Communications Surveys Tutorials*, 10(4):78–93, Fourth 2008.

[2] A. Alrawais, A. Alhothaily, C. Hu, and X. Cheng. Fog computing for the internet of things: Security and privacy issues. *IEEE Internet Computing*, 21(2):34–42, 2017.

[3] S. Bitam, A. Mellouk, and S. Zeadally. Bio-inspired routing algorithms survey for vehicular ad hoc networks. *IEEE Communications Surveys Tutorials*, 17(2):843–867, Second quarter 2015.

[4] S. Bitam and A. Mellouk. *Bio-inspired Routing Protocols for Vehicular Ad-Hoc Networks*. Wiley-ISTE, London, UK and Hoboken, NJ, 2014.

[5] Y. Cao and Z. Sun. Routing in delay/disruption tolerant networks: A taxonomy, survey and challenges. *IEEE Communications Surveys Tutorials*, 15(2):654–677, Second 2013.

[6] S. Chen, J. Hu, Y. Shi, and L. Zhao. Lte-v: A td-lte-based v2x solution for future vehicular network. *IEEE Internet of Things Journal*, 3(6):997–1005, 2016.

[7] F. Chiti, R. Fantacci, D. Giuli, F. Paganelli, and G. Rigazzi. *Communications Protocol Design for 5G Vehicular Networks*, pp. 625–649. Springer International Publishing, Cham, 2017.

[8] F. Chiti, R. Fantacci, Y. Gu, and Z. Han. Content sharing in internet of vehicles: Two matching-based user-association approaches. *Vehicular Communications*, 8:35–44, 2017.

[9] F. Chiti, R. Fantacci, R. Mastandrea, G. Rigazzi, Á. S. Sarmiento, and E. M. M. López. A distributed clustering scheme with self nomination: Proposal and application to critical monitoring. *Wireless Networks*, 21(1):329–345, 2015.

[10] F. Dressler and O. B. Akan. Bio-inspired networking: From theory to practice. *IEEE Communications Magazine*, 48(11):176–183, 2010.

[11] F. Dressler and O. B. Akan. A survey on bio-inspired networking. *Computer Networks*, 54(6):881–900, 2010.

[12] M. Gerla and L. Kleinrock. Vehicular networks and the future of the mobile internet. *Computer Networks*, 55(2):457–469, 2011.

[13] S. H. Sun, J. L. Hu, Y. Peng, X. M. Pan, L. Zhao, and J. Y. Fang. Support for vehicle-to-everything services based on LTE. *IEEE Wireless Communications*, 23(3):4–8, 2016.

[14] L. Hanzo and R. Tafazolli. A survey of qos routing solutions for mobile ad hoc networks. *IEEE Communications Surveys Tutorials*, 9(2):50–70, Second 2007.

[15] H. Hartenstein and L. P. Laberteaux. A tutorial survey on vehicular ad hoc networks. *IEEE Communications Magazine*, 46(6):164–171, 2008.

[16] X. Hong, K. Xu, and M. Gerla. Scalable routing protocols for mobile ad hoc networks. *IEEE Network*, 16(4):11–21, 2002.

[17] X. Hou, Y. Li, M. Chen, D. Wu, D. Jin, and S. Chen. Vehicular fog computing: A viewpoint of vehicles as the infrastructures. *IEEE Transactions on Vehicular Technology*, 65(6):3860–3873, 2016.

[18] A. Iwata, C.-C. Chiang, G. Pei, M. Gerla, and T.-W. Chen. Scalable routing strategies for ad hoc wireless networks. *IEEE Journal on Selected Areas in Communications*, 17(8):1369–1379, 1999.

[19] L. Junhai, Y. Danxia, X. Liu, and F. Mingyu. A survey of multicast routing protocols for mobile ad-hoc networks. *IEEE Communications Surveys Tutorials*, 11(1):78–91, First 2009.

[20] E. Lee, E. K. Lee, M. Gerla, and S. Y. Oh. Vehicular cloud networking: Architecture and design principles. *IEEE Communications Magazine*, 52(2):148–155, 2014.

[21] T. Li, M. Zhao, A. Liu, and C. Huang. On selecting vehicles as recommenders for vehicular social networks. *IEEE Access*, 5:5539–5555, 2017.

[22] Y.-H. Li and X. Tian. Quorum sensing and bacterial social interactions in biofilms. *Sensors*, 12(3):2519–2538, 2012.

[23] C. R. Lin and M. Gerla. Adaptive clustering for mobile wireless networks. *IEEE Journal on Selected Areas in Communications*, 15(7):1265–1275, 1997.

[24] K. Manjappa and R. M. R. Guddeti. Mobility aware-termite: A novel bio inspired routing protocol for mobile ad-hoc networks. *IET Networks*, 2(4):188–195, 2013.

[25] M. Meisel, V. Pappas, and L. Zhang. A taxonomy of biologically inspired research in computer networking. *Computer Networks*, 54(6):901–916, 2010.

[26] K. Mershad, H. Artail, and M. Gerla. We can deliver messages to far vehicles. *IEEE Transactions on Intelligent Transportation Systems*, 13(3):1099–1115, 2012.

[27] F. Mezghani, R. Dhaou, M. Nogueira, and A.-L. Beylot. Content dissemination in vehicular social networks: Taxonomy and user satisfaction. *IEEE Communications Magazine,* 52(12):34–40, 2014.

[28] J. Nzouonta, N. Rajgure, G. Wang, and C. Borcea. Vanet routing on city roads using real-time vehicular traffic information. *IEEE Transactions on Vehicular Technology*, 58(7):3609–3626, 2009.

[29] L. Pelusi, A. Passarella, and M. Conti. Opportunistic networking: Data forwarding in disconnected mobile ad hoc networks. *IEEE Communications Magazine*, 44(11):134–141, 2006.

[30] A. Prindle, J. Liu, M. Asally, S. Ly, J. Garcia-Ojalvo, and G. M. Suel. Ion channels enable electrical communication in bacterial communities. *Nature*, 527(7576):59–63, 2015.

[31] H. Seo, K. D. Lee, S. Yasukawa, Y. Peng, and P. Sartori. LTE evolution for vehicle-to-everything services. *IEEE Communications Magazine*, 54(6):22–28, 2016.

[32] D. Soldani and A. Manzalini. Horizon 2020 and beyond: On the 5G operating system for a true digital society. *IEEE Vehicular Technology Magazine*, 10(1):32–42, 2015.

[33] P. Stoodley, K. Sauer, D. G. Davies, and J. W. Costerton. Biofilms as complex differentiated communities. *Annual Review of Microbiology*, 56:187–209, 2002.

[34] D. Tacconi, D. Miorandi, I. Carreras, F. Chiti, and R. Fantacci. Using wireless sensor networks to support intelligent transportation systems. *Ad Hoc Networks*, 8(5):462–473, 2010.

[35] M. Tao, K. Ota, and M. Dong. Foud: Integrating fog and cloud for 5G-enabled v2g networks. *IEEE Network*, 31(2):8–13, 2017.

[36] S. M. Tornell, C. T. Calafate, J. C. Cano, and P. Manzoni. DTN protocols for vehicular networks: An application oriented overview. *IEEE Communications Surveys Tutorials*, 17(2):868–887, Secondquarter 2015.

[37] L. J. G. Villalba, D. R. Canas, and A. L. S. Orozco. Bio-inspired routing protocol for mobile ad hoc networks. *IET Communications*, 4(18):2187–2195, 2010.

[38] W. Wang, F. Xie, and M. Chatterjee. Small-scale and large-scale routing in vehicular ad hoc networks. *IEEE Transactions on Vehicular Technology*, 58(9):5200–5213, 2009.

[39] D. J. Watts and S. H. Strogatz. Collective dynamics of small-world networks. *Nature*, 393(6684):440–442, 1998.

[40] N. Wisitpongphan, F. Bai, P. Mudalige, V. Sadekar, and O. Tonguz. Routing in sparse vehicular ad hoc wireless networks. *IEEE Journal on Selected Areas in Communications*, 25(8):1538–1556, 2007.

[41] K. D. Wong, K. Tepe, W. Chen, and M. Gerla. Inter-vehicular communications. *IEEE Wireless Communications*, 13(5):6–7, 2006.

[42] F. Yang, S. Wang, J. Li, Z. Liu, and Q. Sun. An overview of internet of vehicles. *China Communications*, 11(10):1–15, 2014.

[43] J. Y. Yu and P. H. J. Chong. A survey of clustering schemes for mobile ad hoc networks. *IEEE Communications Surveys Tutorials*, 7(1):32–48, First 2005.

Chapter 3

Application of Nature Inspired Algorithms for Wireless Multi-hop Ad Hoc Network Optimization Problems in Disaster Response Scenarios

Jesús Sánchez-García, José Manuel García-Campos, Daniel Gutiérrez Reina, Sergio Luis Toral, and Federico Barrero

University of Seville

CONTENTS

3.1   Introduction

The use of algorithms inspired in nature is an active research topic with applications in many scientific areas such as computer science [1], communications [2], and many others. Nature inspired algorithms resemble nature strategies to solve optimization problems aimed at finding a pseudo-optimal solution and spending moderate resources (time and computing capacity). Examples of these algorithms are Hill Climbing, Simulated Annealing, Tabu Search, Evolutionary Algorithms, Particle Swarm Optimization, and Ant Colony Optimization, among others. Nature inspired algorithms have been used in the development of wireless multi-hop ad hoc networks [3] and, in particular, when these networks are used for establishing communications in disaster scenarios [4].

Wireless multi-hop ad hoc networks[1] consist of a set of wireless electronic devices, usually called nodes, which can communicate with each other forming a network [5]. The idea behind these networks is that there is not a predefined topology, and there is neither any infrastructure nor a specific hierarchy among

---

[1]In this chapter, the term *ad hoc network* is used as a shorter form to refer to *wireless multi-hop ad hoc networks*.

nodes. Therefore, there are not some nodes acting as routers and others as terminals. On the contrary, every node is responsible for routing data packets in the network and at the same time is a terminal. The communication between two nodes that are out of range is carried out through other nodes via multi-hop paths.

Disaster scenarios have a random, unpredictable, and changing nature. Due to this, setting up a wireless multi-hop ad hoc network in disaster scenarios is not a trivial task. However, the problems related to the deployment and maintenance of these types of networks in a disaster scenario can be addressed as optimization problems. The main optimization problems that appear when using wireless multi-hop ad hoc networks in disaster scenarios are the placement of the nodes, the routing of packets, and the nodes' mobility. Metaheuristics, and in particular nature inspired algorithms, have been widely used to solve these types of problems [6–8].

This chapter is intended for gathering and organizing the information about current works in the field of nature inspired algorithms for wireless multi-hop ad hoc networks used in disaster scenarios. This chapter continues as follows: Section 3.2 introduces the concept of wireless multi-hop ad hoc networks and the main problems that appear when they are used in disaster scenarios. Section 3.3 describes the main concepts of several nature inspired algorithms. Section 3.4 describes specific works that used nature inspired algorithms in disaster scenarios. Section 3.5 describes the lessons learned and remaining challenges that result from an analysis of the literature. Finally, Section 3.6 presents the conclusions.

3.2 Introduction to Wireless Multi-hop Ad Hoc Networks and Disaster Scenarios

With the latest advances in wireless communications, new communication paradigms have been born. Wireless multi-hop ad hoc networks [4] are a communication paradigm based upon the following characteristics: (1) the network consists of a set of nodes with no predefined infrastructure; and (2) the nodes play a twofold role; first, they relay and route other nodes' messages, and second, they also are receivers of the messages directed to them. Wireless multi-hop ad hoc networks can be considered a big family of smaller communication paradigms such as Mobile Ad Hoc Networks (MANETs), Vehicular Ad Hoc Networks (VANETs), Wireless Sensor Networks (WSNs), Wireless Mesh Networks (WMNs), and Delay Tolerant Networks (DTNs). Wireless multi-hop ad hoc networks have been envisioned as a flexible and fast-deploying communication infrastructure for disaster scenarios [4]. Normally, when a disaster occurs, typical communication infrastructure, like the cellular network, is likely to be malfunctioning or destroyed. Disasters can be due to natural causes (e.g. hurricanes, floods, etc.) or man-made disasters such as terrorist attacks.

In the case that nodes can move, these networks receive the name of MANETs [9]. The mobility of nodes in a MANET may range from low to medium speeds. Usually, MANET nodes are related to the communication devices carried by walking people (e.g. smartphones). Thus, it can be said that MANET nodes represent individuals' movements. For this reason, their mobility is normally unexpected as it depends on people's behavior. MANETs performance is strongly affected by the nodes' mobility, and consequently, broadcasting schemes and specific protocols are used for routing packets over the network. Common wireless technologies used are WiFi (IEEE 802.11 b/g/n/s) and Bluetooth (IEEE 802.15.1).

VANETs [10] are an evolution of MANETs in which the communication devices (transceivers) are integrated within terrestrial vehicles. The speed of VANET nodes is higher than in typical MANET's scenarios. Also, VANETs mobility is predictable to a certain extent, as cars only move over roads. Due to this fact, there may be some patterns depending on the roads' layout, such as urban, highways, and others scenarios. VANETs wireless technologies are evolutions of other well-known technologies with some improvements for the mobility conditions of cars. These technologies are mainly WAVE (IEEE 802.11p) and CALM architecture [11].

With the advancements of electronics developed in the last decade, small aerial vehicles have appeared as candidates to carry wireless communication devices. These are called Unmanned Aerial Vehicles (UAVs) or drones. UAVs have been used recently as nodes to build wireless multi-hop ad hoc networks, which are called Aerial Ad Hoc Networks (AANETs) [12]. AANETs have the ability to perform fast network deployments over the air because they do not have to avoid the typical ground obstacles. Also, some types of UAVs can hover on specific positions if needed to provide coverage to certain areas. The wireless technology that is usually used in AANETs is WiFi (IEEE 802.11a/p/s).

DTNs [13] are defined as a low density network, typically with fewer nodes than a MANET. DTN nodes follow the store-carry-forward strategy in order to deliver messages from source to destination. This strategy means that when a node receives a message, it keeps it in its memory until it finds a different node to exchange the message with. For this reason, a higher mobility of nodes increases the probability of node encounters, and thus the probability of delivering messages. This type of network is useful for delivering non-time-sensitive data. The wireless technologies used in DTNs are normally the same as the ones used for MANETs.

WSNs [2] are another type of multi-hop ad hoc networks. Nodes in WSNs are usually sensing devices capable of communicating wirelessly. The most used wireless technology is ZigBee (IEEE 802.15.4), but others, such as WiFi, are used as well. WSNs are usually considered static networks, but when sensors are embedded in mobile robots they can have certain mobility capabilities. However, WSNs nodes' mobility is usually considered to be low. Normally, they are organized following a fixed topology. The most common architecture in WSNs consists in a few nodes, called sinks, which collect the information from the rest of the sensing nodes.

Finally, WMNs are similar to MANETs. The difference with MANETs is that in WMNs the mobility is very low or nonexistent. The common scenario for WMNs is not related to connecting people's smartphones to each other but creating a low cost communication infrastructure. Mainly, WMNs are used to extend the Internet connectivity in wireless local networks. The wireless technologies that are usually used are WiFi (802.11b/g/n) and WiMAX (IEEE 802.16).

The main characteristics of each of the wireless multi-hop ad hoc networks mentioned in this section are shown and compared in Table 3.1.

3.2.1 Disaster scenario mobility models

The most feasible way to analyze the performance of different wireless multi-hop ad hoc networks in disaster situations is by simulating the communications over a scenario of interest. For this reason, a model of the scenario is a fundamental tool. In this subsection we review some of the most frequently used disaster scenario mobility models. These models can be classified according to different aspects. However, the main features to consider for organizing them are: (1) the individuals that are modelled (victims, first responders[2], or both) and (2) whether the individuals' mobility is random or presents a certain level of tactics.

The Random Waypoint model (RWP) [14] has been used for a long time as the reference of nodes mobility in MANETs. This is a simple model in which nodes move randomly over a predefined area. In 2003, it was demonstrated that this model has some characteristics that can provide misleading results [15] with regard to the speed and the distribution of nodes. However, both its simplicity and the lack of alternative mobility models resulted in the extensive use of RWP. Nevertheless, a disaster scenario is complex by nature, and so is the mobility of victims and first responders in it. For this reason, the RWP mobility model was substituted by more detailed and realistic mobility models.

The Disaster Area (DA) model [16] is based on a method called separation of the room. In this mobility model, a disaster scenario is divided into different context-based regions. These regions are: incident site, casualty treatment area, transport zone, and technical operation command zone. The incident site corresponds to the area affected by the disaster; usually, the victims are located all over the area and the first responders, who are patrolling this area randomly, carry the discovered victims to other regions. The other regions correspond to areas in which rescue workers provide first aid care and medical transport to victims. The DA model is integrated with the mobility generator BonnMotion [17]. In DA, the first responders' movements are based in the RWP mobility model, although

---

[2]In this chapter we use the terms *rescue workers* and *first responders* indistinctly to refer to emergency response teams.

Table 3.1 Multi-hop ad hoc networks family for disaster scenarios

Paradigm	Nodes	Node speed	Mobility	Wireless technology	Topology
MANET	Hand-held devices	Low–medium	Unexpected	WiFi, bluetooth	Dynamic and no hierarchy
VANET	Terrestrial vehicles	High	Restricted to roads. With patterns.	WiFi, WAVE, CALM	Dynamic and no hierarchy
AANET	Aerial vehicles, UAVs	Low to high (depends on the application)	With no restrictions. Controlled depending on the service.	WiFi, WAVE	Dynamic and no hierarchy
DTN	Hand-held devices or vehicles	Medium	Unexpected	WiFi, Bluetooth	Dynamic and no hierarchy
WSN	Sensors	Low	Usually fixed to a location	ZigBee, WiFi	Fixed and hierarchical
WMN	Communications' devices	Low	Usually fixed to a location	WiFi, WiMAX	Fixed and hierarchical

it is obvious in reality their movements are not random; on the contrary, they are organized [18]. This model focuses on first responders' mobility and only considers the victims nodes within the incident site.

The Composite Mobility (CoM) described by Pomportes, Tomasik and Vque [19] presents a mobility model for disaster scenarios consisting of the combination of more basic mobility models. This approach creates a model with more details and is closer to human mobility. The CoM model highlights the importance of three factors for modelling human mobility; these are: (1) realistic human movements, (2) group mobility, and (3) obstacle avoidance. The proposed solution to include these factors is to use the Levy-Walk model [20] to represent a more realistic human mobility. Also, the CoM model uses the Reference Point Group Mobility model [21] for representing the rescue workers' collaboration when moving in groups. Moreover, a modified version of the Voronoi diagram [22] is used as an obstacle avoidance strategy. The CoM model is realistic; however, it assumes that the scenario characteristics are known in advance, which may not be possible in real disaster scenarios.

In Ref. [23] the CORPS (Cooperation, Organization and Responsiveness in Public Safety) mobility model is presented. In CORPS the concept of events is introduced. An event is a location which may attract or cause first responders to move away. Specific event types can be attended by specific first responders' roles (e.g., a fire has to be assisted by a firefighter). The first responders are modelled as moving randomly in search mode. When they find an event, they provide assistance to the event if they have the required role (e.g. a firefighter acts on fires). If they do not have the required role they stay in search mode. When a rescue worker finds an event, the location and type of the event is broadcast so other first responders can come and collaborate. CORPS does not model the mobility of victims. CORPS' mobility model is integrated with a mobility generator tool called *MobiGen*. However, the fact of first responders moving randomly when they are in search mode is not very close to reality because they are organized even when performing search-for-victims operations [24].

In Ref. [18], the Role-Based Urban Post-Disaster (RBUPD) mobility model is presented. The RBUPD model presents many detailed characteristics from disaster areas and is based on real disaster data. The individuals are called *agents* and may be: (1) victims; (2) first responders, such as firefighters or people near the scenario area; or (3) vehicles, such as ambulances or privately held vehicles. The concept of *events* is also used, which represent different conditions in a disaster scenario affecting the scenario and the agents (e.g. a fire, a car accident, etc.). The mobility of the agents is considered semi-random, but when an event attracts an agent the mobility is based on path planning models towards the event location. The composition of simpler and independent mobility models depends on the role represented by the type of the agent and the area characteristics (obstacles, open areas, streets, etc.).

In Ref. [12] an urban disaster scenario area (UDS) is modelled. Different regions are defined according to typical characteristics that appear when a disaster occurs in an urban area. Some of the regions defined are: (1) streets and buildings; (2) blocked streets due to wreckage; (3) open areas, such as parks; (4) collapsed buildings, in which the wreckage may have trapped people inside; and (5) prohibited areas that represent potential dangers and threats. Each of these regions presents specific mobility characteristics. For example, in the case of a blocked street the victims will move along it until they reach a point blocked with wreckage, or in the case of an open area, the victims will move around all over the area. The victims' movement in each region is based on the RWP as defined in Ref. [17]. The UDS model is focused on victims' mobility and is very detailed.

Table 3.2 summarizes the main characteristics of the mobility models used in disaster scenarios simulations.

3.2.2 Optimization problems of wireless multi-hop ad hoc networks for disaster scenarios

The main aim of the networks deployed in disaster scenarios is to enable wireless communication services among first responders and, in some cases, also among victims. Wireless multi-hop ad hoc networks are envisioned as one of the network models that can respond better to the communication requirements of disaster scenarios. However, disaster scenarios present several problems in terms of communications. These problems can be treated as optimization problems and nature inspired algorithms have been used to solve them. Here we describe the main optimization problems of wireless multi-hop ad hoc networks that can be found in disaster scenarios.

3.2.2.1 Nodes placement

In disaster scenario areas, rescue members need to communicate among themselves and also with the officer in command. For this reason, rescue workers usually place communication devices acting as access points or relaying nodes in multiple fixed locations within the scenario. Typical networks for this purpose are WMNs or MANETs with some nodes fixed in specific locations.

The nodes' placement problem refers to finding the specific locations to place fixed nodes, with the aim of increasing network connectivity. Usually, the location of nodes is decided based on the context information, such as the victims' density in the area. It is possible that the nodes to be deployed are carried by some kind of vehicles, such as a firefighter's truck. Despite being transported by vehicles, the network is not a VANET or an ad hoc network with high mobility. The node or the access point will be static at a specific location. Each access point can be positioned in a different location and the combination of all the locations

Table 3.2 Disaster scenarios mobility models

Mobility model	RWP	DA	CoM	CORPS	RBUPD	UDS
Regions	*No*	*Yes*	*Yes*	*No*	*No*	*Yes*
Movement	Random	Random (within-regions)	Levy-Walk and RPGM	Random (search-mode) and target-oriented	Random and target-oriented	Random (within-regions)
Victims model	Yes	Yes	No	No	Yes	Yes
First responders model	Yes	Yes	Yes	Yes	Yes	No
Tactics	No	No	Yes	Through events	Through events	No
Scenario	Not specified	Not urban (mainly)	Urban	Urban and not urban	Urban	Urban

will improve the quality of communication services, depending on several factors such as coverage range, signal propagation, interference cancellation, connected devices density, and others. Thus, this is a combinatorial optimization problem.

Finding the best positions for placing relaying nodes for a network is not a trivial task. The aim is to avoid or minimize the blind zones, which are the areas in which a first responder or a victim is unable to connect to the network. There are also constraints in terms of the number of relaying nodes available to deploy and the time spent in the deployment process. As an example, the deployment time is very important because the success of the rescue operations is directly related to it. Therefore, using brute force or exhaustive algorithms is not an option because they spend lots of time checking all the possible positions in order to find the optimum. On the contrary, nature inspired algorithms perform quite well by providing quasi-optimal positions in a shorter time. Some works addressing this specific problem are [25–27].

3.2.2.2 Routing protocols

In MANETs, the packets' transmission from a source node to other nodes in the network is severely affected by the nodes' mobility [28]. Routing protocols, which are responsible for finding a communication path between the source and destination nodes (e.g. between two first responders), play a very important role. Discovering a route is the step before initiating the transmission of application packets. If the route is not found, the application packets will not be transmitted properly to their destinations. Fixed networks' routing protocols are not directly applicable to wireless multi-hop ad hoc networks. Moreover, in the few cases that they can be adapted to ad hoc networks, their performance is not as good as desired.

Routing in ad hoc networks differs from conventional routing in fixed networks (such as the Internet) in various ways: (1) there is not infrastructure, (2) wireless links are unreliable, (3) nodes may fail, and (4) routing protocols have to meet strict energy saving requirements. There are two main approaches [5] for routing packets through an ad hoc network: (1) Topology-based protocols, which use the information about the available nodes' links in the network in order to establish routing routes; and (2) Position-based protocols, which forward packets to the neighbors that are closer to the destination node, making these decisions only based on the nodes' physical position. The most well-known position-based protocol is the Greedy Geographic Forwarding [29].

Within the topology-based approach, there are two main categories of routing algorithms for Ad Hoc Networks: (1) reactive and (2) proactive protocols. On the one hand, in reactive routing protocols (also called *demand routing protocols*) a node initiates a route discovery throughout the network only when it wants to send packets to one or several destinations. On the other hand, in proactive routing, each node continuously maintains up-to-date routes to every other

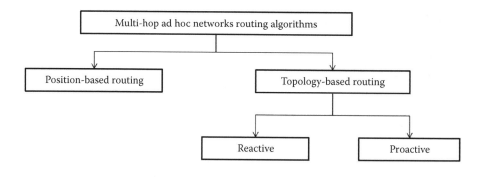

Figure 3.1 Multi-hop ad hoc routing algorithms classification.

node in the network. In proactive protocols the routing information is periodically transmitted throughout the network in order to maintain the routing tables' consistency. A detailed comparison between reactive and proactive protocols is available in Ref. [30]. Figure 3.1 shows the main categories of routing protocols for wireless multi-hop ad hoc networks.

It is important to highlight that there is not a perfect routing protocol for all types of scenarios. Therefore, it is a crucial task to develop adaptive, robust, and self-healing routing protocols. Nature inspired approaches meet these requirements, and for this reason they are a research area of importance for ad hoc routing protocols. Some works on this area are [4,7,31].

3.2.2.3 Nodes' mobility

Nodes' mobility is one of the most important parameters affecting the performance of wireless multi-hop ad hoc networks [28]. Typically, by creating a MANET among first responders and also using complementary relaying nodes (as described in the nodes' placement problem), there are still chances that some first responders could be isolated in specific zones of the scenario. When the communication devices are terrestrial (ground-to-ground) there are many aspects that can affect the quality of the service such as obstacles, fading effects, multi-trajectory propagation, and others. Current efforts are trying to create wireless multi-hop ad hoc networks for disaster scenarios by embedding wireless communication devices in small-sized aerial vehicles.

The aim is to create an AANET made of UAVs as a support network for providing communication services to the nodes in the ground. There are three main aspects to address in this type of network: (1) providing communications coverage in the disaster area, (2) exploring the scenario in order to look for new victims and unknown regions, and (3) tracking victims, first responders, or even specific events.

Providing coverage missions refers to deploying a set of mobile nodes, acting as wireless access points over the disaster scenario, and providing communications services to the maximum area possible. Air-to-air communication links
are more energy efficient than ground-to-ground links. As the nature of disaster
scenarios is highly dynamic, the UAVs usually have a strategy to adapt to the
changing conditions in the ground (i.e., victims and first responders density and
mobility). The initial stage of the coverage problem is commonly finding the best
positions for the UAVs in order to provide communication services to the maximum area within the scenario. This initial stage is, in some cases, similar to the
nodes' placement problem described in Section 2.2.1. After this stage, the UAVs
recalculate their positions periodically in order to adapt to the mobility of nodes
on the ground. The mobility in this situation ranges from low to medium mobility, as the UAVs providing coverage usually adapt to the scenario with gradual
movements. There are works dealing with the coverage problem, such as [13,32].
This problem is similar to the topology control problem in general-purpose ad
hoc networks.

Exploration missions are based on a set of aerial vehicles in search of potential victims or first responders that are located in an unknown location. These
missions can be carried out with some known-in-advance data about the scenario
and the target (e.g. the scenario area, the obstacles, the target's previous locations, etc.) or can be performed completely blind, with no other information than
the area to scan. When a target, such as a group of victims, is discovered, the
UAVs may transmit the victims' locations and aerial photos of the area to the
rescue teams so they can plan the operations accordingly. Exploration missions
are a very common topic in the literature of disaster scenarios [8,33,34].

Tracking missions are those in which a set of UAVs are responsible for flying close to specific mobile targets. The main aim of these missions is to provide communication support to the target or to gather data about it and send this
information to first responders while the target is moving. Examples of tracking
missions can be found in works such as [35,36].

3.3   Nature Inspired Optimization Algorithms

The majority of metaheuristics optimization algorithms can be considered
nature-inspired. There are many ways to classify these optimization algorithms.
Rather than trying to create a new classification, we will follow the one suggested in Ref. [37], in which the algorithms are grouped into trajectory-based
and population-based algorithms.

On the one hand, in trajectory-based optimization, only one state of the search
space is visited in each iteration of the algorithm. As the algorithms execute, different states are visited, and it can be considered that the algorithm describes a

trajectory. It can be said that there is a single particle moving from one state to another. Examples of trajectory-based algorithms are Hill Climbing, Simulated Annealing, or Tabu Search, among others. These algorithms are easy to implement and do not consume many computational resources, which make them suitable for mobile wireless devices powered by batteries. Moreover, they are able to find quasi-optimal solutions in a relatively short time. On the other hand, population-based algorithms are based on a set of particles or individuals that are exploring in parallel different regions of the search space. Examples of these algorithms are evolutionary algorithms and swarm algorithms, among others. Although these algorithms are more resource-consuming than trajectory-based algorithms, they are capable of finding good solutions in hard optimization problems. Both trajectory-based and population-based algorithms are able to solve complex problems like nondeterministic polynomial (NP) time, also called NP problems [37].

3.3.1   Trajectory-based algorithms

Trajectory-based algorithms often are referred to as algorithms where there is only one agent. We may define the term *agent* (or also *particle*) in this context as an entity that perceives a specific region of the search space and evaluates the quality solution that may be found on that region [38]. The agent moves throughout different states of the search space describing a trajectory. This trajectory may be kept in the memory or not, depending on the final aim of the algorithm. Usually, in each algorithm's iteration, the agent visits potential candidates for the solution and selects the best one. The candidate solutions could be called neighbors, as they are states that are close to the current state in which the agent is in (in terms of the search space). Visiting only the neighbors of the current state is because the agent can perceive only a limited part of the search space. For this reason, these algorithms are also called *local search algorithms*. These algorithms also receive the name of *single-solution based metaheuristics* [39] because they only provide one solution.

As is usual in optimization problems, a specific objective and a scale is needed in order to select the best candidate solution. Normally, this scale is modelled as a *fitness function* that associates a value to each state of the search space. This function can receive the name of *objective function* as well. By using a fitness function, the algorithm can compare the candidate solutions before selecting the most appropriate one. Usually, the objective of the optimization is to find either the global maximum or minimum value of the fitness function (depending on the application). It is a common problem of these algorithms to find the global maximum (or minimum), avoiding getting stuck in local maxima (minima), plateaus, or ridges [38]. The definition of the fitness function is part of the design of the optimization problem and greatly depends on the application (Figure 3.2).

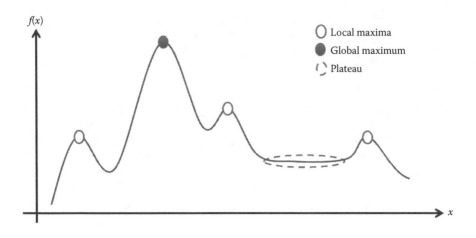

Figure 3.2 Generic fitness or objective function.

Trajectory-based algorithms iteratively evaluate potential candidate solutions. The trajectory described by the algorithm depends on the shape of the fitness function and the specific algorithm used. There are two main approaches to finish the optimization process. On the one hand, the algorithm can iterate until there is no improvement in the average quality of the solutions (average fitness value). On the other hand, the algorithm can also be implemented with a maximum number of iterations, stopping at this number whether there is still room for improvements or not. The approach selected depends on the type of the application for which the algorithm is used. We refer to both approaches as *stop criterion*. In the following sections we describe common trajectory-algorithms used in wireless multi-hop ad hoc networks problems in disaster scenarios.

3.3.1.1 Hill climbing

Hill climbing [38] is one of the simplest trajectory-based algorithms. The hill climbing algorithm is analogous to the behavior of a person climbing a mountain (in its uphill version); also it can resemble the behavior of the water that falls from the peak of a mountain (downhill version). In both situations, the agent (the person or the water) is trying to take the greatest slope in order to reach the highest (or the lowest) height, respectively, in the minimum amount of time. By "taking the greatest slope" we refer to select a neighbor of the current state of the problem in which the increase (or decrease) with respect to the parameter evaluated in the fitness function is the greatest. In Figure 3.3 a generic hill climbing algorithm pseudocode is shown in its uphill version (which seeks the maximum value of the fitness function).

Usually, the hill climbing algorithm is run several times using different initial conditions for each run (the initial state of the search space). These initial

```
Hill Climbing (HC)
1: Current_Position ← Start_Position
2: Best_Position ← Current_Position
3: While (StopCriterion == False):
4:     For (Neighbor_of(Current_position)):
5:         New_Position ← Evaluate(Neighbor_of(Current_Position))
6:         If (New_Position > Best_Position):
7:             Best_Position ← New_Position
8:         End if
9:     End For
10:    Iterations++
11: End While
```

Figure 3.3 Hill Climbing algorithm.

conditions are chosen randomly. In a single run, the initial conditions may lead the algorithm to get stuck in a local optimum. By choosing random initial conditions the algorithm is able to explore the search space and have higher probability of reaching the global optimum. This combination of local search and randomness, the two main components of metaheuristics [37], allows the algorithm to avoid getting stuck in local optima. The best candidate obtained among all the runs is selected as the optimal value.

3.3.1.2 Simulated annealing

Simulated annealing [37] is an abstraction of the metallurgy process known as the *annealing of metals*. In this process, a metal which has been shaped at high temperature is cooled gradually, spending a specific amount of time at different temperatures. This allows the material's internal structure to reach states with minimum energy and thus, having materials with fewer flaws and better performance. This process was adapted in 1953, from the mathematical point of view, in order to create an algorithm that is able to solve optimization problems. Originally it was known as the Metropolis algorithm [40].

Simulated annealing consists of an agent that combines local search operations within a specific area of the search space together with random jumps to other areas. This behavior allows the algorithm to escape from local optima. This is performed by using the Boltzmann probability, which is determined by Equation 3.1.

$$p = e^{-\Delta E / k_B T} \tag{3.1}$$

where

ΔE is the difference between the energy levels of the current and the previous state,

k_B is the Boltzmann's constant, and

T is the temperature that controls the annealing process.

For each iteration of the algorithm, the agent performing the search evaluates the objective function f; if the candidate solution evaluated is better than the previous one it is accepted. In contrast, if the agent finds a worse candidate solution then a random value is calculated and, if this random value is smaller than the Boltzmann probability, the candidate is accepted. On the contrary, in the case that the value is greater than the Boltzmann probability, the candidate is rejected. Accepting worse candidates as a solution represents the fact of jumping to other areas of the search space for escaping from local optima.

Another important feature is that with the decrease of the temperature T, the Boltzmann probability becomes smaller, and thus with the passing of the time (and the iterations), the chances for a worse candidate to become an accepted solution are fewer.

The initial values selection of the algorithm parameters is of paramount importance. Selecting proper values for the temperature T, the cooling rate, and the cooling schedule will affect the algorithm performance. Depending on these parameter values the algorithm may reach the final iteration with a Boltzmann probability of zero or not. Also, these parameters will determine if the algorithm has time to stabilize at each temperature T. In Figure 3.4 the pseudocode of a generic simulated annealing algorithm is shown.

Simulated Annealing (SA)
1: Current_Position ← Start_Position
2: Best_Position ← Current_Position
3: Temperature ← Initial_Temperature
4: While (StopCriterion == False):
5:     Repeat
6:        New_Position ← Evaluate(Current_Position, Random_Movement)
7:        If (New_Positon > Best_Position):
8:           Best_Position ← New_Position
9:        Else:
10:          Boltzmann_Probability ← Calculate_Boltzmann(Temperature, ΔE, k_B)
11:          Random_Number ← Generate_Random
12:          If (Boltzmann_Probability > Random_Number):
13:            Best_Position ← New_Position
14:          End if
15:        End if
16:     Until the Thermodynamic equilibrium is reached
17:     Temperature ← Update_Temperature
18:     Iterations++
19: End While

Figure 3.4 Simulated Annealing algorithm.

3.3.1.3 Tabu search

Tabu search [41] is a powerful metaheuristic that was first used by Fred Glover back in 1986. This algorithm has the property of using a memory structure for storing the already visited states. By checking this memory the algorithm avoids visiting again an already evaluated candidate state of the search space. It is based on the concepts of the human memory, ranging from short-term memory up to long-term memory.

The most common memory used in this algorithm is a short-term memory that stores a finite number of the most recently visited states of the search space. This avoids losing the latest improvements of the algorithm on the recent iterations. It is also common to include an intermediate-term memory structure with the aim of intensifying the exploration in specific areas of the search space. Also, a long-term memory can be considered in order to leverage a broader exploration of the search space [41]. This is similar to the approach of the simulated annealing algorithm for escaping the local optima.

Usually, the idea behind the Tabu search is to integrate it with other heuristics. This hybrid approach provides other algorithms, such as hill climbing algorithm or simulated annealing, the ability to avoid revisiting already explored areas. Figure 3.5 shows an example of the Tabu search algorithm integrated with hill climbing.

3.3.2 Population-based algorithms

Population-based algorithms are characterized by a set of particles or agents that explore the search space at the same time. In these algorithms there is not a single

Tabu Search (TS)
1: Current_Position ← Start_Position
2: Best_Position ← Current_Position
3: Tabu_list ← Current_Position
3: While (StopCriterion == False):
4:     For (Neighbor_of(Current_position)):
5:         If Neighbor is not in Tabu_list:
6:             New_Position ← Evaluate(Neighbor_of(Current_Position))
7:             If (New_Position > Best_Position):
8:                 Best_Position ← New_Position
9:             End if
10:            Tabu_list ← New_Position
11:        End if
12:    End For
13:    Iterations++
14: End While

Figure 3.5 Tabu Search algorithm.

trajectory, but there are a set of agents and each one describes a different trajectory in parallel. As it occurred with the trajectory-based methods, the population-based algorithms normally use a fitness or objective function in order to measure the quality of the candidate solutions. Population-based algorithms also use similar *stop criteria* as the ones described for trajectory-based algorithms.

As we have several agents, we can have different approaches in order to find the optimal solution to the problem that we are dealing with. A possible approach would be to consider that each agent looks for its best local solution, not considering the quality of the other agents. The drawback to this approach is if the agents act independently from each other they are not making the most of the potential of the population knowledge (e.g., sharing the best quality solutions with other agents).

Another approach would be to let each agent seek its local best but being influenced by the agents that had the best quality in the previous iterations of the algorithms; this is the basic idea of the Evolutionary algorithms (EAs) that we describe in the following section. Another option is to make the agents follow simple rules of interaction between each other and let them organize according to these interactions in order to find the best quality solutions; we can include in this category the swarming or swarm intelligence algorithms.

3.3.2.1 Evolutionary algorithms

EAs propose to use nature-inspired evolving strategies for solving complex problems. EAs are based on the Darwinian theory of evolution [42], which describes the capacity of biological systems to modify their genetic material to adapt to a changing environment and ensure their survival. EAs are iterative heuristics that evolve a set of candidate solutions, represented as individuals that are grouped in a population [43]. Figure 3.6 represents a basic layout of an evolutionary algorithm implementation.

Evolutionay Algorithm (EA)
1: InitPopulation(Population)
2: Evaluate(Population)
3: While (StopCriterion == False) do
4:       Parents ← ParentSelection(Population)
5:       NewPopulation← GeneticOperators(Parents, Population)
6:       Evaluate(NewPopulation)
7:       Population ← NewPopulation
8: End While
9: BestIndividual← FindBest(Population)
10: Return BestIndividual

Figure 3.6 Evolutionary algorithm.

According to Ref. [39] Genetic algorithms are one of the most used evolutionary strategies. The basic idea of a Genetic algorithm is that a population composed of potential solutions, namely individuals, evolves over time trying to generate the best solutions based on the previous solutions. Figure 3.7 illustrates the idea of the evolution of different generations (i.e., a Genetic algorithm that has *g* generations, and each generation is composed of *n* individuals, where *g* and *n* are design parameters. After the *g* generation, the resulting population is composed of the best individuals (solutions) found along the execution of the Genetic algorithm).

For example, in a scenario in which we seek to deploy several nodes in a disaster area to play the role of communications access points, we can think of the solutions such as the positions of the access points (nodes) in the disaster scenario *P*. The Genetic algorithm begins with an initial population that is selected randomly. Each individual of the initial population represents a different potential solution. In the target optimization problem, a potential solution is given by the nodes' coordinates in the disaster area. This is called chromosome structure, containing the genetic information of the individual or potential solution. For example, for *N* nodes, a potential solution representation is like the one shown in Figure 3.8, where *Xi*, *Yi* represent the *x* and *y* coordinates of an access point *i*.

As a summary, we can say that: (1) a chromosome is usually represented as a string of bits and is divided into a series of genes, and (2) a gene represents a value for one or more features in the problem.

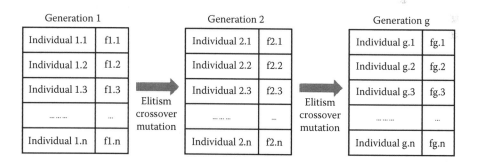

Figure 3.7 Evolution of Genetic algorithms.

| *X*1, *Y*1 | *X*2, *Y*2 | *X*3, *Y*3 | . . . | *Xn*, *Yn* |

Figure 3.8 Genetic information of an individual or potential solution.

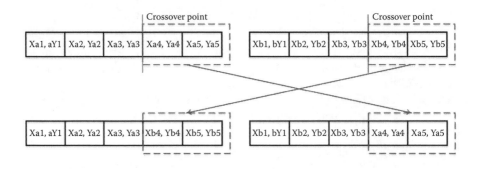

Figure 3.9 Single crossover operation.

Once we have generated the initial population, it is evaluated using a fitness function that determines the quality of a given solution. It means that each individual is evaluated by a fitness function. The fitness function is specifically defined for each optimization problem.

Then, a number of parents are chosen by a selection procedure. The main genetic operators to create new individuals are applied over the selected parents such as crossover and mutation. By crossover operation, the genetic information of the two parents (solutions) is swapped. There are many different crossover operators such as one-point crossover, two-points crossover, uniform, and others. Figure 3.9 illustrates a single crossover operation between two selected parents for an initial optimization problem with potential solutions composed of five nodes of the aforementioned problem. The crossover point is selected randomly.

The mutation operator consists of modifying the genetic information of an individual. Again, different mutation operators can be applied such as Gaussian and shuffle indexes, among others [44]. Figure 3.10 illustrates the mutation operation over a selected parent. Only part of the individual is modified to generate a new one.

Both crossover and mutation operators are applied according to certain probabilities p_c and p_b. The main objective of crossover is to combine the genetic information of two individuals to determine whether the solution is improved or not based on the existing solution. On the contrary, the main goal of mutation is to try to explore new locations in the search space, avoiding the evolution of the population to get stuck in local optimal values. Once the composition of the new population is determined by applying the genetic operators, the new population is evaluated. After that, if the stopping criterion is not reached, a new iteration is started.

3.3.2.2   Swarm algorithms

The terms *swarming, swarm intelligence,* or *swarm behavior* are normally used to refer to algorithms that mimic the behavior of groups of animals such as birds,

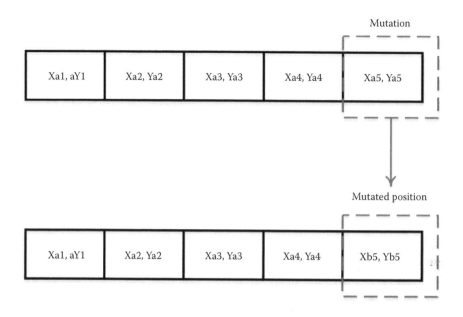

Figure 3.10 Mutation operation.

insects, or others. Thus, these algorithms consist of a group of agents. The behavior of each agent is usually represented by a set of simple rules. These rules define how each agent interacts locally with its closest neighbors in the swarm. Usually a single agent cannot perform complex tasks. However, the addition of all agents' interactions normally yields an efficient collective behavior able to carry out complex activities. This collaboration among agents is the main potential of the swarm intelligence.

There are many algorithms that can be classified as swarm algorithms, almost as many as types of animals that live in groups. Some of the best known are Particle Swarm Optimization (PSO) and Ant Colony Optimization (ACO), among others.

Particle Swarm Optimization

The PSO algorithm [37] was first presented by Kennedy and Eberhart in 1995 [45]. The PSO algorithm is not related to a specific type of animal; on the contrary, it resembles the common behavior of different animals such as bird flocks or fish schools. In the PSO algorithm, each of the agents is called *particle*. Each particle is a candidate solution of the optimization problem and it updates its position in each of the algorithm iterations. The position of each particle is updated according to Equation 3.2.

$$x_i(t+1) = x_i(t) + v_i(t+1) \tag{3.2}$$

In Equation 3.2, the position of particle i is calculated according to its current position, denoted by $x_i(t)$, and a velocity component $v_i(t+1)$. The particles' velocity is ruled by several factors. These usually are: (1) local best, (2) global best, and (3) random movements. Due to the *local best factor* each particle is attracted towards its own best value within the search space. This local best value is selected among the past values of the fitness function during the previous iterations of the algorithm. This factor favors visiting local optima in a limited region of the search space.

The *global best factor* can be defined as the best value of all the local best values. It also represents an attraction point for each particle movement. In order to be able to know the global best value, the PSO algorithm assumes that there is global communication among all the particles. In some implementations of the algorithm, this global best can be selected from the entire population of particles, or otherwise it can be selected among the group formed by each particle and its neighbors.

The *random movements factor* provides a random component to the movement vector of each particle, which is different for each of the iterations. This component enables the exploration of undiscovered areas and allows the algorithm's escape from local optima.

According to the following terms, the velocity equation for each particle is as shown in Equation 3.3:

$$v_i(t+1) = v_i(t) + C_1 \varphi_1 (P_i(t) - X_i(t)) + C_2 \varphi_2 (P_{gi}(t) - X_i(t)) \qquad (3.3)$$

where the terms are as follows:

$v_i(t)$: Current velocity.

C_1 and C_2: Constants representing the intensity of the attraction of a particle towards its local best (C_1) or the global best (C_2).

φ_1 and φ_2: Random values that represent the exploration and diversity component of the algorithm. They usually follow a uniform distribution within the range [0,1]

$P_i(t)$: The local best of particle i.

$P_{gi}(t)$: The global best of particle i.

Both Equations 3.2 and 3.3 are defined for one dimension. Usually the types of problems encountered in wireless multi-hop ad hoc networks for disaster scenarios have at least two dimensions. Thus, there will be a pair of Equations 3.2 and 3.3 for each dimension. In Figure 3.11 a generic PSO algorithm is shown (in this case with the aim of minimizing the fitness function).

Ant colony optimization

The ACO algorithm was first proposed by Marco Dorigo as part of his PhD dissertation back in 1992 [46]. ACO is based on the behavior of ants when they are looking for food. Real ants use positive feedback (pheromones) to mark the path they followed from the nest up to a food source and also from the food source

Particle Swarm Optimization Algorithm (PSO)
1: InitPopulation(Population)
2: While (StopCriterion == False) do
3:    for each particle i do
4:        Velocity ← CalculateVelocity(particle i)
5:        NewPosition ← CalculatePosition(CurrentPosition, Velocity)
6:        EvaluateFitness(NewPosition)
7:        Find current best for particle i
8:    end
9:    Find global best
10: end While
11: Output(Current best for each particle, Global best)

Figure 3.11 Particle Swarm Optimization Algorithm.

back to the nest. After some time passes since an ant has laid its pheromones, these start to evaporate. Initially, when no pheromones are laid on any route, the ants start looking for food selecting random directions. When the ants start finding food, several paths could be marked with pheromones as different ants might have followed different routes towards the food source. Over time, some paths become the preferred ones with a more intense pheromone trail.

Usually this algorithm is based on a graph $G = (V, E)$ where the vertices V represent the possible positions of the ants and the edges E the paths between positions. When an ant is located at a position i, this ant usually can choose among a set of possible positions to move to, represented by n_d, for its next movement. The probability of an ant to move to a position j when it is located at position i (i.e., to select a specific route) is defined by Equation 3.4.

$$p_{ij} = \frac{\phi_{ij}^{\alpha} d_{ij}^{\beta}}{\sum_{i,j=1}^{n_d} \phi_{ij}^{\alpha} d_{ij}^{\beta}} \tag{3.4}$$

where the terms are defined as follows:

α, β: These are the influence parameters ($\alpha > 0$, $\beta > 0$) and are usually given a typical value $\alpha \approx \beta \approx 2$; in the case of $\alpha = \beta = 1$ the ants will follow the routes with higher concentration of pheromones,

ϕ_{ij}: The pheromone concentration on the route between the position i to position j,

d_{ij}: The desirability to take the route between the position i to position j; a common representation of this is usually based on the distance between the positions i and j. This implies that selecting shorter routes will have priority over selecting longer ones (assuming that the distances between the positions are known in advance).

As mentioned before, the less-traveled paths will gradually decrease their pheromone levels due to the evaporation effect. This fact favors the routes that

Ant Colony Optimization (ACO)
1: Init_Pheromones
2: Define_Pheromone_Evaporation_Rate
3: While (StopCriterion == False) do
4:    for each ant k do
5:        Generate_New_Solutions (ant k path)
6:        Evaluate_New_Solutions
7:        Update_Pheromone
8:    end
9: Find the current best
10: end While
11: Output(Best_Results, Pheromone_Distribution)

Figure 3.12 Ant Colony Optimization Algorithm.

are more efficient or shorter for finding food. The evaporation effect is given by Equation 3.5.

$$\phi(t) = \phi_0 e^{-\gamma t} \tag{3.5}$$

where the terms are defined as follows:

ϕ_0: The initial concentration of pheromones,

γ: A constant representing the pheromone decay rate or evaporation rate,

t: The time.

A possible simplification of the pheromone evaporation formula can be applied when, $\gamma t \ll 1$, as the Equation 3.5 can be substituted by Equation 3.6.

$$\phi(t) \approx (1 - \gamma t)\phi_0 \tag{3.6}$$

In Equation 3.6, if the time increment is considered to be unitary (i.e., $\Delta t = 1$) then the pheromone update formula can be represented by Equation 3.7.

$$\phi_{ij}^{t+1} \approx (1 - \gamma)\phi_{ij}^t + \delta\phi_{ij}^t \tag{3.7}$$

where the terms are defined as follows:

$\gamma \in [0, 1]$: The constant representing the evaporation rate,

$\delta\phi_{ij}^t$: An increment representing the amount of pheromone deposited at time t, when an ant travels a distance L when moving from the position i to position j; usually this increment is $\delta\phi_{ij}^t \propto 1/L$.

A generic implementation of the ACO algorithm is shown in Figure 3.12.

3.4 Nature Inspired Algorithms in Disaster Scenarios

Nature inspired algorithms have been applied to optimize the different problems described in Section 2.2 for wireless multi-hop ad hoc networks in disaster scenarios: (1) the nodes' placement, (2) the packets routing problem, and (3) the nodes' mobility to adapt to changing conditions of the scenario. In this section, we review different works found in the literature, grouped by the type of nature inspired algorithm used.

3.4.1 Trajectory-based algorithms applications

In Ref. [12] a group of UAVs is used to provide communication services to victims and first responders (ground nodes) over a disaster scenario area. The UAVs in the AANET self-deploy in order to have as many ground nodes as possible under their coverage areas. UAVs movements follow a hybrid strategy based on GPS and a dissimilarity metric (i.e., the Jaccard distance). The Jaccard distance $J_d \in [0, 1]$ measures the dissimilarity of two sets. A set is represented by a UAV and the ground nodes under its coverage area (i.e., its serviced ground nodes). When $J_d \approx 0$, it means that two UAVs are very similar (which means that they are servicing to almost the same group of ground nodes). When $J_d \approx 1$, it means that two UAVs are very dissimilar (they share only a few or no ground nodes). The UAV movement strategy is based on a *virtual spring* model. If two UAVs are very similar they try to separate from each other with a force proportional to the current Jaccard distance that they have. On the contrary, if they are very dissimilar they try not to separate more from each other so they do not lose the connectivity with each other. The aim is to find the optimum Jaccard distance (to be maintained between each couple of UAVs) that maximizes the total number of serviced ground nodes. Two trajectory-based algorithms are used for this purpose, hill climbing and simulated annealing. There is one constraint; the algorithms have to keep each UAV connected at least to another UAV of the AANET (i.e., they have to maintain a connected network). The UAVs adapt their movements to the scenario changing conditions (i.e., the victims' mobility). Thus, [12] can be also classified as a nodes' mobility problem.

In Ref. [47] several trajectory based algorithms are used to optimally place a set of autonomous robots. The autonomous robots form a support network that intends to improve the connectivity of an ad hoc network. The nodes of the generic ad hoc network are called *general nodes*. The autonomous robots are called *support nodes* and act as relay nodes between the general nodes that have suffered from a communication link failure. A static objective function is defined that evaluates the maximum number of covered broken links. By using a static objective function, the nodes are considered static. Thus, this problem can be considered as a node's placement problem. Also, a dynamic objective function is considered, which takes into account the mobility of the general nodes. Thus, the support nodes also move in order to adapt to general nodes' mobility. Hill climbing and Tabu search with a short-term memory are used for this optimization problem. The results show that both algorithms perform similarly. However, hill climbing was slightly faster than Tabu search. These two trajectory-based algorithms are also compared against a population-based algorithm, the PSO. PSO reaches solutions closer to the global optimum than trajectory-based approaches, but it is slower in terms of time.

WMNs can be used to improve Internet connectivity in disaster scenarios. The wireless mesh routers' positions have to be calculated carefully to maximize the network connectivity and performance. This type of problem can be considered a nodes' placement problem. In order to solve this problem, works such as [26,48,49] use Tabu search. The works [27,50,51] use simulated annealing. Hill climbing is used in Refs [49,52,53]. With regard to the packets routing in MANETs, other works such as [31,54] apply the Tabu search algorithm. In Ref. [31], Tabu search showed better results than simulated annealing and Genetic algorithms.

3.4.2 Population-based algorithms applications

3.4.2.1 Evolutionary algorithms applications

In Ref. [9] a MANET built among first responders is considered to be deployed over a disaster scenario. The easiest routing protocol for MANETs is based on the flooding strategy in which each node relays packets to all its neighbors. In practice, this strategy presents severe collision problems, high traffic, and high energy consumption. Probabilistic broadcasting strategies are proposed as a better alternative than flooding. In probabilistic broadcasting, a node will broadcast a received packet to all its neighbors according to a given forwarding probability. This probability can be fixed or dynamic. If it is dynamic it allows the routing technique to adapt to network requirements or to optimize some communications aspects. In Ref. [9], the aim is to optimize several metrics of a MANET by using probabilistic broadcasting, namely: (1) the reachability (which is the ratio of nodes receiving the broadcast packet to the total number of nodes in the network), (2) the total number of messages transmitted in the network, and (3) the delay or the total time elapsed since a packet is originated at the source node until the last receiving node receives it. Therefore, a multi-objective optimization is proposed in order to optimize the three metrics. Genetic algorithms and the Pareto front are used together in order to determine the optimal combinations of the metrics. This work addresses the routing problem.

In Ref. [55] Genetic algorithms are used in order to evolve the controllers of an AANET. The main objective is to connect several first responders who are performing rescue operations. Specific constraints of this work are that: (1) the vehicles used are fixed-wing micro aerial vehicles (MAVs) and have to stay in motion; (2) the MAVs do not have GPS sensors and only rely on heading sensors and communications with other MAVs to find their positions in the network; and (3) the first responders' location is not known in advance, only the location of the base where the UAVs are launched from and the disaster area. This work proposes Genetic algorithms for evolving the neural controllers that define the behavior of the UAVs. The aim is to find the controller that performs best. However, the solutions produced by the Genetic algorithms are not the final controllers implemented in the MAVs. On the contrary, the solution is reverse-engineered in order to understand the controller behavior and to capture its main functionalities. Then, these functionalities are transformed in simpler swarming rules that are implemented in the MAVs. This procedure allows using the benefits of Genetic algorithms for evolving controllers and then, the transformation into simple swarming rules enables using the controllers in broader types of scenarios and applications. This work can be classified as a nodes mobility problem.

In Ref. [25] the aim is to improve the connectivity of a MANET. First responders are MANET nodes coordinating rescue tasks. When all the nodes are ground nodes, the connectivity is severely affected by problems such as multi-path propagation, fading effects, noise, and others. Also, the nodes' mobility causes connectivity losses. Therefore, the deployment of auxiliary nodes for improving the MANET connectivity is proposed. Thus, this work can be classified as a nodes placement problem. These auxiliary nodes will be deployed at the first stages of the rescue operations on fixed locations and will keep the first responders connected to each other even if they are moving. The disaster scenario area is divided in a regular grid. Each of the vertices of this grid is a possible location for an auxiliary node. The number of auxiliary nodes is limited. Genetic algorithms are used to calculate the best auxiliary nodes' locations for maximizing the network connectivity. This strategy increased the MANET connectivity between 11% and 27% when considering different disaster scenarios' conditions.

Also, the nodes placement problem is addressed in Refs [6,52,56,57]. In Ref. [56], Genetic algorithms are used in order to reach a uniform distribution of MANET nodes in an unknown area. The aim is to maximize the area coverage with the nodes distribution. In Ref. [57], the initial deployment of a group of robots forming a MANET is implemented by using a hybrid version of PSO and evolutionary algorithms. In Refs [6,52], the routers placement problem related to a WMN is addressed by using Genetic algorithms. With regard to the ad hoc network packets routing, in Refs [58,59], evolutionary algorithms are used in order to solve the multicast routing problem in MANETs. Finally, the nodes mobility problem is addressed in Refs [60,61]. In Ref. [60], Genetic algorithms are used to adapt the MANET nodes' movements in unknown environments using only limited information from the nodes' neighbors. In Ref. [61], evolutionary algorithms are used in order to control the UAVs movements in search and rescue scenarios.

3.4.2.2 *Swarming algorithms applications*

In Ref. [36], a swarm of Micro Unmanned Aerial Vehicles (MUAVs) is used as an AANET. The MUAVs are given the mission to detect and track a plume of toxic gases. This is a node's mobility problem as described in Section 2.2, with exploration and coverage provision features. Different swarming algorithms are compared in this work. These algorithms aim to reach two main objectives: (1) from a macro perspective, reaching the greatest coverage, and (2) from a micro perspective, maintaining a distance between each pair of UAVs that guarantees the quality of the communications links. Thus, these swarming strategies can be defined as communication-aware mobility algorithms. The macro perspective aims to fulfill the primary mission target (e.g. detect and track the plume). The micro perspective aims to maintain the connectivity among MUAVs and solve connectivity loss problems. Although there is not a reference to a specific swarm algorithm, this consideration of macro and micro objectives is similar to the PSO approach, where the individuals follow the global and also the individual best solutions.

In Ref. [33] a set of 10–20 MAVs is used to create a Swarm of MAVs Network (SMAVNET). The MAVs do not rely on GPS systems, but they rely only on proprioceptive sensors and communications with other MAVs in order to calculate their positions. The ACO algorithm is used in order to explore the scenario and create a multi-hop communication link between the first responders and the command base. A set of MAVs is classified as node-MAVs, which position themselves creating a graph that is used to represent the different routes between the base and the first responders. These node-MAVs are also the holders of virtual pheromones. The rest of the nodes are ant-MAVs that rely on the node-MAVs to choose different routes, basing their decisions on the virtual pheromone levels. The results show that ACO allows the SMAVNET to create the communication links among the base and first responders effectively. We can classify [33] as a nodes placement problem, specifically when the SMAVNET is deploying. Also, as the SMAVNET nodes are continuously moving in order to enable the discovery of the first responders' location and connect with the base, this can be considered a nodes mobility problem with coverage and exploration features.

In Ref. [7] the AntHocNet algorithm performance is analyzed. The AntHocNet was first presented in Ref. [62], and is a routing protocol based on the ACO algorithm. AntHocNet has an adaptive approach that combines the characteristics of reactive and proactive protocols for MANETs. On its reactive side, AntHocNet has agents, called *reactive forward ants*, that perform on-demand paths setup to specific destinations, mostly at the time of starting a communication session. These routes are stored in tables that associate a specific pheromone level (quality) to each path. On the proactive side, a set of agents, called *proactive forward ants*, performs, several tasks for monitoring, maintaining, and improving the paths stored in the pheromone tables. These proactive tasks are done only during the time that a session is open. This hybrid approach is a specific characteristic of AntHocNet because other nature inspired routing solutions were focusing only on a proactive scheme. The simulation results demonstrate that AntHocNet provides better performance than Ad hoc On Demand Vector routing protocol

(AODV) in several metrics such as delay, throughput, and packet delivery ratio, among others.

Also, the nodes placement problem is addressed by using PSO in works such as [47,57,63]. In Ref. [57], the initial deployment of a group of robots forming a MANET is implemented in a complex scenario (e.g., search and rescue operations). In Ref. [47], the aim is to place optimally a set of autonomous robots. These robots create a support network for improving the connectivity in an ad hoc network in which nodes are considered static under certain conditions. In Ref. [63], the wireless routers' positions of a WMN are calculated in order to optimize the network performance. The nodes mobility problem is addressed by Refs [35,47,64]. In Ref. [47], PSO is used in order to adapt a set of autonomous robots positions for improving the connectivity in a MANET. In Ref. [35], a group of UAVs is used to detect and track a chemical plume. A set of simple swarm rules is used in order to control the UAVs' movements. In Ref. [64], a swarm strategy based on bird flock behavior is implemented. The aim is to use a set of intelligent swarm agents in order to improve a generic MANET connectivity and avoid communication bottlenecks. Finally, works such as [65–68] address the packets routing problem in MANETs. In Ref. [66], a clustering and routing algorithm based on bird flocking behavior is proposed for enhancing MANET performance, scalability, and energy consumption. In Ref. [67], the Termite Hill Building algorithm is used in order to improve several network parameters like the throughput and the delay. The Termite algorithm is similar to the ACO as the termites also use pheromone trails. In Ref. [68], different routing techniques are proposed for solving the routing problem in MANETs. These techniques are ACO, Bee Colony Optimization, and Termite Hill Building algorithm. In other works swarm algorithms are used to improve several characteristics of well-known MANETs routing protocols such as AODV. An example is the Multi-Route AODV Ant routing [65], which uses ACO in order to reduce the routing overhead, buffer overflow, and end-to-end delay.

As we have seen in Section 3.4, several nature inspired algorithms have been applied to wireless multi-hop ad hoc networks for disaster scenarios. In Table 3.3 the works described in this section are included, organized according to the problem that they aim to solve, and the type of nature inspired algorithm used.

3.5   Lessons Learned and Open Challenges

As a result of the study of the literature, several lessons learned have been gathered and are described in this section. Wireless multi-hop ad hoc networks is a multidisciplinary research area covering topics from network theory, robotics, and computer science, among others. Several topics of these scientific fields have been optimized, as has been shown in this chapter. However, there are other aspects that can be optimized by using nature inspired algorithms. It is important

Table 3.3 **Nature inspired algorithms works on wireless multi-hop ad hoc networks for disaster scenarios**

	Hill climbing	Simulated annealing	Tabu search	Evolutionary	Swarm
Nodes deployment	[47,49,52,53]	[27,50,51]	[26,47–49]	[6,25,52,56,57]	[33,47,57,63]
Routing		[31]	[31,54]	[9,32,58,59]	[7,65–68]
Nodes mobility	[12,47]	[12]	[47]	[55,60,61]	[33,35,36,47,64]

to notice that wireless multi-hop ad hoc networks is a relatively young research area. For this reason, we also include in this section the open challenges that are yet to be addressed. These open challenges are potential research areas to be explored with the help of nature inspired algorithms.

3.5.1   *Lessons learned*

We classify the lessons learned in several categories: those related to (1) the nodes placement problem, (2) the routing problem, (3) the nodes mobility problem, and (4) others related to the disaster scenario models. With respect to the nodes deployment problem, there are works using trajectory-based, evolutionary, and swarm algorithms. It is more common to find trajectory-based algorithms used for placing the nodes in WMNs, as for example [47,50]. Evolutionary and swarm strategies also have been used for placing wireless mesh routers. Also, population-based algorithms have been used for optimizing the initial deployment of MANET's nodes in unknown scenarios, as in Ref. [33]. Sometimes, the population-based algorithms that address this initial deployment are followed by a second stage in which they implement an adaptive behavior to the scenario conditions (e.g., first responders' mobility). This can be considered as a node mobility problem. Also, some of the trajectory-based approaches are used as a reference for comparing newer evolutionary and swarm approaches against them.

Regarding the routing problem, it is unusual to find trajectory-based algorithms by themselves. Instead, the trajectory-based algorithms are usually part of other strategies for optimizing specific aspects of a bigger algorithm. An example is described in Ref. [69] for the Multiphase Discrete Particle Swarm Optimization routing protocol. Other works use trajectory-based algorithms as a reference for comparing them against newer algorithms [31]. Despite this, there are some recent works that use pure trajectory-based algorithms in routing problems such as in Ref. [31], but the majority of them date from about a decade ago. As shown in Table 3.3, there are plenty of works in the literature addressing the routing problem by using evolutionary or swarm algorithms.

In the case of the nodes mobility problem, the majority of the works make use of swarm strategies, which is quite logical as normally these are based in the movement of groups of animals. Evolutionary approaches are also used, but the number of works is fewer than those using swarm algorithms. Although there are some works using trajectory-based algorithms for addressing the nodes mobility problem, such as [12], the majority used them as a reference for comparison purposes as well.

With respect to the disaster scenario mobility models, the works described show that there is a tendency of increasing the level of detail for mimicking the human behavior in rescue and emergency operations. The majority of works focus on modelling the human mobility of rescue workers, and some of them include a little bit on the mobility of victims. This is a logical approach, since in the early years of this research area the rescue workers were the only ones with wireless communications devices. However, nowadays almost every person has a smartphone that can be used for communications in disaster scenarios [34].

3.5.2 Open challenges

Regarding the open challenges, we mention in this section several aspects that we consider unexplored areas. First, most of the works described in this chapter have treated the three problems mentioned in Section 2.2 as single-objective optimization problems. Single-objective optimization aims to find the optimal value of only one variable. However, in wireless multi-hop ad hoc networks there are usually several parameters that can be optimized jointly in order to understand the effects of one parameter onto another. Only a few works, such as [9], have addressed multi-objective optimization problems. Therefore, there is yet an open research area with respect to multi-objective optimization for ad hoc networks.

Some of the works that we already described implement a centralized-like optimization. These works can be applied to realistic situations only under certain conditions; for example, if the disaster scenario is known in advance, the communication between nodes is ideal, and other considerations. However, it is very difficult to guarantee that these conditions will be met if a real disaster scenario occurs. Due to that, these conditions are not realistic. Thus, adapting these centralized strategies to more distributed nature inspired algorithms, making use of local, rather than global, information, is yet an area to exploit.

Moreover, other nature inspired techniques have not been commonly applied to wireless multi-hop ad hoc networks. Some of these techniques are Cooperative Coevolution [70] and Genetic Programming [71]. There is the challenge in checking if these techniques can outperform the ones presented in this chapter.

Also, the majority of the works have been carried out by using computer simulations. This has permitted the development of lots of cost-effective research works in this research field. This is logical as it is not feasible to emulate a real disaster scenario with all of its details and uncertainty. Also, the costs of building significant testbeds, with real wireless communications devices and mobility conditions, are high. However, the latest advancements in electronics are making it possible to build tinier wireless devices and to embed them in small robots. For this reason, in a few years it will not be strange to build realistic testbeds with very small and real communications. Building these testbeds and testing all the nature inspired algorithms in more realistic testbeds is also an open challenge.

With respect to the disaster scenario models, only a few of the works have modelled the victims' mobility such as in [12,18]. A complete and detailed victims' mobility model, emulating the behavior of victims in a disaster scenario, has not been proposed yet. Victims' movements could be modeled by using swarm rules adapted to the human behavior when facing an accident. Thus, there is an important open challenge in the application of nature inspired algorithms for generating the victims' behavior in disaster scenarios. There are very few works in this area (maybe none) and the scalability advantages and local interactions of nature inspired algorithms may provide to victims' mobility models the realism that they do not have yet.

3.6 Conclusions

Disaster scenarios present complex problems for wireless multi-hop ad hoc networks that can be treated as optimization problems. The main problems addressed in this field can be classified in three categories: (1) the nodes placement problem, (2) the packets routing problem, and (3) the nodes' mobility problem. In order to solve these problems, different nature inspired strategies have been used. These strategies can be classified as trajectory-based and population-based. The literature reviewed has shown that there is not an algorithm that can be classified as the best one over the others. The performance of an algorithm strongly depends on the type of problem, the scenario conditions, and the variables to be optimized, among other considerations. However, nature inspired algorithms have shown many advantages for solving the types of problems described in this chapter. Wireless multi-hop ad hoc networks are decentralized systems by nature and also consist of many nodes. For this reason, nature inspired algorithms are a great candidate for these types of problems and their potential in this area has not been fully realized yet. There are new nature inspired algorithms to apply in this research field, other aspects to optimize, and other ways to address these problems (e.g., such as multi-objective optimization problems). Thus, we expect that this chapter serves as an incentive for leveraging more research in the field of nature inspired algorithms applied to wireless multi-hop ad hoc networks for disaster scenarios.

Bibliography

[1] P. C. Vinh, Concurrency of self-* in autonomic systems, *Future Generation Computer Systems*, vol. 56, pp. 140–152, 2016.

[2] F. Dressler, I. Dietricha, R. Germana and B. Krger, A rule-based system for programming self-organized sensor and actors networks, *Computer Networks*, vol. 53, no. 10, pp. 1737–1750, 2009.

[3] D. G. Reina, P. Ruiz, R. Ciobanu, S. L. Toral, B. Dorronsoro and C. Dobre, A survey on the application of evolutionary algorithms for mobile multi-hop ad hoc network optimization problems, *International Journal of Distributed Sensor Networks*, vol. 2016, Article ID 2082496, 13 p, 2016.

[4] D. G. Reina, M. Askalani, S. L. Toral, F. Barrero, E. Asimakopoulou and N. Bessis, A survey on multi-hop ad hoc networks for disaster response

scenarios, *International Journal of Distributed Sensor Networks*, vol. 2015, Article ID 647037, 16 p, 2015.

[5] M. Conti and S. Giordano, Multihop ad hoc networking: the theory, *IEEE Communications Magazine*, vol. 45, no. 4, pp. 78–86, 2007.

[6] A. Barolli, T. Oda, M. Ikeda, L. Barolli, F. Xhafa and V. Loia, Node placement for wireless mesh networks: analysis of WMN-GA system simulation results for different parameters and distributions, *Journal of Computer and System Sciences*, vol. 81, no. 8, pp. 1496–1507, 2015.

[7] A. P. Patil, K. Rajanikant and H. P. Rakshith, Analyzing the performance of AntHocNet protocol for MANETs, *International Journal of Computer Applications*, vol. 57, no. 5, pp. 20–25, 2012.

[8] S. Hauert, S. Leven, J. C. Zufferey and D. Floreano, Communication-based swarming for flying robots, *IEEE International Conference on Robotics and Automation, Workshop on Network Science and Systems Issues in Multi-Robot Autonomy*, Anchorage, AK, 2010.

[9] D. G. Reina, J. M. León-Coca, S. L. Toral, E. Asimakopoulou, F. Barrero, P. Norrington and N. Bessis, Multi-objective performance optimization of a probabilistic similarity/dissimilarity-based broadcasting scheme for mobile ad hoc networks in disaster response scenarios, *Soft Computing*, vol. 18, no. 9, pp. 1745–1756, 2014.

[10] B. T. Sharef, R. A. Alsaqou and M. Ismail, Vehicular communication ad hoc routing protocols: a survey, *Journal of Network and Computer Applications*, vol. 40, pp. 363–396, 2014.

[11] J. M. León-Coca, D. G. Reina, S. L. Toral, F. Barrero and N. Bessis, Intelligent transportation systems and wireless access in vehicular environment technology for developing smart cities, In: Nik Bessis and Ciprian Dobre (eds). *Big Data and Internet of Things: A Roadmap for Smart Environments*, Springer, Cham, 2014, pp. 285–313.

[12] J. Sánchez-García, J. García-Campos, S. Toral, D. G. Reina and F. Barrero, An intelligent strategy for tactical movements of UAVs in disaster scenarios, *International Journal of Distributed Sensor Networks*, vol. 2016, Article ID 8132812, 20 p, 2016.

[13] R. Ciobanu, D. G. Reina, C. Dobre, S. Toral and P. Johnson, A history-based forwarding scheme for delay tolerant networks using Jaccard distance and encountered ration, *Journal of Network and Computer Applications*, vol. 40, pp. 279–291, 2014.

[14] D. Johnson and D. Maltz, Dynamic source routing in ad hoc wireless networks, *Mobile Computing*, vol. 1996, pp. 153–181, 1996.

[15] J. Yoon, M. Liu and B. Noble, Random waypoint considered harmful, *22th Annual Joint Conference of the IEEE Computer and Communications (INFOCOM)*, San Francisco, CA, 2003.

[16] N. Aschenbruck, E. Gerhards-Padilla and P. Martini, Modeling mobility in disaster area scenarios, *Performance Evaluation*, 66, pp. 77–790, 2009.

[17] N. Aschenbruck, R. Ernst, E. Gerhards-Padilla and M. Schwamborn, Bonn-Motion: a mobility scenario generation and analysis tool, *Proceedings of the 3rd International ICST Conference on Simulation Tools and Techniques (SIMUTools '10)*, Torremolinos, Malaga, Spain, 2010.

[18] D. Costantini, M. Munch, A. Leonardi, V. Rocha, P. Mogre and R. Steinmetz, Role-based urban post-disaster mobility model for search and rescue operations, *IEEE 37th Conference on Local Computer Networks Workshops*, Clearwater, FL, 2012.

[19] S. Pomportes, J. Tomasik and V. Vque, A composite mobility model for ad hoc networks in disaster areas, *Journal on Electronics and Communications*, vol. 1, no. 1, pp. 62–68, 2011.

[20] I. Rhee, M. Shin, S. Hong, K. Lee and S. Chong, On the Levy-Walk nature of human mobility, *The 27th IEEE Conference on Computer Communications (INFOCOM)*, Phoenix, AZ, 2008.

[21] X. Hong, M. Gerla, G. Pei and C.-C. Chiang, A group mobility model for ad hoc wireless networks, *Proceedings of the 2nd ACM International Workshop on Modeling, Analysis and Simulation of Wireless and Mobile Systems (MSWiM '99)*, New York, 1999.

[22] A. Jardosh, E. Belding-Royer, K. Almeroth and S. Suri, Towards realistic mobility models for mobile ad hoc networks, *9th Annual International Conference on Mobile Computing and Networking*, San Diego, CA, 2003.

[23] Y. Huang, W. He, K. Nahrstedt and W. Lee, CORPS: event-driven mobility model for first responders in incident scene, *IEEE Military Communications Conference*, San Diego, CA, 2008.

[24] L. Conceição and M. Curado, Modelling mobility based on human behaviour in disaster areas, *Wired/Wireless Internet Communication: Proceedings of the 11th International Conference of Wired/Wireless Internet Communication (WWIC)*, St. Petersburg, Russia, Springer Berlin Heidelberg, 2013, pp. 56–59.

[25] D. G. Reina, S. L. Toral, N. Bessis, F. Barrero and E. Asimakopoulou, An evolutionary computation approach for optimizing connectivity in disaster response scenarios, *Applied Soft Computing*, vol. 13, no. 2, pp. 833–845, 2013.

[26] F. Xhafa, C. Sánchez, A. Barolli and M. Takizawa, Solving mesh router nodes placement problem in wireless mesh networks by Tabu Search algorithm, *Journal of Computer and System Sciences*, vol. 81, no. 8, pp. 1417–1428, 2015.

[27] S. Sakamoto, A. Lala, T. Oda, V. Kolici, L. Barolli and F. Xhafa, Application of WMN-SA simulation system for node placement in wireless mesh networks: a case study for a realistic scenario, *International Journal of Mobile Computing and Multimedia Communications (IJMCMC)*, vol. 6, no. 2, pp. 13–21, 2014.

[28] A. Munjal, T. Camp and N. Aschenbruck, Changing trends in modeling mobility, *Journal of Electrical and Computer Engineering*, vol. 2012, Article ID 372572, 16 p, 2012.

[29] X. Jin, R. Zhang, J. Sun and Y. Zhang, TIGHT: a geographic routing protocol for cognitive radio mobile ad hoc networks, *IEEE Transactions on Wireless Communications*, vol. 13, no. 8, pp. 4670–4681, 2014.

[30] S. Mohseni, R. Hassan, A. Patel and R. Razali, Comparative review study of reactive and proactive routing protocols in MANETs, *IEEE International Conference on Digital Ecosystems and Technologies (DEST)*, Dubai, United Arab Emirates, 2010.

[31] K.-W. Jang, A Tabu Search algorithm for routing optimization in mobile ad-hoc networks, *Telecommunication Systems*, vol. 51, no. 2, pp. 177–191, 2011.

[32] M. Di Felice, A. Trotta, L. Bedogni, K. Chowdhury and L. Bononi, Self-organizing aerial mesh networks for emergency communication, *IEEE 25th Annual International Symposium on Personal, Indoor, and Mobile Radio Communication*, Washington DC, 2014.

[33] S. Hauert, L. Winkler, J.-C. Zufferey and D. Floreano, Ant-based swarming with positionless micro air vehicles, *Swarm Intelligence*, vol. 2, no. 2, pp. 167–188, 2008.

[34] J. Munoz-Castaner, P. Counago Soto, F. Gil-Castineira, F. Gonzalez-Castano, I. Ballesteros, A. di Giovanni and P. Colodron Villar, Your phone as a personal emergency beacon: a portable GSM base station to locate lost persons, *IEEE Industrial Electronics Magazine*, vol. 9, no. 4, pp. 49–57, 2015.

[35] M. Scheutz, P. Schermerhorn and P. Bauer, The utility of heterogeneous swarms of simple UAVs with limited sensory capacity in detection and tracking tasks, *Proceedings of IEEE Swarm Intelligence Symposium (SIS)*, Pasadena, CA, 2005.

[36] K. Daniel, S. Rohde, N. Goddemeier and C. Wietfeld, Cognitive agent mobility for aerial sensor networks, *IEEE Sensors Journal*, vol. 11, no. 11, pp. 2671–2682, 2011.

[37] X. Yang, Engineering optimization, *Engineering Optimization: An Introduction with Metaheuristic Applications*, John Wiley & Sons, Hoboken, NJ, 2010, pp. 19–22. http://eu.wiley.com/WileyCDA/WileyTitle/productCd-0470582464.html.

[38] S. J. Russel and P. Norvig, *Artificial Intelligence: A Modern Approach*, Pearson, Upper Saddle River, NJ, 2010.

[39] I. Boussada, J. Lepagnotb and P. Siarry, A survey on optimization metaheuristics, *Information Sciences*, vol. 237, pp. 82–117, 2013.

[40] N. Metropolis, A. W. Rosenbluth, M. N. Rosenbluth, A. H. Teller and E. Teller, Equation of state calculations by fast computing machines, *The Journal of Chemical Physics*, vol. 21, no. 6, pp. 1087–1092, 1953.

[41] J. Brownlee, *Cleveer Algorithms: Nature-Inspired Programming Recipes*, Lulu, Melbourne, Australia, 2011.

[42] C. Darwin, *On the Origin of Species by Means of Natural Selection, or the Preservation of Favoured Races in the Struggle for Life*, John Murray, London, 1859.

[43] B. Dorronsoro, P. Ruiz, G. Danoy, Y. Pigne and P. Bouvry, *Evolutionary Algorithms for Mobile Ad Hoc Networks*, John Wiley & Sons, Hoboken, NJ, 2014.

[44] DEAP Project, DEAP: distributed evolutionary algorithms in Python, 2014. Available: http://deap.gel.ulaval.ca/doc/default/api/tools.html#operators.

[45] J. Kennedy and R. Eberhart, Particle swarm optimization, *IEEE International Conference on Neural Networks*, Perth, WA, Australia, 1995.

[46] M. Dorigo, Optimization, learning and natural algorithms, PhD Thesis, Politecnico di Milano, 1992.

[47] G. Martins, M. Rutherford and K. Valavanis, Search methodologies for node recovery in robotic swarms, *Mediterranean Conference on Control and Automation (MED)*, Corfu, Greece, 2011.

[48] F. Xhafa, C. Sánchez, A. Barolli and M. Takizawa, A Tabu Search algorithm for efficient node placement in wireless mesh networks, *International Conference on Intelligent Networking and Collaborative Systems (INCoS)*, Fukuoka, Japan, 2011.

[49] T. Oda, D. Elmazi, A. Lala, V. Kolici, L. Barolli and F. Xhafa, Analysis of node placement in wireless mesh networks using Friedman test: a comparison study for Tabu Search and hill climbing, *International Conference on Innovative Mobile and Internet Services in Ubiquitous Computing (IMIS)*, Blumenau, Brazil, 2015.

[50] S. Sakamoto, E. Kulla, T. Oda, M. Ikeda, L. Barolli and F. Xhafa, Performance evaluation considering iterations per phase and SA temperature in WMN-SA system, *Mobile Information Systems*, vol. 10, no. 3, pp. 321–330, 2014.

[51] V. Kolici, T. Oda, A. Barolli, A. Lala, L. Barolli and F. Xhafa, Application of WMN-SA web interface and NS-3 for optimization and analysis in WMNs considering different number of mesh routers and architectures, *International Conference on Intelligent Networking and Collaborative Systems (INCOS)*, Taipei, Taiwan, 2015.

[52] D. Elmazi, T. Oda, A. Lala, V. Kolici, L. Barolli and F. Xhafa, Analysis of node placement in wireless mesh networks using Friedman test: a comparison study for genetic algorithms and hill climbing, *International Conference on Complex, Intelligent, and Software Intensive Systems (CISIS)*, Blumenau, Brazil, 2015.

[53] S. Sakamoto, A. Lala, T. Oda, V. Kolici, L. Barolli and F. Xhafa, Analysis of WMN-HC simulation system data using Friedman test, *International Conference on Complex, Intelligent, and Software Intensive Systems (CISIS)*, Blumenau, Brazil, 2015.

[54] P. Dahiya and R. Johari, VAST: volume adaptive searching technique for optimized routing in mobile ad-hoc networks, *IEEE International Advance Computing Conference (IACC)*, Gurgaon, India, 2014.

[55] S. Hauert, J.-C. Zufferey and D. Floreano, Evolved swarming without positioning information: an application in aerial communication relay, *Autonomous Robots*, vol. 26, no. 1, pp. 21–32, 2009.

[56] C. Sahin, M. Umit Uyar, S. Gundry and E. Urrea, Self organization for area coverage maximization and energy conservation in mobile ad hoc networks, *Transactions on Computational Science XV*, pp. 49–73, 2012.

[57] M. S. Couceiro, R. P. Rocha and N. M. F. Ferreira, Ensuring ad hoc connectivity in distributed search with Robotic Darwinian Particle Swarms, *IEEE International Symposium on Safety, Security, and Rescue Robotics (SSRR)*, Kyoto, Japan, 2011.

[58] Y. Liu and J. Huang, A novel fast multi-objective evolutionary algorithm for QoS multicast routing in MANET, *International Journal of Computational Intelligence Systems*, vol. 2, no. 3, pp. 288–297, 2009.

[59] J. Huang and Y. Liu, MOEAQ: a QoS-aware multicast routing algorithm for MANET, *Expert Systems with Applications*, vol. 37, no. 2, pp. 1391–1399, 2010.

[60] S. Gundry, J. Zou, E. Urrea, C. Sahin, J. Kusyk and M. Umit Uyar, Analysis of emergent behavior for GA-based topology control mechanism for self-spreading nodes in MANETs, *Advances in Intelligent Modelling and Simulation*, Berlin, Germany, Springer Berlin Heidelberg, 2012, pp. 155–183.

[61] G. Varela, P. Caamaño, F. Orjales, A. Deibe, F. López-Peña and R. J. Duro, Autonomous UAV based search operations using constrained sampling evolutionary algorithms, *Neurocomputing*, vol. 132, pp. 54–67, 2014.

[62] G. Caro, F. Ducatelle and L. Gambardella, AntHocNet: an ant-based hybrid routing algorithm for mobile ad hoc networks, *International Conference on Parallel Problem Solving from Nature (PPSN)*, Birmingham, UK, 2004.

[63] S. Sakamoto, T. Oda, M. Ikeda, L. Barolli and F. Xhafa, A PSO-based simulation system for node placement in wireless mesh networks: evaluation results for different replacement methods, *International Conference on Broadband and Wireless Computing, Communication and Applications (BWCCA)*, Kraków, Poland, 2015.

[64] A. Konak, G. E. Buchert and J. Juro, A flocking-based approach to maintain connectivity in mobile wireless ad hoc networks, *Applied Soft Computing*, vol. 13, no. 2, pp. 1284–1291, 2013.

[65] A. Abd Elmoniem, H. Ibrahim, M. Mohamed and A.-R. Hedar, Ant colony and load balancing optimizations for AODV routing, *International Journal of Sensor Networks and Data Communications*, vol. 1 (2012), Article ID X110203, 14 p, 2011.

[66] M. Tiwari and S. Varma, Bird flight inspired clustering based routing protocol for mobile ad hoc networks, *International Journal of Computer Science and Network Security*, vol. 10, no. 3, pp. 119–128, 2010.

[67] G. Sharvani, N. Cauvery and T. Rangaswamy, Adaptive routing algorithm for MANET: TERMITE, *International Journal of Next-Generation Networks*, vol. 1, no. 1, pp. 38–43, 2009.

[68] H. N. Saha, A. Chattopadhyay and D. Sarkar, Review on intelligent routing in MANET, *International Conference and Workshop on Computing and Communication (IEMCON)*, Vancouver, BC, Canada, 2015.

[69] Z. Ali and W. Shahzad, Critical analysis of swarm intelligence based routing protocols in adhoc and sensor wireless networks, *International Conference on Computer Networks and Information Technology (ICCNIT)*, Abbottabad, Pakistan, 2011.

[70] F. Wei, Y. Wang and T. Zong, A novel cooperative coevolution for large scale global optimization, *IEEE International Conference on Systems, Man, and Cybernetics (SMC)*, San Diego, CA, 2014.

[71] D. Lynch, M. Fenton, S. Kucera, H. Claussen and M. O'Neill, Scheduling in heterogeneous networks using grammar-based genetic programming, In: Malcolm I. Heywood, James McDermott, Mauro Castelli, Ernesto Costa, and Kevin Sim (eds). *Genetic Programming*, Springer, Cham, 2016, pp. 83–98.

Chapter 4

Introducing NovaGenesis as a Novel Distributed System-Based Convergent Information Architecture

Antonio Marcos Alberti and Marco Aurelio Favoreto Casaroli

Instituto Nacional de Telecomunicações (INATEL)

Rodrigo da Rosa Righi

Universidade do Vale do Rio dos Sinos

Dhananjay Singh

Hankuk University of Foreign Studies

CONTENTS

4.1 Introduction

The Internet is no longer a network of fixed computers among governmental institutions. It has invaded most aspects of our life and society, changing lifestyles, business practices, work relations between employers and employees, communication channels, and even social interactions. Technically, the Internet can be seen as a single communication subsystem providing communication among all of the hosts that are connected to it [1,2]. Modern Internet's facet concerns a vast collection of interconnected computer networks of many different types, including, for example, a wide range of wireless communication technologies such as wireless-fidelity (Wi-Fi), worldwide interoperability for microwave access (WiMAX), bluetooth, and fourth-generation mobile phone networks. Differences among the networks are masked by the fact that all of the computers attached to them use the Internet protocols to communicate with one another. While the usage of standardized and widespread adopted protocols is useful for connecting devices, the price for accomplishing portability is high [3–5]. The protocols of the TCP/IP reference model are organized in a single, monolithic stack, in which most of the layers are processed in software [6]. Although this model originally presents four layters, Figure 4.1 illustrates the five-layered common vision of both physical and virtual communication paths when using TCP/IP. TCP/IP layers and the Internet itself were designed in an era where technological development was completely different from today [7]. The computing and communications capacities were much more limited.

The current Internet is also a very large distributed system. It enables users, wherever they are, to make use of services such as the worldwide web (WWW), email, virtual conferences, and file transfer. We define a distributed system as one

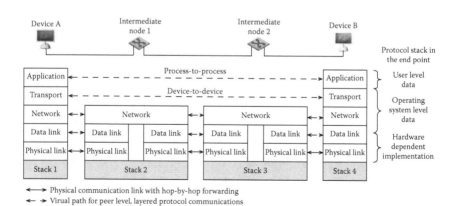

Figure 4.1 TCP/IP layered protocol communication between two end point devices. Functionally, intermediate nodes only require the bottom three layers of the reference model.

in which hardware or software components located at networked computers communicate and coordinate their actions only by passing messages [8,9]. They can incorporate large numbers of nodes and provide distributed system services for global organizations. Building on this foundation, larger-scale distributed systems started to emerge in the 1990s in response to the dramatic growth of the Internet during this time. More recently, it is possible to observe some trends such as the extensive use of both cloud computing tools and mobile commerce environments, as well as the diffusion of big data as a mandatory science for successful business companies [10–12]. The level of heterogeneity in such systems is significant in terms of networks, computer architecture, operating systems, programming languages, and the development teams involved. This has led to an increasing emphasis on open standards and associated middleware technologies. Furthermore, performance and the notion of transparency (at location, mobility, scaling, and access levels) are more and more crucial for adopting or not adopting newer distributed systems technologies [8,13].

As the Internet was opened to the general public and began to be used for an increasingly diverse mix of applications, a complex agglomerate of patchwork solutions was applied to extend its scope – creating some inconsistencies that are now questioned. Thus, the Internet scales and roles have changed considerably from its original purposes [1,14]. Many researchers started to wonder whether the current stack can support the multifaceted exponential growths we are experiencing on a number of devices, mobility, interactivity, content, and traffic. Concerned with this situation, several initiatives emerged worldwide to reshape the Internet under the banner of the so called *"future Internet"* (FI) design [15–17]. The term was adopted in the first initiatives aimed at rethinking the Internet, including the 4D project [18], the future Internet design initiative [19], the global environment for network innovations [20], and the European future Internet assembly (FIA) initiative [21]. By FI, we mean any Internet-like network that could emerge in the future. This includes evolutionary approaches, where the fundamental protocols of the current Internet are maintained and where new ideas are introduced incrementally, or revolutionary approaches, where the architecture is redesigned from scratch.

The revolutionary approaches are usually called "clean-slate" architectures [22]. For many researchers, the current Internet is a complex agglomerate of protocols that inherits the grown legacies of decades of patchwork solutions. New layers have been added to support novel features, significantly changing the original purposes. Protocols were added to extend the architecture towards new usages, creating unexpected interactions with other protocols. Additionally, the Internet was designed in an era where technological development was completely different from today. Computing and communication hardware capacities were orders of magnitude less capable than today. In this scenario, many researchers have questioned the adequacy of the current Internet architecture to meet our society's needs. They defend a more deep redesign—a revolution in Internet architecture. The argument for a more deep redesign is that some limitations on the Internet are fundamentally architectural. They can not be definitely solved by incremental patches. Thus, clean slate architectures have the potential to disrupt current technologies and their limitations, providing new environments for technology evolution. Since every technology has a saturation point, clean slate proponents believe that original Internet design has reached its saturation point and it is time for renovation.

The main benefits one can expect from a clean slate Internet are: (1) better efficiency, since new protocols could adapt themselves to fit exactly on the service requirements; (2) more flexibility, mainly because of the improved role of software in architectures; (3) simplicity, by eliminating the edges left by decades of incremental work; (4) evolvability, by designing for evolvable architectures rather than for a fixed set of protocols; and (5) the cohesive integration of essential abstractions, eliminating unnecessary duplications and generalizing solutions for problems that repeat themselves at different levels. The current Internet was itself a clean slate approach in the beginning, running over telephonic networks. Now, the large majority of telephony is Internet-based. This is a good argument that favors clean slate designs: they can accommodate previous designs in a more flexible, adaptable, and essential way, reducing costs and accelerating technology evolution.

In this chapter, we propose the concept of a clean slate *convergent information architecture*, which covers information processing, storage, exchanging, and visualization. Its scope is not limited to the original Internet scope: a robust network with flexible applications to exchange information among computers. Instead, the convergent information architecture embraces not only networking aspects, but also computing, storage, and visualization. Thus, it looks to redesign both communication mechanisms and protocol implementations for enabling a program running anywhere to address messages to programs anywhere else with acceptable performance and portability levels. Also, it looks for the convergence of computing and communications, merging technologies like cloud computing, mobile computing, Internet of things (IoT), service frameworks, and distributed systems. Its scope is so broad as the notion of where information can be processed, exchanged, stored, or visualized. Therefore, to design a convergent information architecture is more broad than designing protocols and their stack. It includes the computing, storage, and services/applications aspects. Ideally, according to the algorithmic information theory [23], the problem is to design an *essential information architecture*, or equivalently a simple integrated information architecture.

In contrast, the Internet today is a global networking infrastructure that supports our services and applications outside the network stack. Examples of applications are e-government, e-commerce, or any kind of data processing [16]. In other words, the scope of a convergent information architecture is not limited to the original Internet scope—it is much broader than networking—covering a huge information- and services-based business ecosystem. With this perspective in mind, we are developing a new architecture called *NovaGenesis* (NG[1]). NG is a clean slate convergent information architecture to be universally applied. It embraces not only content exchange and distribution (e.g., Internet, transport networks, or content-centric networks) [24], but also content processing and storage (e.g., cloud computing services and applications). NG has addressed the full information architectures convergence by building a generic information processing interface wherein the information can come in to infrastructure as a shared data buffer. Depending upon the legacy networks, or hardware product, an information interface solution shall be imparted. For software interfaces, a software solution shall be decided to support the data exchange from future Internet platforms to their respective platforms. A significant challenge will be the development of a generic message that will minimize the overheads and maximize the data throughput.

Considering NG concepts, networking protocols are implemented as services (or processes) that can be acquired by other communicating services following a publish/subscribe semantic. In this way, NG proposes the concept of *protocols-implemented-as-a-service* (PIaaS), where protocols are dynamically combined with upper level services. Thus, protocol implementations dynamically work with self-organizing in accordance with the user and communication semantic needs. For instance, only the necessary protocols for a particular purpose are combined, passing the idea that the protocol stack is rewritte on-the-fly as needed. The PIaaS opens a new world of possibilities in protocol design, since the communication stack can be redefined at runtime. Besides network protocols, this principle is also applied to any architectures' services or applications. This combination allows us to define NG as a large self-organizing distributed system. In addition, it is also correct to consider NG as a future Internet architecture (FIA), since a convergent information architecture has a broader scope than an Internet. Figure 4.2 illustrates some of the most important information architectures we have and their relative scopes, as well as their relationship with the distributed systems concept.

Considering FIA architectures, another interesting point refers to the increment of software layers on their definitions. Historically, with the exponential growth of computing power, more and more architectural aspects are being addressed at the software level. Examples include the advent of software-defined systems, things, radio, networks, infrastructure, everything—the so called software-defined everything paradigm, as well as the virtualization technologies that are being applied in cloud computing [25]. Even the IoT is strengthening the relationship among the physical and virtual worlds, creating software to represent ordinary things. Thus, all these architectures (Internet, computing clouds, future Internet, and convergent information) are becoming strongly

---

[1]http://www.inatel.br/novagenesis/

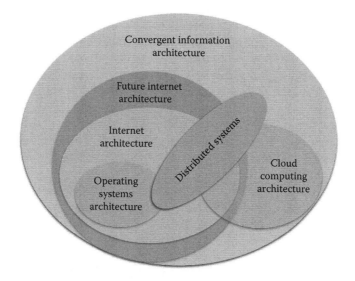

Figure 4.2 Information architecture and their relative scopes.

software-based, overlapping one another, and becoming deeply related to distributed systems, since in essence they are all distributed systems, and most often their implementations depend on other distributed systems. Thus, the quest for future architectures is not of a matter of mutually exclusive efforts, but instead a matter of essentially and synergistically combining all of them. Therefore, considering the aforementioned context, this chapter presents some core concepts of distributed systems under the vision of the future Internet, including the design choices adopted on the NG proposal.

The motivation for the work is clear: *when you think of a new model for the Internet or even for a convergent information architecture, it is impossible to not consider its impact over the functioning of existing distributed systems. Moreover, a new Internet may benefit from the state-of-the-art of underlying distributed systems.* Consequently, this chapter focuses on exploring the synergies, dependencies, and common requirements between the future Internet and distributed systems, discussing the relationships between both research areas. It provides a detailed description of NG and newer challenges in distributed systems (evolutionary and revolutionary). Furthermore, it addresses communication models, message passing, naming, identification, addressing, and locating issues, as well as approaches for modeling distributed systems and communication protocols under the concepts of NG. With this scope, this chapter will naturally discuss other research areas, including cloud-based storage and networking, network management, virtualization, IoT, operating systems, and even hardware solutions.

The remainder of this chapter is organized as follows. Section 4.2 presents the NG architecture in detail. Section 4.3 discusses the newer challenges in the distributed systems area. Section 4.4 represents the core of the chapter, discussing how NG can address DSs challenges. Finally, Section 4.6 presents the final considerations, emphasizing both the main contributions of the chapter and future work.

4.2 NG Convergent Information Architecture

The main question of FI design is: considering the current state of the art computing and communications technologies, is it possible to redesign the Internet to best meet our information society needs? Dozens of initiatives worldwide believe that it is possible [26–29]. Among them is the NG initiative being developed in Brazil. NG design started with a concept-driven survey on FI requirements, technologies, and challenges [15]. Based on this background, a set of promising ingredients to solve the identified requirements and challenges were chosen. In 2012, there was a search for synergies among selected ingredients for future convergent information architectures [30]. The results of this study fueled NG design, by defining a set of fundamental distributed systems where any information treatment or exchanging is seen as a service. The subsections below present the fundamental concepts, architectural design, and implementation details of the current NG version.

4.2.1 Fundamental concepts

4.2.1.1 Existences

The physical world in the Earth is full of physical entities (e.g., cars, houses, cables, and satellites). Furthermore, the human mind is capable of creating virtual realities, full of virtual entities (e.g., computer programs, files, and distributed systems). Adopting these definitions, a convergent information architecture can be seen as the combination of physical and virtual existences aimed at dealing with information at a universal scale. In this context, what things (physical or virtual) may belong or be connected to this universal information infrastructure? Considering the idea of the IoT, in which virtually anything can belong or be connected to the Internet, possible answers to this question will certainly step into the philosophical domain. The search for an appropriate term to define entities that can inhabit these realities led us to the concept of *individual existences*. An individual existence is anything that can be classified as independent or separated from others. Examples are people, cars, houses, software instances, equipment, virtual content, virtual information objects, or any other individual thing that simply exists. Figure 4.3 shows a set of physical and virtual individual

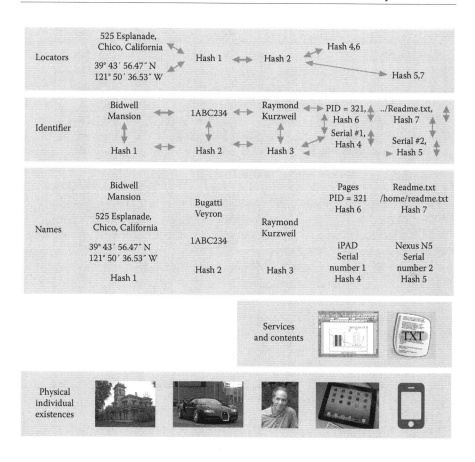

Figure 4.3 People attribute "weak semantics" and meaningful names to physical (e.g. a car or a house) and virtual existences (a computer program or a file). If they are unique in some scope, they can be used as identifiers and locators. Therefore, bindings among names (or name-bindings) can capture all sorts of relationships among virtual and physical existences. They can represent semantic relationships like "contains," "is contained," or "close to." In this example scenario, the car is "close to" the house, and "contains" the tablet and smartphone. Also, the person "is contained" in the car.

existences. Physical existences are at the bottom of the figure, while the virtual ones are at the top. A physical existence can be a *substrate* or the host for one or more virtual existences. To illustrate some individual existence relationships consider a Readme.txt file that is stored in a physical device—let's say an iPad. This file can be processed by a person—let's say Raymond Kurzweil—using the computer program Pages running on an iPad. In this case, the iPad is the substrate where the Pages instance exists. In the next subsections, we will explore these individual existences' relationships, covering several other concepts required to understand NG.

4.2.1.2 Contents, services, contracts, distributed systems, and protocols

Content can be defined as a piece of information, composed by chunks of binary data. For example, a textual file as illustrated in Figure 4.3. A *service* is a data (or information) processor. Thus, a textual file can be the input or the output of a word processor service (e.g. the Pages in Figure 4.3). In addition, a networking information processor, such as a frame forwarder, is a service like any other, since it receives information at the input, adds some control data, and forwards the processed information as an output. In this context, a process that is instantiated in an operating system in essence provides information processing, which in turn can be defined as a service. Therefore, in this chapter we propose the following definition: a service is an existence aimed at processing, exchanging, or storing information. Services negotiate their job using contracts, which are contents that describe service requirements, clauses to be respected, evaluation, and finalization criteria, among other data. All the information required to describe and regulate a service offer can be included in a contract. In service-oriented architecture a contract is usually called service-level agreement (SLA).

Continuing, a *distributed system* can be seen as a set of collaborating services (virtual existences), inhabiting geographically distributed substrates (physical existences). NG services form a distributed system aimed at exchanging, processing, and storage contents.

Finally, there is the notion of a protocol. A protocol is a set of rules, actions, and procedures agreed upon among peers to facilitate communication. In NG's point of view, even protocol implementations provide services, since everything they do is exposed to other protocol implementations as services. Historically, protocols and services were regarded as different concepts. The international organization for standardization open systems interconnection (OSI) model [31] defines a set of rules for exchanging messages between peer entities (in the same layer). Tanenbaum [32] points out that in the OSI model entities use protocols to implement their services. The protocols establish how communication will proceed between peer entities. In this chapter, no distinction is made between a protocol implementation and a service. A protocol implementation is a service that processes, stores, and exchanges information in order to build networks. Thus, services use other services indefinitely, including the ones required to implement a network. This chapter proposes the original concept of *protocols-implemented-as-a-service* (PIaaS).

4.2.1.3 Names and identifiers

Names inhabit the human mind. People like to denote things by names. *Names* are symbols used to denote one or more individual existences. In this case, *to denote* means to represent something by signals. By definition, names denote meaning and sense. However, there are names that are almost randomly generated, having "weak semantics." One can call a car by "xyzwertyu"; however, this name makes sense only for the owner or other closely related people who have been introduced to it. In another example, one can denote a car by a sequence of symbols that typically carry "weak semantics" [i.e., the numbers and letters on its license plate (e.g., 1ABC234 in Figure 4.3)]. Or yet, one can call a car by its brand, for instance "Buggati Veyron," which has more "strong semantics." A final example is the binary word obtained at the output of a hashing algorithm. This binary word (also called hash code) can be used as a name—a self-certifying name (SCN). In this case, the binary input of the hash function can be the existence itself (e.g. computer program executable, source code, or information files) or other binary input related to the existence being named (e.g. existences' immutable attributes). In the first case, the name is said to be self-certifiable, because at any time the existence's binary words can be hashed again to get exactly the same name. In the second case, the perennial physical existence attributes can be digitalized again to certify the name. Figure 4.3 illustrates some hash names calculated for physical and virtual existences. "Hash 1" for example, can be obtained from perennial attributes of the "Bidwell Mansion" in the United States, such as its physical proportions. "Hash 2" can be obtained from car attributes, like its chassis number or serial numbers. "Hash 3" could be based on biometrics of the human body. "Hash 4" can be obtained from computer serial number or processor unique IDs. "Hash 6" can be generated from entire executable binaries.

Names can be used as identifiers if they are unique in some scope. The scope can be a domain, a city, or a country, etc. Therefore, to be used as an identifier in some scope, a name must be unique in that scope. For example, in a certain small city, the name "John Smith" can be used as an identifier, while in other major cities more than one person could have this name. Thus, we can define an *identifier* as symbols used to unambiguously identify some individual existence from others in some extent. The name "Raymond Kurzweil" identifies the famous entrepreneur and inventor worldwide.

4.2.1.4 Locators, bindings, and name-bindings

A *locator* denotes the current position where an individual existence is inhabiting or attached to in some space. A space is the set of all possible positions where some individual existence can inhabit or be attached. Therefore, from a certain space definition, one can determine how close or far two existences are. Interestingly, a name can be a locator if it is possible to derive notions of distance from its interpretation. For instance, geographic coordinate systems are composed by three names: latitude, longitude, and elevation. In Figure 4.3, consider the famous "Bidwell Mansion" in the United States. The name "39°43″56.47′N 121°50″36.53′W" can be used as a locator for this physical existence, as well as the address "525 Esplanade, Chico, California." Even the identifier "Bidwell Mansion" can be used as locator if it is unique nation wide. Interestingly, using some mapping systems it is possible to derive the notion of distance between the entities named as "Bidwell Mansion," "39°43″56.47′N," and "1ABC234."

Frequently, names are related to each other to capture relationships between underlying existences. In other words, the relationships among named-existences are represented by the relationships among their names. An adjacent existence is another individual existence close to or that contains some existence. For example, the name of a street ("Esplanade") can be related to the name of a city ("Chico"). Relationships like "Esplanade" "is contained by" "Chico," or "Esplanade" "contains" the car "identified by" "1ABC234" can be used to represent physical relationships among named existences.

In this context, we can define a *name-binding* (NB) as an additional existence that represents relationships among two or more existences by linking their names. Again, considering Figure 4.3, a name-binding between the identifiers "39°43″56.47′N 121°50″36.53′W" and "Hash 1" can represent that the physical existence named by "Hash 1" is at the aforementioned geographical position. In addition, note that it is possible to create name-bindings exclusively between "weak semantic" names (i.e., an SCN). As an example, consider a name-binding between "Hash 1" and "Hash 2."

4.2.1.5   Publish, subscribe, and rendezvous

The publish/subscribe communication model is an alternative for the "receiver accepts all" model used on the current Internet. *To publish* means to make available some NB with or without attached information to possible peers, while *to subscribe* means to query for some published NB. The *rendezvous* is the process of authenticating and authorizing the subscriber according to the publisher's instructions.

4.2.2   Current design

The key idea of NG can be summarized in the following statement: *services (including protocol implementations) that organize themselves based on names, name-bindings, and contracts to meet semantically rich goals and policies.* Every component of the architecture is offered to others by publishing several name-bindings. NG accommodates two types of names: natural language names (NLNs) and SCNs. NLNs are expressed using people's language, like English or Portuguese. For example, "John Smith" is a natural English name. SCNs are obtained as the output of selected hash functions, as previously stated. The architecture adopts name-bindings to relate one name to several other related names (i.e., the NBs express relationships among individual, named existences).

An NB between both can represent that the host inhabits such domain. The NBs are structured in a [key, value(s)] format.

4.2.2.1 Services structure

NG named-services are virtual existences aimed at exchanging, processing, and storing information in any computational substrate, which can be physical or virtual (overlay or simulated). Therefore, named-services can be implemented as components inside a simulator, as object instances inside a physical or virtual computer operating system, or even as parallel processes in a physical hardware platform.

In the current design, we adopted an object oriented design approach. Figure 4.4 represents our current design model. It presents two boxes, demonstrating the interaction between two computers hosting NG. Each NG service has several block objects to implement functions. They can be classified as two types: common or specialized. There are two common blocks that are instantiated in all services, as can be seen in Figure 4.4: gateway (GW) and hash table (HT). The GW provides: inter block communication (IBC) inside a service and inter service communication. The IBC is a typical form of communication among dynamically instantiated classes inside a service. The communication between services instantiated at different substrate resources (real world hosts, virtual hosts, or simulated hosts) is done by a specialized block called Proxy/Gateway (PG), which will be further discussed. The GW employs an event-driven discipline to process NG messages according to scheduled times. It removes a message from the head of a priority queue and calls back the message's destination block. If a message is destined to a block outside a service, the GW forwards the message to the proper destination service.

4.2.2.2 Publish/subscribe services

The HT block stores the NBs of a service locally. These NBs can be published to other services through a publish/subscribe service (PSS). In other words, a service can expose its NBs to other services. This is done by publishing them to a PSS. Thus, other services can subscribe to a service's NBs. The PSS does the rendezvous between publisher and subscriber services, enabling them to discover how published names are related each other in a secure way. Eventually, services can successively subscribe NBs to identify and locate other existences, storing these NBs in their local HTs, and routing information based on them.

Instead of storing the NBs at its own HT, the PSS forwards them to a generic indirection resolution service (GIRS). This service selects an instance of another service called a hash table system (HTS) where finally the NB is stored. Thus, the PSS does not store any published NB. It forwards them to a GIRS which selects the proper HTS where the NB is stored. Typically, one domain can have several PSS, GIRS, and HTS instances. The HTS provides a mechanism to retrieve

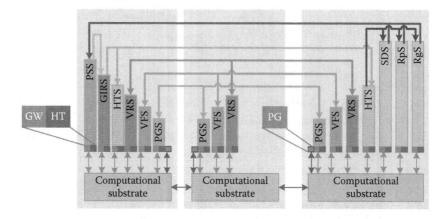

Figure 4.4 NG architecture overview. The arrows in the figure illustrate core services interrelationships and their message exchanging. A PGS represents hosting resources, which are negotiated to other components, enabling the emergence of distributed forwarding and routing services. PGSs start operation by sending a "hello" message to other peers. After the "hello," they can transmit messages destined to other PGSs. VFSs establish virtual links over PGSs. Continuing, VRSs can contract VFSs to build their routing tables. If they don't know how to route a message, they subscribe the names on the header of the NG message until they find out a path. These three *networking services* (PGS, VFS, and VRS) follow a bottom-up paradigm to create a communication network, where protocols are implemented as services (named here as PIaaS). The communication network can be a simulated overlay network, an experimental overlay network, or a real exclusive NG network, with NG hardware. NBs are published/subscribed via PSS. The GIRS selects the appropriate HTS instance to store a certain NB. These three *pub/sub services* (PSS, GIRS, and HTS) provide pub/sub functionality to other services. Finally, the SDS, RpS, and RgS prepare the environment for semantics rich service orchestration, aimed at high level application development.

already published NBs and deliver them directly to an authorized subscriber. To improve the scalability of the design, NBs are categorized, and the HTS implements a separated hash table to each NB category. Also, published NBs and content can be removed to avoid scalability issues. In other words, the PSS provides different *time to live* policies, enabling the deletion of unused bindings and content. In addition, load balancing and scalability on pub/sub demand can be provided by additional PSSs, GIRSs, and HTSs. This solution also provides flexibility on services placement and mobility. Depending on the experimented delay, core services can be migrated or duplicated to the service clients. In the worst case, core services like PSS, GIRS, and HTS could be deployed in every host. NG flexibility enables human operators to define policies to optimize the placement and redundancy of core processes. In addition, the core programs must be implemented to be as lightweight as possible, similar to the kernel of operating systems. Efficient memory access and concurrent programming is also a requirement. We are going to discuss NG requirements in more depth in Section 4.4.

4.2.2.3 Networking services

Aiming at to facilitate the migration to the NG approach, a proxy/gateway service (PGS) was designed. The PGS accumulates the role of a proxy and a gateway. It is a proxy for physical or virtual hardware resources and services inside an NG domain, representing those physical hardware/software existences inside the architecture. It is also a gateway, since it encapsulates NG traffic over some physical or virtual networking technology. The PGS can encapsulate messages over operating system sockets or simulated sockets, in the case of a network simulator implementation. This is done by a specialized block called Proxy/Gateway (PG). Every PGS works with one or more peer PGS instantiated at different substrate computers. For example, a PGS could be instantiated at one real world computer, while the peer PGS could be in a disjointed computer. Or a PGS could be instantiated at one virtual computer (a virtual machine), while the peer PGS could be in a disjointed virtual machine in the same real world computer. Or a PGS could be instantiated at one simulated computer, while the peer PGS could be at a different simulated computer. All possibilities can be implemented. However, for bootstrapping purposes, all PGSs need to be initialized with the peer PGS addresses. The addresses should make sense in the real world, virtual, or emulated network. They also could be NG names in a pure NG model.

Since an address is a name that denotes the position where an existence can inhabit or be attached, the PGS relays on its local HT block or the domain HTS where to store name-bindings between already established address formats (e.g., a real world or a simulated MAC Ethernet) or NG addresses. Independently of the address format used to connect PGSs, inside NG all communication is name-oriented.

In addition to the gateway service, the PGS also performs initialization operations for the NG services inside some computer. It can publish the existence of some PSS, HTS, or GIRS inside a host to other PGSs in the same domain, but at different hosts.

The architecture also envisions a set of services aimed at forwarding and routing information. The networking interfaces available are represented by PGSs that "sell" their free capacities to instances of a virtual forwarding service (VFS). Every VFS instance needs to negotiate with a link proxy (a PGS) in order to obtain a slice of its resources (i.e., a fraction of the link capacity in bits per second). After a successful negotiation, the VFS publishes frames to the peer PGS, which subscribes the frames and forwards them to the interface of the physical link. An instance of a PGS and one or more instances of the VFS could be embedded at some network equipment model or operating system (e.g., an access point) representing its resources and exposing its forwarding capabilities. In this case, these VFS instances can negotiate the forwarding capabilities with the peer VFSs at the other side of the link, as illustrated in Figure 4.4.

Additionally, the NG architecture foresees a virtual routing service (VRS) aimed at offering *routing-as-a-service*. VRS instances negotiate with VFS instances to route packets over established forwarding contracts. Thus, a route consists of selecting the best set of hops based on established SLAs. All the operations are SLA-based, which can cover detailed QoS support. Since all the entities are uniquely named globally, the SCNs can be used as identifiers (IDs), creating a distributed, hierarchical, self-certified service access point scheme. Finally, the architecture proposes a virtual network service (VNS), which can represent a cluster of VRS and VFS services.

4.2.2.4   Search, discovery, reputation, and regulation services

The NG name-oriented publish/subscribe scheme enables entities to publish entire ontologies as NBs. To explore the potential behind this feature, NG proposes a search and discovery service (SDS), which can help other entities to recursively search over published bindings. The idea is that the SDS will do a rank of best matches, helping entities to select appropriate bindings. The architecture also includes a genesis service to help with the semantics rich instantiation of entities, a reputation service (RpS) to store the reputation services have according to previous established contracts, and a regulation service (RgS) to monitor entities' social behavior.

4.2.2.5   Summary of terminology

A summary of NG terminology is shown on Table 4.1 and illustrated in Figure 4.5. The proposed services can be classified into four categories.

- *Networking services*—Aimed at providing a distributed message forwarding and routing service. PGSs represent real world or virtual hardware and provide gateway capacity. VFSs negotiate link capacity and configurations. VRSs create routing tables over VFS links.

- *Pub/sub services*—Aimed at providing a distributed publish/subscribe service. HTSs store the published NBs and content. GIRSs determine the best HTS to store some NB or content. PSSs do the rendezvous among publishers and subscribers, forwarding NBs and content to a peer GIRS.

- *Search and discovery services*—Aimed at providing a distributed semantics-rich search and discovery service. SDSs iteratively subscribe published NB and content to reduce candidate names and entities to a rank of selected possible peers.

Table 4.1 **NovaGenesis terminology**

Term	Meaning
Existence	Anything that inhabits the real or virtual worlds.
Name	A set of natural language or engineered symbols attributed to existences.
Identifier	A name that is unique in some scope.
Locator	A name that offers the notion of distance in a space.
Name binding (NB)	A binding among names.
Block	The smallest named structure inside a process.
Service	Virtual existences aimed at exchanging, processing, and storing information in any computational substrate.
Hash table (HT)	A block that implements a hash table to store name bindings.
Gateway (GW)	A block that implements inter block and inter service communication.
Proxy/gateway (PG)	A block inside the PGS that implements message forwarding to other peer PGSs.
Hash table service (HTS)	A service that stores and retrieves name bindings at domain level.
Generic indirection resolution service (GIRS)	A service that selects an HTS to store a name binding.
Publish/subscribe service (PSS)	A service that does the rendezvous among name bindings or content publishers/subscribers.
Proxy/gateway service (PGS)	A service that represents some core services in a certain computer. Also, it is a gateway for other PGSs.
Virtual forwarding service (VFS)	A service capable of exposing, negotiating, and establishing virtual links over the PGS exposed resources.
Virtual routing service (VRS)	A service capable of exposing, negotiating, and routing messages over virtual links.
Virtual network service (VNS)	A service capable of exposing, negotiating, and establishing virtual networks using virtual routers and links.
Search and discovery service (SDS)	A service that does iterative searching and ranking of the results according to other services queries.
Reputation service (RpS)	A service that ranks other services regarding SLA success.
Regulation service (RgS)	A service that verifies other services' SLAs regarding policies, regulations, and etiquettes.

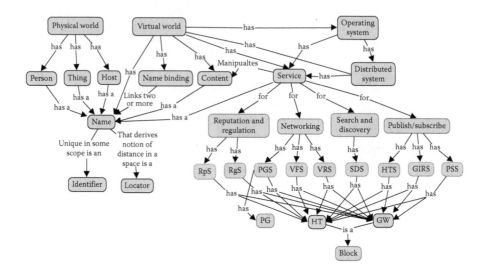

Figure 4.5 Map of NG terminology and relationships. Things are classified into physical or virtual. All individual existences can have attributed names, which can be used as identifiers and/or locators. An NB links two or more names. The virtual world encompasses content and services, which are used to implement NG. The services form a distributed system, which has instances into several operating systems.

- *Reputation and regulation services*—RpSs are contracted by other services to manage reputation, while RgSs verify ongoing SLAs regarding policies, regulations, and etiquettes.

4.2.3 Current implementation

A proof-of-concept of NG architecture was implemented in Linux OS. Figure 4.6 illustrates the implementation scenario with two desktop computers. Inter service communication inside a host was implemented using shared memory (usually know as interprocess communication or IPC), while service communication among hosts was implemented using traditional Linux OS sockets. There are three options: TCP, UDP, or raw sockets, which enable direct transportation of NG messages inside Ethernet or Wi-Fi frames. Inside the NG services, all communication is based on self-certifying names. However, during bootstrapping the processes use a well-known key to generate a shared memory ID (shmid). After a first "hello" message, every service proposes a new key, which is linked to the process SCN and stored on the local HTs of the peer services. Every service initializes the communication to the local OS PGS. The inter service communication among hosts starts similarly. Every PGS is initialized with the real world MAC address of peer PGSs and sends a "hello" message for them. A binding

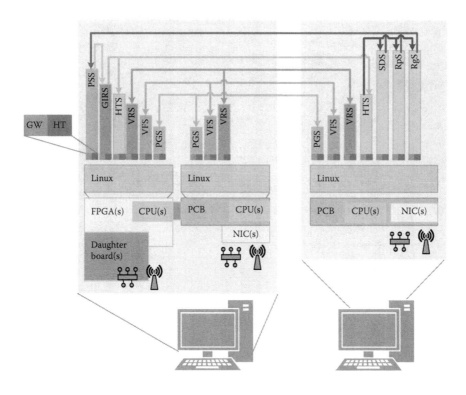

Figure 4.6 Example of NG architecture implementation on a Linux OS.

between the PGS SCN and the legacy address is stored at the local HTs of the peer PGSs. Future versions will dispense legacy addresses, since all the required services will have SCNs. Thus, the PGSs would store only SCN-to-SCN bindings at initialization procedure.

Figure 4.7 provides a simplified sequence diagram of the actions taken during the bootstrapping of the NG services illustrated in Figure 4.6. The bootstrap procedure is divided into four phases: "hello," exposition, discovery, and contracting. At the end of this bootstrap sequence, the core is ready for publishing/subscribing messages. Services that are instantiated after this bootstrap sequence follow a similar procedure. The exposition and discovery steps are periodic, enabling the core to self-organize—in the event of services release or instantiation. In this figure, we adopted a real Wi-Fi network, without TCP/IP. NG messages are encapsulated by the PGS using a *raw socket* in Linux OS. The procedure starts with a "hello" phase. In this case, every PGS is initiated with the MAC addresses of the peer PGSs. Thus, every PGS sends a "hello" message (transaction *a* in figure) to their peer PGSs, which contains several initialization NBs. The peers answer with an "OK" message (*b*). These messages

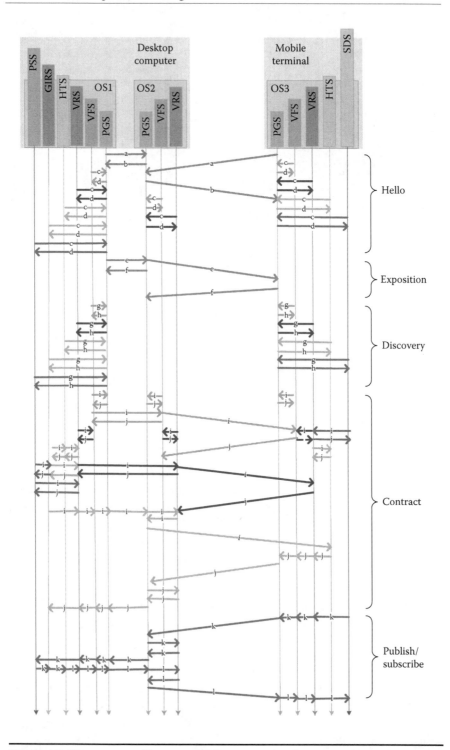

Figure 4.7 The sequence diagram of NG core services bootstrapping in the current implementation.

are encapsulated in Wi-Fi frames, configuring the inter node IPC. Also, every NG service sends a "hello" message (c) to the PGS in its OS. These "hello" messages configure the intra node IPC parameters. They are acknowledged by the PGS on transactions d. After the "hello" phase, the PGSs exchange the NBs they have (e and f). The core services can now discover possible peers (g and h) by consulting the PGS hash table (HT) on their local OS. If some service matches the natural language queries used, peer contracting proceeds by sending SLAs to the candidates (i). If the SLAs are acceptable, services agree on them (j). Observe that: (1) The peer PGSs establish contact each other; (2) the VFSs contact peer VFSs to form a virtual link; (3) the VFSs contact peer PGSs, which represent NICs; (4) the VRSs contract one or more VFSs to forward their traffic; (5) the VRSs contact peer VRSs to create a virtual network; and (6) this GIRS instance has two peer HTS candidates in this example and it selects both. The final phase in this example is the pub/sub case, which is allowed after all core processes have finished contracting. The SDS publishes some NB to the PSS (k). The message is routed from the mobile terminal to the desktop computer using the entities' self-certifying names in a connectionless communication model. In the figure, the NB is stored in the HTS instance at OS1. If the SDS is publishing some NB (or a content), the HTS acknowledges the successful storage in transaction l. If it is subscribing some NB, then the transaction l carries the desired information. The authentication and authorization steps, which are related to all activities, are not shown in the figure for the sake of simplicity.

Another interesting approach to deploy NG is to assume flexible hardware, such as a field-programmable gate array (FPGA). In this case, NG can be run inside the FPGA—as a set of processes at the embedded OS. A PGS can expose the resources available at the FPGA, enabling the creation of virtual links over universal serial bus, Ethernet, peripheral component interconnect express, or any other digital communication technology. Note that from the point of view of the proposed architecture, the internal computer bus is treated identically to any physical link between two equipments. Another approach is to embed NG services on a Linux kernel for mobile devices, similarly to what is done for the Android OS. In this solution, the NG architecture can be seen as a set of distributed systems that extends the Linux OS, enabling rich contract- and name-oriented self-organization. However, this approach will be the target of a future work.

4.3 Distributed Systems Challenges

Today, we cannot talk about computing without mentioning its distributed facet. The growing access to the Internet together with the low cost of network and computing devices, such as network interface cards, cables, switching elements, and wireless devices, favored the widespread use of networked environments for computing either simple or complex tasks [9]. Text editing, video and audio conferences, chat, and remote banking are tasks that employ distributed systems in a seamless and effective way. Now it is possible to look at novel tasks including integration, dynamism, synchronization, and consistency that will be present in the next-generation of distributed systems, such as those used to control smart grid production, IoT, and large metro subway infrastructures [33–35]. In this way, the next subsections describe some research points and challenges in the state of the art on distributed systems, linking them to the needs of the FI.

4.3.1 Support for the emerging cyber-physical systems (CPS)

A cyber-physical system (CPS) is a system of collaborating computational elements controlling physical entities [36]. Embedded computers and networks monitor and control the physical processes, with feedback loops, where physical processes affect computations and vice versa. In this way, CPS integrates the dynamics of the physical processes with those of the software and networking, providing abstractions and modeling, design, and analysis techniques for the integrated whole.

Mobile cyber physical systems, in which the physical system in question has inherent mobility, are a prominent subcategory of CPSs. Examples of mobile physical systems include mobile robotics and electronics transported by humans or animals. For tasks that require more resources than are locally available, one common mechanism for rapid implementation of smartphone-based mobile CPS nodes utilizes the network connectivity to link the mobile system with either a server or a cloud environment. This approach enables the system to perform complex processing tasks outside limited resources nodes.

Emerging CPSs will be coordinated, distributed, and connected, and must be robust and responsive [37]. The CPS of tomorrow will need to greatly exceed the systems of today in capability, adaptability, resiliency, safety, security, and usability. Examples of the many CPS application areas include the smart electric grid, smart transportation, smart buildings, smart medical technologies, next-generation air traffic management, and advanced manufacturing. CPS will transform the way people interact with engineered systems, just as the Internet transformed the way people interact with information.

4.3.2 Heterogeneity and integration at global scale

The next-generation of distributed systems needs to integrate a wide range of devices, ranging from traditional micro-controllers, to the latest generation of smartphone and tablet devices, while dealing with varying protocols network and bus interconnects, software platforms, and processor types and capabilities [38].

Particularly at the network viewpoint, the FI simultaneously can deal with local-area gigabit networks, wireless connections, and long-haul, high-latency connections present on wide-area switched connections.

Modern Internet applications have giant scale, so that even within a single data center there are various generations of equipment. They may involve specialized nodes, from well-provisioned servers down to nodes whose specialty is fitting in pockets. Participating nodes are often owned and administered by many different entities, and they are deployed globally on top of the largest distributed system of them all, the Internet, which exemplifies all of these characteristics.

Therefore, new and existing components must be compatible with existing systems to provide cost-effective system maintenance and upgrades. Finally, the FI must deal with integration and openness properly, aiming at making the distributed system independent from the heterogeneity of the underlying environment: hardware, platforms, protocols, and languages.

4.3.3 Mobility at different levels: user, host, and code

The support for host mobility makes it possible for a mobile terminal to disconnect from one wireless network and to connect to another seamlessly, without requiring any intervention from the user. This happens thanks to the employment of standard interfaces for building the communication stack of mobile and stationary devices. Considering the mobile computing context, the portability of many of these devices, together with their ability to connect conveniently to networks in different places, makes this branch of computing feasible [13]. Portability also appears at the middleware level by defining standard application programs interfaces (APIs) for developing distributed applications. In this way, the message passing interface (MPI) is a standard interface for cluster and grid programming. For instance, it is possible to write a single MPI program and compile it against a lot of MPI implementations for taking profit of system area networks or optimized scheduling algorithms [12,39].

Mobile computing concerns the execution of computing tasks while the user is on the move, or visiting places other than their usual environment (roaming) [39]. The main idea is to offer both location and access transparencies by providing the access to the Internet, or Intranet, anywhere at any time. Nowadays, we can observe the exponential growth of tablet and smartphone usage, which normally use a wireless connection (i.e., IEEE 802.11) or third/fourth generation of mobile telephony technology. The presence of computers everywhere only becomes useful when they can communicate with one another. In this way, the network core infrastructure must deal with this growth of devices by providing multihoming, simultaneous connectivity, and mobility support, as well as by offering efficient communication among them.

4.3.4 Efficient collaboration model

A distributed system, in a general view, can be implemented by following one of these strategies: (1) client-server or (2) peer-to-peer. The client-server is the architecture that is most often cited when distributed systems are discussed. It is historically the most important and remains the most widely employed. In particular, client processes interact with individual server processes in potentially separate host computers, in order to access the shared resources that they manage. Servers may in turn be clients of other servers. In the peer-to-peer architecture, all of the processes involved in a task or activity play similar roles, interacting cooperatively as peers without any distinction between client and server processes or the computers on which they run. In practical terms, all participating processes run the same program and offer the same set of interfaces to each other.

While the client-server model offers a direct and relatively simple approach to the sharing of data and other resources, it scales poorly. The centralization of service provision and management implied by placing a service at a single address does not scale well beyond the capacity of the computer that hosts the service and the bandwidth of its network connections. Thus, the FI must design efficient communications between processes by providing them some guarantees; for example, quality of service.

4.3.5 Adaptation on communication protocols

Considering the traditional Internet, many frequently used services run over transmission control protocol/Internet protocol (TCP/ITP) stack, with reserved port numbers. These include the following: (1) The hypertext transfer protocol (HTTP) is used for communication between web browsers and web servers; (2) The file transfer protocol allows directories on a remote computer to be browsed and files to be transferred from one computer to another over a connection; (3) Telnet provides access by means of a terminal session to a remote computer; and (4) The simple mail transfer protocol is used to send emails between computers. TCP offers end-to-end reliable communication between the pairs, offering control for lost and duplicated messages, message ordering, and flow control.

The TCP is normally used when reliability is the main communication requirement. On the other hand, for some applications, it is acceptable to use a service that is liable to occasional omission failures. For example, the domain name system (DNS), which looks up DNS names on the Internet, is implemented over user datagram protocol (UDP). Voice over IP also runs over UDP. UDP datagrams are sometimes an attractive choice because they do not suffer from the overhead associated with the guaranteed message delivery mechanisms. Here, the challenge encompasses adaptation in order to employ protocols in accordance to the communication context (local or wide-area network, need for strong reliability, performance levels, QoS, geographical areas and administrative domains, number of hops, people involved, and so on).

4.3.6 Supporting the requirements for the IoT

Another challenge for large-scale distributed systems is dealing with what is known as the IoT: the pervasive presence of a multitude of Internet-enabled things, ranging from tags on products to mobile devices to services, and so forth. From a distributed-systems perspective, the challenge is to move away from the network- and things-oriented views and provide a view in which the collaboration of all these Internet-enabled things form, indeed, a coherent distributed system. We are witnessing the integration of wireless sensor and actuator networks as a distributed computing platform with more traditional cloud-based systems to which specific computations and storage facilities are offloaded.

Again, one of the challenges consists of addressing the inherent heterogeneity of the modern large-scale distributed system composed by IoT devices. Nowadays, solutions for IoT are already available, among them: message queue telemetry transport (MQTT) [40], extensible messaging and presence protocol (XMPP) [41], advanced message queueing protocol (AMQP) [42], and data distribution service (DDS) [43]. The MQTT is typically used for machine-to-server (M2S) scenarios, while the XMPP and AMQP are typically employed in server-to-server (S2S) scenarios. DDS is focused on machine-to-machine (M2M) communications. While the MQTT and XMPP do not support real time operation, the DDS is focused on supporting timely data distribution among devices. Concerning the relationship with the current Internet, the MQTT, XMPP, and AMPQ relay on the TCP/IP sockets, while the DDS specifies a reliable real-time publish-subscribe wire protocol DDS interoperability wire protocol specification (DDS-RTPS) aimed at UDP/IP stack or other data encapsulations (TCP/IP or direct shared memory inside a node). Thus, timeliness of data distribution, reliability, efficiency, flexibility, and QoS are important requirements for M2M communications, but can also be important for M2S and M2M scenarios.

Another challenge is dealing with the large amount of devices expected on the IoT. This "army" of devices will challenge established process communication and collaboration models, process addressing, and communication protocols. The MQTT and XMPP depend on the deployed IP addressing scheme. The DDS-RTPS protocol was standardized to be fundamentally transported by UDP messages, although it may be encapsulated by other technologies.

Given the potential size of the data generated by sensors and related devices, a trade-off will need to be found between in-network processing and aggregation techniques versus streaming data to the external support system. This trade-off is not an easy one. It depends on the capabilities of the distributed sensor network, the communication channel between sensor network and support system, and the support system itself. In some cases, the networking delay or the intermittent connectivity could prohibit external support, requiring more computing power at the IoT nodes.

A commonality between MQTT, AMQP, and DDS is the adoption of the publish/subscribe communication paradigm. However, the interoperability between publishers and subscribers can be a problem in the MQTT, since the message payload format needs to be agreed upon among the peers before any data transfer. A challenge is to deal with a large number of publishing/subscribing nodes.

Finally, we can envision the challenge related to spontaneous interactions among the elements and the support of the distributed system needed for that. For example, associations between devices are routinely created and destroyed by identifying and locating a device, such as a printer. According to Presser et al., the idea of "social devices" requires not only the unique identification of nodes, but also the nodes' capacity to discover peers and establish trustable relations. While peer discovery is not in the scope of MQTT and XMPP, the DDS standard provides mechanisms to expose node interests, as well as the kind of information each node can provide. DDS automatically connects the subscribers to the topic-related publishers. Also, the nodes (or a support system) must be capable of semantically interpreting information, allowing them to collaborate with each other towards a common objective. The idea of "social devices" also requires the ability to be aware of the environment situation (situation-awareness) and the capacity to securely exchange data and learned knowledge. While MQTT and XMPP consider security to be other protocols' concern, the DDS standardizes a framework for M2M secure communication. A final requirement related to such situations is to make interoperation fast and convenient even though the users are in an environment they may never have visited before. Different vendor support is fundamental.

4.3.7 Service orientation and utility computing

With the increasing maturity of distributed systems infrastructure, a number of companies are promoting the view of distributed resources as a commodity or a utility, drawing the analogy between distributed resources and other utilities, such as water or electricity. Thus, some companies buy large datacenters and rent processing power and storage capacities to end users. Furthermore, small and mid-sized enterprises do not need to buy an entire set of machines, but they can use outsourced resources. These actions present an improvement over one of the most important metrics when developing a new product: time to market. The term cloud computing is used to capture this vision.

Technically, a cloud can be defined as a set of Internet-based applications, storage, and computing services sufficient to support most users' needs, thus enabling them to largely or totally dispense local data storage and application software. Nevertheless, the main drawback concerns the migration of existing applications to the cloud (i.e., the requirement of presenting a standardized interface independent of the operating system or programming language for the migration of standard client-server applications to the cloud). Because people do not feel comfortable uploading top secret data to cloud databases, sometimes the applications must be rewritten to take advantage of distributed resources.

4.3.8 Context-aware computing and recommendation systems

Concerning the research on distributed systems, we observe the challenge to develop recommender systems enhanced with the information from other users. Finding out which information is needed, and from whom is not obvious. For certain recommender systems it does not really matter whether data is used from a group of similar peers or whether the data is from a randomly selected group. Note that such findings may have a huge impact on the design of a distributed system: whereas the first approach may require intricate distributed algorithms, the second approach may be easy to implement using fairly straightforward techniques, such as gossiping and randomized peer selection.

4.3.9 Naming, addressing, identification, and mapping

Considering the Internet protocols, a sender process identifies a receiver as a tuple containing both Internet address and local port. A local port is a message destination within a computer, specified as an integer. A port has exactly one receiver (multicast ports are an exception) but can have many senders. Processes may use multiple ports to receive messages. Any process that knows the number of a port can send a message to it. Servers generally publicize their port numbers for use by clients. If the client uses a fixed Internet address to refer to a service, then that service must always run on the same computer for its address to remain valid.

Both forms of Internet communication (UDP and TCP) use the socket abstraction, which provides an endpoint for communication between processes. Sockets originate from BSD UNIX, but are also present in most other versions of UNIX, including Linux and OS X, as well as in Windows. Interprocess communication consists of transmitting a message between a socket in one process to a socket in another process. In addition, a receiving process can implement a receive method from any strategy. The receive method does not specify an origin for messages. Instead, an invocation of receive gets a message addressed to its socket from any origin. The receive method returns the Internet address and local port of the sender, allowing the recipient to check where the message came from. It is possible to connect a datagram socket to a particular remote port and Internet address, in which case the socket is only able to send messages to and receive messages from that address.

The current Internet employs a hierarchical naming scheme, where names are attributed to represent hosts and networks. For example, the name host.somedomain.com.br denotes a host that can be found inside a network named somedomain.com.br. These names are mapped to IP addresses by the DNS. In this case, hierarchical meaningful names are used to denote hosts and networks. In the DDS IoT standard, a "global name space" is defined, where named topics are used to provide the rendezvous among data writers and readers. The DDS names enable IoT nodes to express their interests to the network, enabling one node to discover possible peers. Both cases are examples of the importance of semantics on distributed systems and networks.

4.3.10 Exposition and orchestration of resources

The joint orchestration of substrate resources, such as memories, disks, and cores, with middleware resources is still an ongoing work. An example is the integration of open multiprocessing (OpenMP) [44] and MPI [45].

Another issue is how to orchestrate the creation of virtual networks among distributed systems and to map the virtual topology to a physical network.

4.3.11 High performance computing and concurrency

The MPI is the standard API for developing high performance applications for cluster and grid computing. MPI was first introduced in 1994 by the MPI Forum[2] as a reaction against the wide variety of proprietary approaches that were in use for message passing in this field. Considering that a lot of applications written with this standard for cluster and grids already exist, MPI is also used for submitting message passing applications on cloud environments. Two popular implementations of MPI standard are OpenMPI [46] and MPICH [47]. Both provide support for TCP/IP sockets, as well as transport over other high speed networks, such as Infiniband [48], Ethernet [49], iWARP [50], and Myrinet [51]. In general, it is possible to bypass the TCP/IP stack or even the operating system's kernel overhead.

A challenge regarding MPI is the amount of unnecessary memory copies when transferring a message between two processes. First, the sender process needs to copy the message from its own memory to a FIFO queue or a shared memory region. Second, the receiver process needs to copy the message to its private memory. This double copy design not only reduces communication performance, but also creates scalability issues. A possible solution was implemented by Friedley et al. The sender process passes a pointer of its private memory to the receiver process, which copies the message only once directly to its internal memory.

---

[2]www.mpi-forum.org

The MPI also faces challenges regarding concurrency, especially on thread-based implementations for multi-core HPC systems. The saturation in the increase of core's clock frequency generated a rush towards the concurrence. However, programming languages and distributed systems frameworks are not aligned to this direction change. MPI faces challenges regarding thread safety, diversity of node parallelism models, and nonblocking operations.

4.3.12 Scalability, consistency, and availability

One of the key principles of distributed systems is their incremental scalability. Typically, there is a large number of servers supporting the distributed system. Servers can join, leave, or even crash. A failure in one machine is compensated for by others. No service has complete system state information—a "divide to conquer" approach is adopted. To increase reliability and performance, information is replicated among several servers. Therefore, an important issue is information consistency (i.e., how to deal with different versions of the same object or how to correctly determine causality of events that affect some information history). Thus, in general, distributed systems require the ability to determine the consistency of a replicated information object, as well as the ability to keep consistent the information object available in case of server failure. If a consistent version of an information object is not available, the system should then wait until it is available. This is one of the performance drawbacks of keeping information consistent. Finally, distributed systems fault tolerance also includes process resilience. They need to be aware of other processes' states in order to achieve a solid behavior.

4.3.13 Security, privacy, and trust

A universal and global distributed system would have the same performance hurdles and security threats as distributed mobile ad-hoc networks [52]. Some common attacks that represent a critical challenge to trust and reputation systems are described by Zhang in [53].

Sensitive information can be used by control systems that can represent a threat to safety or malicious nodes may attempt to disseminate false or corrupted information. The system should be designed to minimize selfish and malicious behaviour, and to support flexible security and privacy mechanisms.

Applications may require a broad range of protocols and security mechanisms. From simple end-to-end secure channels through public key infrastructure (PKI) to distributed reputation and voting systems, ingenious protocols can be employed using symmetric and asymmetric cryptosystems, cryptographic key management, authentication, authorization and accounting systems, threshold cryptography, etc.

On the other hand, security mechanisms to verify authenticity, integrity, and reputation features may be required for the operation and management of the network itself. Several distributed cryptographic, trust, reputation and currency systems can be combined to promote an integral trust solution making them ideal to be employed in applications built by service composition.

Sophisticated trust and reputation systems have been proposed, some of them even providing distributed reputation and quality assurance of any node, message, or information. Some models designed include a data-centric trust establishment [54] framework and distributed emergent cooperation through adaptive evolution [55].

4.4   Addressing Distributed Systems Challenges with NG

This section provides a discussion on the relationship among distributed systems and NG. Our analysis considers two complementary perspectives: (1) NG as an architecture to address distributed systems challenges and (2) the distributed systems challenges from NG's point of view.

4.4.1   *Interprocess communication*

Strictly speaking, the NG scope also covers computer operating systems, since they can perform tasks like information processing, exchange, storage, and visualization. One of the NG design principles is to employ the same design choices at different levels of the architecture in a hierarchical modular way, by means of recursive structures. However, the proposal can run as a set of applications in the current operating systems, interacting with them through PGS implementations. In the current version, NG processes rely on a gateway block to perform Linux shared memory IPC. NG's "philosophy" can be employed to design self-similar distributed operating systems that use the same design choices on several hierarchical levels, from machines to local networks, domains, and even to the entire Internet.

The idea is to find the essence of the communication process irrespective of the level at which it occurs (i.e., inter process, inter OS, inter host, inter domain, or inter network). Regardless of the level, digital messages are sent and received through unidirectional or bidirectional communication. They are temporarily stored in digital memory. The transmitter and the receiver can be blocked or not during operation, as well as needing binary message synchronization (i.e., a mechanism to detect the beginning/ending of messages in the communication channel binary stream). Digital information and controls need to be delineated according to the "digital edges" of each transmission. Message sequence numbering is another prerequisite, which is necessary if messages are fragmented.

Access control to shared resources is another issue; for example, the access to memory or data structures in multiprocess/multithread environments. Flow control is another concern, since the sender and receiver need to signal when new messages are being serialized, as well as when messages have been successfully received. Finally, there is the identification and localization aspect. Every resource needs to be identified and localized, not only on a network level, but also inside a machine (e.g., a memory address or a resource ID). These prerequirements are similar no matter the level of the communication system. Therefore, the same abstractions can be used at all levels, reusing functionalities at several scales. NG advocates for a self-similar approach for the communication functionalities regardless of the hierarchy level.

4.4.2 Indirect communication

NG adopts the publish/subscribe paradigm. The NG components can provide space/time decoupling between senders and receivers. Name-bindings can be published to some entity even though the target entity is not running at a given time. The HTS provides a temporary cache for the information. When the receiver is instantiated, it can discover the publisher and subscribe the information. The HTS envisions a periodic cleaning of name-bindings to avoid excessive memory consumption. Thus, the publisher needs to specify for how long an NB will remain at the HTS. Additionally, the publisher could be off when the subscriber downloads the binding. The NG persistent names allow representation of the relationships among a new instance of some service to the previous ones. Therefore, prior published information can be related to a new instance of a source service, enabling contents to be traced to the original service despite the instance changes. NG pub/sub solution enables efficient content dissemination from multiple sources to multiple destinations. The HTS temporary cache can be considered a distributed message queue, where senders can publish named contents to several possible interested receivers. When compared to tuple spaces, NG allows services to publish any piece of named data to their peers. Also, NBs can be searched and discovered according to services interests. However, in the NG approach, only the publishers can specify how long content will be stored in the temporary cache.

4.4.3 Addressing

NG uses natural language and self-certifying names to identify a service in some scope. When a service initializes, it publishes bindings among several NLNs and SCNs. Thus, every service can be addressed by subscribing these bindings recursively. As the combination of a port number and an IP number addresses a port uniquely in the current Internet, SCN tuples allow for the same for NG services. For example, consider a proxy service that inhabits some OS. This service can generate an SCN—let's say A1—and assume that this name is an address for this service inside of the local OS. Likewise, the OS can generate an SCN— let's say B1—and assume that this name is an address for this OS inside a certain host. The host can also have an SCN—let's say C1—which can be used to address this host inside a domain (D1), and so on. The resultant tuple, A1-B1-C1-D1, enables any other service to address a message to this proxy service globally. Additionally, natural language names linked to this tuple facilitate search and discovery of service access points.

SCNs are verifiable. They can be checked anytime for integrity. SCNs are flat, since they do not depend on network hierarchy. Only the bindings among SCNs change according to network hierarchy. SCNs can be unique globally, avoiding the current lack of addresses on the IPv4 Internet. In case of host mobility, the services' addresses (tuples) change, but their identifiers (i.e., SCNs) remain the same. In other words, only the NBs change. The SCNs remain the same while entities do not change their binary patterns or attributes.

In the current Internet, when a host moves from an autonomous system to another, its IP address could change, causing a change in the identity of the host. This results in an undesirable loss of traceability, as well as possible connection losses. In the NG approach, there is no loss of traceability, since the host remains with the same SCN after movement. Suppose the host of the aforementioned proxy service moves to a new domain—let's say D2. The SCN tuple changes to A1-B1-C1-D2, while the host continues with the C1 SCN despite the movement. Therefore, the mobility of a host on the NG approach requires removing the first NB (C1-D1) from HTS and the publication of a new NB between C1 and D2. This solution is self-similar, since it could be applied to the mobility of any existences, including content, services, hosts, etc.

Some emerging concerns regarding this proposed addressing scheme are: (1) the loop problem, since an SCN tuple could return to the same original point, making the message undeliverable. Some mechanism to avoid addressing loops will be required; (2) the overhead of a long addressing chain. Long chains can be avoided by replacing a large quantity of existences by a few key existences in the path; and (3) scalability of name resolution. The larger the number of existences, the larger the number of NBs to be resolved. The distributed nature of pub/sub services helps with load sharing. Also, replicas of pub/sub services can be instantiated to deal with a large number of name resolution requests. From a vertical point of view, the most important NBs are locally stored inside every service at the HT block. In a domain level, several PSS/GIRS/HTS instances can store NBs in a distributed fashion. Also, by hierarchically structuring domains the number of hierarchical pub/sub services can be largely supported.

4.4.4   Peer collaboration model

The NG architecture enables SLA-based peer-to-peer collaboration, as well as the traditional client-server model. Services can establish collaboration in a distributed way, sharing the available resources. Service "clones" (or replicas) can run on different machines to improve scalability, performance, and robustness. In this case, the pub/sub services are a solution at hand to deal with NBs update and coherence. The optimal equilibrium among substrate resources, load, and performance can be obtained by optimizing the SLAs, number of distributed peers, and positioning. Since NG proxy/gateways represent the required substrate resources, SLAs can be established among service peers and proxies. This "trust network" can be extended to include QoS monitoring, reputation, and regulation systems. The idea is to create an entire digital business ecosystem (DBE) for integrated systems evolution, covering all aspects of services' life-cycling. In this DBE, clusters of collaborating services will emerge, compete, fragment, disappear, and evolve under the evolutionary pressures of humans and machines. The peer collaboration models can self-optimize according to these evolutionary pressures (e.g., SLAs, trust networks, QoE, reputation, etc.).

4.4.5 Communication protocols

These protocols are implemented as services in the NG proposal. They are similar to any other service; therefore services and applications need to establish dynamic contracts with them. The idea is to implement network protocols as clusters of collaborating services, possibly fragmenting them in smaller "utilities" that can be composed dynamically to form on demand protocol stacks. For example, transmission control can be achieved by a cluster of distributed services, including the ones for retransmission, error check, flow control, etc. This enables multisource, multipath, and multidestination transmission control, a feature desired to improve content distribution performance in an FI. The transport services can be customized to specific needs of the applications. Every feature can be included or not, creating a wide range of possibilities that do not exist in the current Internet. Another important benefit is that network protocols will evolve like any other service. Thus, the service provided by some protocol implementation, as well as the dynamic combination of PIaaS or the logical and geographical placement of these PIaaS can be based on the prerequirements of client services or available substrate resources. NG enables the creation of trend- or social-driven networks by adapting PIaaS according to user, service, or XI substrate resources needs. The notion of PIaaS was presented in Section 4.2.1.2.

4.4.6 Internet of things

The recent development in the FI architecture shows that IP based Internet architecture has limitations when it comes to interconnection of devices in the world of IoT objects and devices. Scalability and portability are two points where NG can score over other Internet architectures. NG architecture will have native support for distributed systems and the ability to evolve its functionality to accommodate new, as yet unforeseen, requests over time for exchanges and distribution of data. A simple NG service is being developed to be embedded at IoT nodes. This service aims at implementing some NG novelties at sensors and actuators, enabling them to exchange name-based messages, as discussed in Section 4.4.3. For those small capacity IoT nodes, proxy/gateway services can represent them at the NG cloud, enabling dynamic contract establishment in the name of the things. The PGS model provides a distributed gateway and interoperability solution adequate for the heterogeneity of IoT platforms, protocols, and devices implementation. The PGS can also be extended to change configurations at the controlled IoT nodes, as well as to detect their status. This model goes on the direction of a software-defined IoT, where nodes are controlled by NG services.

The Internet architecture follows a "narrow waist" design, which has a great impact in the success of the present Internet. It forces the applications and protocols to be made above the waist and supporting the physical media, physical layers, and access technologies below the waist. But this has a drawback, especially regarding the dual semantics of IP addresses and obsolete fields in headers. As the technology is evolving we are talking about the interconnection of billions of devices. This is where the present Internet architecture faces problems. This is where IPv6 comes in to play, but it also faces the same problems because of the same "narrow waist" and the large size of datagram headers. The NG pub/sub "narrow waist" resembles DDS link protocol, but with the advantage of integrating several FIA ingredients, like naming structure, binding resolution, software-defined, mobility-friendly, self-organizing, service-oriented designs. In fact, NG extends DDS in many ways, including "semantics rich" integrated orchestration of contents, services, and IoT resources. NG provides a renewed naming scheme with dynamic messaging, where identifiers are decoupled from locators, supporting mobility by rebuilding NBs. NG protocols are implemented as services, enabling dynamic protocol orchestration, self-adaptation, and evolution. It enables the emergence of more efficient and modern protocols to IoT, which can operate aware of several issues, like energy, delay, communication opportunities, etc.

4.4.7 Scalability, consistency, and availability

NG encompasses services targeted to expose computer and communication resources to other services. These PGS facilitate distributed systems to react to changes in the substrate level. Substrate resource awareness is created by PGS instances that publish relevant computer and communication resource information to peer distributed systems. Therefore, client distributed systems can negotiate contracts with PGS instance and subscribe their valuable information about machine status and other peer service hosts. NG design already provides the required "information bus" implemented using core pub/sub services. Changes on substrate resources can be announced to distributed system peers, making them aware of information and communication technologies (ICT) environment.

Regarding the consistency and availability of replicated information objects, NG already employs distributed HT structures. However, the pub/sub services do not follow a classic ring topology used in DHTs. The GIRS can implement replica strategies for information objects and NBs, naturally optimizing their placement in order to achieve good performance and robustness. The consistency of NBs and information objects (IOs) is granted by the SCNs employed at the DHT key values. New versions will naturally have new names and therefore avoid the consistency and unavailability problem typical of distributed content replicas. Additionally, NG NBs can be used to represent a new version of previously published contents. NG pub/sub services already incorporate the mechanisms to deal with causality of events and version control. History of IOs can be determined throughout recursively navigating over [key, value] pairs published at pub/sub services. Also, NG protocols-implemented-as-a-service approach offers the appropriate tools to implement several protocols already used in distributed systems, like quorum protocols for IO consistency, election algorithms for services coordination, checkpointing for state synchronization and gossip protocols for changes on membership. Gossip protocols can be implemented using these pub/sub services to announce crashes or new servers joining into the system.

4.4.8   Security, privacy, and trust

NG SCNs are generated from existences' immutable patterns. As long as an existence maintains its immutable attributes, its SCN will be the same. Therefore, even in ephemeral ad hoc networks, existences can preserve their SCNs, while opportunistically connecting and communicating. NG services maintain contracts that are bound to existences' SCNs. Thus, an existence's reputation can be determined based on contract analysis. Additionally, NG pub/sub services can support new techniques like data-driven trust [54].

Services can form trust networks, where every service has a reputation, like in the online e-commerce websites we have today. Every service has a reputation and this reputation is verified before establishing service contracts. Thus, services are evaluated regarding possible threats and risks. Secure services with good quality will prosper, while bad services, suspected of being unsafe, will have their reputation reduced, naturally forcing them to improve or disappear.

Services may form agreements with other services to evaluate the reputation of their mutual SLAs. These reputation services (RpSs) can distributively provide reputation and quality assurance for any node, message, or information. Information is secured per se and its dissemination depends on contracting established and traditional secrecy and integrity mechanisms. NG published NBs form a distributed web of relationships, enabling authorized services to navigate among contents, services, and hardware relationships. Authorized services can derive complete graphs of relationships, clearly determining provenance, non-repudiation, and other security properties.

Traditional security services, like PKI, distributed reputation, and voting systems need to be mapped to NG abstractions. Novel approaches can emerge when combining the NG service framework with contemporary techniques. However, this subject is still in its beginning and intensive research is required.

4.5   NG Related Work

The relationship to the distributed systems is not exclusive to NG architecture. Other future Internet proposals also employ DSs in their implementations or aim at addressing open problems in DS landscape. In this section, we compare NG with some of these FI approaches, namely Scalable and Adaptive Internet Solutions (SAIL) [56], eXpressive Internet Architecture (XIA) [57], and Recursive Internet Architecture (RINA) [58]. Since NG is a "clean slate" convergent information architecture, we selected architectures that have a similar origin, scope, and purpose.

4.5.1   SAIL

The SAIL project is funded by the European Commission's 7th Framework Program (FP7) and brings together 25 partners (operators, vendors, and research institutions) aimed at designing a future network architecture in Europe. It encompasses three key technologies, namely: network of information (NetInf) [59], cloud networking (CloNe), and open connectivity services (OConS). The NetInf defines an information-centric network where named data objects (NDOs) can be accessed without a priori localization (i.e., the data is localized only during final transfer phase). Network caching is used to store NDOs closely to the interested peers, maximizing content distribution efficiency and robustness. The NetInf nodes provide two services: (1) the forwarding of NDOs requests to the closest caching/storage nodes and then the transfer back of the NDOs to the requester and (2) a name resolution service (NRS) that can be used alternatively to the name-based routing. The NetInf protocol covers publishing, searching, and subscription for NDOs.

The NetInf [59] relies on a convergence layer (CL) to transport its messages among NetInf nodes. Current realizations of CL include support for HTTP over TCP, UDP, disruption/delay tolerant network, bundle protocol, and Ethernet. In OpenNetInf [60] implementation, nodes communicate via HTTP or TCP. They can also use Google protocol buffers (GPBs) [61] or XML [62] for the serialization of the messages. OpenNetInf nodes are implemented using *Guice* [63], a Java framework developed by Google. Nodes perform search by using a semantic web query language called SPARQL [64]. An SDB database is used to store data on resource description framework [65] format. The NRS is implemented in Java using FreePastry [66] and Multi-level Distributed Hash Table (MDHT) [67].

The OpenNetInf nodes have an event-driven service that can be used to publish, notify, and subscribe NDOs. The aim is to update applications regarding events related to NDOs of interest. The implementation has an event service framework (ESF) that can use one or more available underlying event-driven open source solutions, like the Java implemented scalable Internet event notification architectures (SIENA) [68] or Hermes [69]. Both SIENA and Hermes were developed for the current Internet protocols. Interestingly, the ESF enables applications to use both event-driven systems for their own purposes.

The OConS covers open connectivity services for heterogeneous networking resources. It provides support for multipath, multilayer, multi-domain, and multi-protocol routing for CloNe and NetInf. It includes services for dynamic and distributed mobility and multihoming management, as well as transport network self-management and autonomic control of data center interconnectivity. The OConS is implemented by three functional entities, namely: information management entity (IE), the decision making entity (DE), and the execution and enforcement entity (EE). The IEs manage information collection, aggregation, and preparing for decision making. The DEs make transport decisions considering rules, goals, policies, and preferences. The EEs execute the decisions, enforcing control and management. The model creates a hierarchy of control loops, where open APIs are employed at every interface, including an open API for the entire OConS component.

The CloNe integrates cloud computing and virtual networking to deal with the complex life-cycle of virtual machines and their networking requirements. It introduces the notion of a flash network slice (FNS), which can provide customized, on demand, virtual networking among different provider clouds. CloNe uses implementation specific virtualization interfaces, such as *libvirt*, cloud data management interface, Amazon simple storage service, Open Cloud Computing Interface (OCCI), Amazon elastic compute cloud, libCloud, etc. CloNe provides an application deployment toolkit, which includes a link driver to handle DHCP and SDN configurations.

From CloNe and OConS point of view, NetInf is just an application that requires specific communication support. Meanwhile, CloNe can support NetInf scalable deployment at data centers, managing NetInf components life-cycle and virtual connectivity by creating specific FNSs. From OConS point of view, both NetInf and CloNe are possible clients, since both create overlay communication topologies over the physical substrate. Both also provide discovery mechanisms, which can be seen as duplicated. Importantly, OConS makes no use of NetInf services.

4.5.2 XIA

The XIA is a clean slate *all in one* Internet architecture funded by the U.S. National Science Foundation. The key XIA concepts are: (1) unique identification, (2) expressing intent, (3) flexible addressing, (4) iterative refinement, and (5) intrinsic security. XIA employs global unique identification of networking principals (i.e., administrative domains, hosts, services, and contents). Self-certifying identifiers (XIDs) are calculated by hashing the public cryptographic keys of domains, hosts, and services or the entire binary pattern of contents. Observe that not only XIA [57], but also NG borrowed this idea from previous work, including [70].

To express intent means to give clues to the network—XIDs—that point clearly the desired communication targets (e.g., content, service, host, and domain). For example, when searching for some content using a content ID (CID), an interested client would provide a path that contains the domain ID (AD), host IDs (HIDs), and service IDs (SIDs) where the content can be found. Considering this information, the network can better address the task of satisfying users' intents.

XIA provides a flexible addressing approach where alternative forms of accessing some communicating target (fallbacks) are provided. In other words, XIA routers can use more than one destination address to forward packets to a desired target. Therefore, if some ID is unknown at some router, alternative addressing can help find the desired communicating target (a CID for example). A fallback allows alternative measures in case a router cannot proceed using a preferred intent. Depending on available XIDs on a router, it can iteratively refine the intent by using other XIDs to determine the next hop. For instance, if some router does not know how to route to a certain CID, it can use the AD provided in the address. If the AD cannot route to the desired CID, the router forwards the packet to the informed HID. By using XIDs on packet headers, XIA provides intrinsic security, since address veracity can be verified anytime.

XIA proposes a principal-independent expressive Internet protocol that covers flexible addressing, packet format, routing based on XIDs, and in-network caching of data chunks. The current XIA prototype provides a socket API—similar to the UDP socket. Also, three transport protocols are provided: X-datagram protocol (XDP), X-stream protocol (XSP), and X-chunk protocol (XChunkP). The XDP is an unreliable connectionless message forwarding protocol. The X-stream is a connection oriented reliable message forwarding protocol. Finally, the XChunkP provides reliable content transfer from in-network caches. XIA authors call this mechanism *on-path interception*. XIA routers and hosts can cache content chunks that are transferred to requesters based on their CIDs. Thus, if a content is available in the path to some content server, XIA provides network stored copies of the data chunks to the interested clients. Interestingly, XIA also provides: (1) an X-address resolution protocol, which is similar to the address resolution protocol in the current Internet; (2) an X-control message protocol, which is similar to the Internet control message protocol; and (3) X-host configuration protocol, which provides network bootstrap similar to the current Internet's dynamic host configuration protocol (DHCP). The current router prototype was implemented using Click modular router.

The XIA socket API uses XIDs instead of port numbers and IP addresses. Nowadays, it is implemented as a user-level library using python and C/C++ programming languages. Sockets have been implemented for XDP, XSP, and XChunkP. A service that wants to interact with others needs to select one of these sockets and provide the target services SIDs. Thus, if a target service moves from one host to another, its SID remains the same, while the destination HID should be transparently changed. To enable application communication, XIA proposes an NRS to enable services to resolve XIDs from human language keywords. Associations among human readable names (e.g. "my web server") and XIDs can be published at the NRS.

4.5.3 RINA

RINA is a clean slate architecture designed considering networking to be IPC, and only IPC (i.e., the only thing that a network must provide is a good IPC service for applications). Nowadays, there are four RINA implementations: Alba, TRIA network systems LLC, Boston University and investigating RINA as an alternative to TCP/IP (IRATI) funded by FP7 in Europe. The Alba and Boston University implementations are Java-based, while the TRIA network systems and IRATI are c-based. However, the IRATI RINA offers support to other programing languages via a simplified wrapper and interface generator (SWIG). The IRATI's key RINA abstractions are: (1) distributed IPC facility (DIF), (2) distributed application facility (DAF), (3) resource information base, (4) inter-DIF directory (IDD), (5) error and flow control protocol (EFCP), (6) common distributed application protocol (CDAP), and (7) common application connection establishment phase (CACEP).

The DIF is a single layer abstraction that can be recursively used to create multilayer networks. All DIFs have the same interface and components, regardless of their architectural level. DIFs are populated by IPC processes, which have a common internal structure. The EFCP is implemented as an internal component of the IPC process and is based on the Delta-*t* protocol from the 1970s. It is divided into two protocols: the data transfer control protocol and the data transfer protocol. Another internal component is the RIB. It is a virtual object database. The IPC processes communicate by exchanging operations over the RIB objects using CDAP. The protocol supports object creation, deletion, reading, writing, starting, and stopping. There is an event driven RIB daemon implemented on Linux kernel to access the RIB either locally or remotely. Objects are referred by names or IDs, although the naming/identification approach was not standardized yet.

A DAF is a set of distributed application processes (DAPs) that exchange objects. The DAF communication model requires application naming, the selection of adequate supporting DIFs, the establishment of data flows (with proper QoS) inside the DIFs, the establishment of a connection with peer DAPs (using

CACEP), and carrying of objects (using CDAP). In the current Internet, applications are not named. In RINA, applications are named. Therefore, DIF's flows connect named applications, instead of addresses or ports. Also, flows are identified inside applications by port IDs. The CACEP implements application connection allowing two applications to discover peer names, as well as to authenticate each other. Application connection occurs after flow establishment on a supporting DIF. After connection establishment, communication continues using CDAP messages, which are encoded using GPB, XML, ASN.1, or JSON.

The IDD is a DAF that contains several DAPs supported by one or more DIFs. This distributed system provides application discovery when applications are supported by different DIFs. In other words, there is no DIF that provides communication between the two applications. The problem requires the expansion of an already existent DIF (to cover both applications) or the creation of a new DIF.

While the IRATI RINA implementation is divided between user and kernel spaces in Linux/Unix, the service provided by a DIF is exposed to upper layers or distributed applications by an IPC API. The API contains six methods: allocate flow, data write, data read, deallocate flow, register application, and unregister application. Upper layers can use this API to access the bottom layers' services. The IRATI RINA also provides two shim (a small code to intercept IPC API and redirect to traditional sockets) to expose the TCP/UDP/IP and Ethernet protocols as a regular DIF to upper layers. Thus, it is possible to forward RINA messages over legacy Internet protocols. Netlink sockets, system calls, and sysFS virtual file system are some of the external software used by IRATI RINA implementation. They are accessed by a a common library called *librina*.

4.5.4 Comparison

Table 4.2 provides a comparison among the aforementioned FIA approaches considering distributed systems point of view. Contrary to current Internet approach, where applications have no name—and must be addressed by socket addresses— all studied FI initiatives provide named-entities APIs to client applications. These names are perennial, therefore enabling DSs to establish communication without depending on mutable sockets. Unique named applications are a better abstraction for current DSs requirements, than the 1970s sockets were.

The investigated FI architectures have different strategies regarding integration and heterogeneity requirements. SAIL adopts three independent components: OConS, CloNe, and NetInf. The building blocks at each component need to respect defined macro interfaces. CloNe and OConS deal with heterogeneity of technologies at transport and cloud networking components, respectively. XIA and RINA integration follows a layered approach. However, they differ on number and structure of layers. XIA proposes static layers, like in the current Internet. RINA proposes a generic layer, which can be instantiated and customized

Table 4.2 Comparison of FIA approaches from the distributed systems point of view

FI initiative	SAIL	XIA	RINA	NovaGenesis
API	NetInf offers a NDO-oriented API. OConS offers an API called orchestration service access point (OSAP) for its applications or to the NetInf or CloNe components. The OSAP enables the clients to specify the required connectivity services, as well as the OConS to report the status of requested services. CloNe uses diverse APIs to access other cloud computer solutions, like *libvirt*, OpenStack, OCCI, etc.	Offers socket APIs for each principal.	IRATI RINA offers an IPC API for applications (DAFs).	Offers a pub/sub API for native application development. All NG services are exposed to upper level services adopting a name-oriented approach. Future versions will have shims for transparent support of legacy applications.
Hetero-geneity and Inte-gration	NetInf, CloNE, and OConS are independent components of SAIL. Therefore, some superposition of functionalities is tolerated in the name of independent deployability. OConS and CloNe have built-in support for heterogeneity on transport and cloud, respectively. OConS supports multi RAT at wireless terminals. network selection is also provided. OpenNetInf implementation is based on controllers, that enable the heterogeneity of underlaying search, resolution, transport, and event-driven services. Programming language independency is not clear - many components are implemented exclusively in Java.	The current prototype adopted Linux/Unix and Python and C/C++ programming languages. All components are integrated in stack modules, similarly to the current Internet.	Current prototype adopted Linux/Unix. RINA implementations diverge in terms of programming languages. IRATI RINA kernel components are in C, while Java was the selected language at the user space. SWIG is used to map user space Java code to C libraries. Future versions could support Perl, Python, and Ruby languages at user space.	All services, including protocols, are available to SLA negotiation. NovaGenesis provides a fully integrated environment, where seamless integration of building blocks can occur using the same naming, pub/sub, and name storage approach. NG proxy/gateways enable to interact with any consolidated technology, including TCP/IP, Ethernet, and Wi-Fi. Operating system and programming language independency is being planed to future versions.
Collabora-tion Model	Services can implement peer-to-peer, client-server models, etc. NetInf in-network caching helps on peer-to-peer support. OConS multi-path routing facilitates peer-to-peer implementation.	Services can implement peer-to-peer, client-server models, etc. In-network caching help on peer-to-peer support. Named-services can discover each other using *NameSrv* component and establish collaboration.	DAFs can implement peer-to-peer or client-server models. However, internal DIF components are opaque for applications, requiring functionality duplication at DAFs. Application discovery can involve IDD component.	Supports SLA-based peer-to-peer, traditional client-server, and other collaboration models. Aims at creating "trust networks" among services.

(Continued)

Table 4.2 (*Continued*) Comparison of FIA approaches from the distributed systems point of view

FI initiative	SAIL	XIA	RINA	NovaGenesis
IPC	The OConS has an intra/inter-node communication (INC) component that forwards messages using traditional IPC, like Unix sockets. NetInf relays on event-driven packages (SIENA and Hermes) to perform IPC.	The current prototype employs GPB messages over TCP/IP sockets for intra node IPC. Inter node IPC is done via Click router.	It is the main abstraction in design. Provides a common IPC API for inter node and intra node communication, as well as shims for TCP/IP, UDP/IP, Ethernet, etc.	Inter node process communication employs a shared memory approach. Intra node IPC employs proxy/gateway systems (PGSs) to encapsulate NG messages over legacy networks using sockets (TCP/IP, Ethernet, Wi-Fi).
IoT	NetInf naming supports IoT. However, it is not clear whether NetInf nodes can run embedded. In general, NetInf, CloNe, and OConS were not designed focusing on IoT scenarios.	Expressiveness by XIDs supports the "social devices" idea, enabling self-organization, name-based routing, and content-distribution. However, XIA stack needs to be evaluated for real IoT scenarios. It is not clear how XIA will scale/support timeliness, heterogeneity, and performance for IoT.	DIFs and DAFs can be embedded on IoT devices that support Linux. However, it is not clear how RINA will scale/support all the ideas behind IoT. For example, IDD could not fit on small devices or device mobility could not fit IoT requirements.	The proxy/gateway approach fits well on the high heterogeneous, mobile, distributed, loosely coupled, opportunistically connected IoT environment. NovaGenesis name-oriented and contract-based self-organization creates the adequate support for IoT devices. NG naming, search and discovery, pub/sub, SLA, and mobility support were designed thinking on IoT scalability issues. The current version readily supports M2S and S2S. However, the current prototype is too heavy to be embedded. Future versions will address this challenge, as well as M2M real-time communication.
Service Orientation	The architecture offers an OConS service orchestration process (SOP). It is not name-based. Neither NetInf provides the "named service abstraction", nor CloNe. NetInf is focused on named content only. CloNe concerns to virtual entities, but lack on named services orchestration as well.	Creates a name-oriented digital business ecosystem. Peer application discovery is done via *NameSrv*. Reputation and negotiation are left to applications.	Creates a name-oriented digital business ecosystem. Peer application discovery is done via IDD. Reputation and negotiation are left to applications.	Creates a SLA-based, name-oriented, digital business ecosystem. Peer application discovery and SLA negotiation are exposed to applications by SDS, RpS, and RgS.

(Continued)

Table 4.2 (*Continued*) Comparison of FIA approaches from the distributed systems point of view

FI initiative	SAIL	XIA	RINA	NovaGenesis
Naming	NetInf employs flat names (public signatures and content hashing). OConS employs 128 bits names for hosts. Services have additional 16 bits, which are added at the less significant portion of the host name.	Supports natural language and flat names to current principals, i.e. domains, hosts, services, and content. XIDs have 160 bits long.	Application natural language and flat naming is supported inside DIFs and DAFs. However, it lacks on content, host, domain, and other entities naming. Application and services names are global unique. A complete naming implementation is missing.	Employs both natural language and flat names, which are related via published name bindings. NG naming can accommodate the DS requirements.
Identification and Routing	Inside NetInf, contents are requested by their names. The closest copies are routed in the reverse request path or using underlay routing. At the OConS, the INC component maps destination host names to locators. Notwithstanding, OConS supports multi-path, multi-layer, multi-domain, multi-protocol routing for CloNe and NetInf.	All principals have unique XIDs. XIA routing employs alternative paths (fallback) to route content requests to some communicating target.	The scope of entities/content identification and routing is restricted to DIFs. Hierarchical or flat routing can be implemented. Routes are sequences of DIF process addresses. A node address at a certain DIF is considered as a node name by the underlaying DIF.	All routing and forwarding is name-based. Content requests and delivery are routed using SCNs in message header.
Mapping	The OpenNetInf provides two NRS implemented in Java: MDHT and Pastry. However, it is not used inside OConS, where the INC component is responsible to map OConS IDs to locators, such as IPv6 addresses. The OpenNetInf NRS can be used by DSs inside NetInf. Name resolution is duplicated at OConS and NetInf.	Provides a NRS able to map from human readable names to XIA IDs. Other services and distributed systems can take advantage of this NRS.	Provides a distributed application and process name resolution service called IDD, which is focused on RINA application search and discovery. DIFs map application names (including DIF process names) to underlaying DIF processes addresses.	Name-to-name or name-to-content bindings are published/subscribed and stored on HTs. The PSS is available to services, including DS.

(*Continued*)

Table 4.2 (*Continued*) Comparison of FIA approaches from the distributed systems point of view

FI initiative	SAIL	XIA	RINA	NovaGenesis
Mobility	NetInf decouples names from location. OConS provides a dynamic and distributed mobility manager (DDMM) component, where mobile terminal can access a number of access networks. The radio access technology (RAT) is selected on a flow basis approach. Handover decision includes available technologies, measurements per interface (throughput, delay, load), per flow QoS, operator and user preferences, etc.	Supports mobility of all entities through rebinding XIDs.	At the lowest levels, DIF process addresses are mapped to specific interfaces. Therefore, host mobility requires rebinding of process address to interface names.	Decouples identifiers (unique names in a scope) from locators (names that allows a notion of distance in a certain space) for all entities and content. Mobility support requires update of published name bindings at local domain PSS.
Resource Exposition	OConS resources are exposed to CloNe and NetInf to enable virtual networks establishment. Computing resources are orchestrated at CloNe. The support for computing hardware exposition is not clear. There is not a multi-path-on-top-of-OConS convergence layer at NetInf.	It is left to the user. The support for communication and computing hardware exposition is out of scope.	An underlaying DIF can perform flow admission control based on a list of QoS parameters passed by the IPC API. However, there is no native support for the exposition of substrate resources to DIFs. The support for computing hardware exposition is out of scope.	Substrate resources are exposed by proxy/gateway services, which represent physical resources during SLA negotiation. The support for computing hardware exposition is possible, but not implemented in current version.
HPC and Concurrency	CloNe merges cloud computing with networking to provide high performance computing and storage.	Current prototype is implemented using Click modular router at user space. Memory copies and concurrency are not addressed.	IRATI RINA provides a "fast execution path" for time sensitive functionalities, like packet read/write. Additionally, these functionalities are implemented in the kernel space to reduce context switches.	The current version is implemented only at user space. IPC is based on traditional shared memory. Future versions will included zero copy, context switching, and thread safety as requirements.
Communication Model	Inside NetInf, it is employed a request/response model, while OConS employs the traditional receiver accepts all model. OConS inter node communication provided by INC can be use multi-protocol technologies, like IPv6, Ethernet, Wi-Fi, WiMAX, etc. Additionally, in OpenNetInf nodes can subscribe for changes on named-data objects. Therefore, there is an event-driven pub/sub service for events that affect NDOs and applications.	Employs a request/response model. Named-content transfer starts after a secure session establishment based on XIDs and public/private key pairs.	To communicate applications need to discover each other and authenticate in a common DIF.	Services use publish/subscribe communication paradigm to securely transfer content. A security association can be established during SLA negotiation. Receiver accepts all model is used only for core services bootstrapping.

indefinitely. Interfaces are defined between consecutive layers. NG proposes the most flexible model, where services can be combined in any possible way (since it respects the pub/sub paradigm). Core services can be contracted by any other service in the architecture. The NG protocols-implemented-as-a-service concept enables FIA designers to implement protocols as any other service. Protocol implementation, services, and any application are implemented using the same service development kit. Additionally, NG employs the same programming interface at all services.

Regarding programming language, the proposed FIA prototypes adopt Java, C/C++, Python, etc. However, all architectures will possibly be translated to other languages as they become more mature. The same is valid for operating systems compatibility. Besides the portability issue, the performance of FIA implementations needs to be more carefully studied and evaluated.

Regarding processes collaboration model, FIA proposals provide APIs that helps support the peer-to-peer model, as well as other traditional models. In-network caching is offered by NetInf, XIA, RINA, and NG. However, the internal building blocks of the studied approaches differ regarding accessibility to processes when helping to support collaboration models. NG building blocks (e.g., PSS, PGS, and HTS) are accessible to applications. In other words, application can establish contracts with core NG processes. Therefore, applications do not need to duplicate functionalities, since they are already available inside NG. OpenNetInf does something similar by giving access to its event-driven service. But, the general solution adopted on XIA, RINA, and SAIL was to prohibit access, which will force FI applications to duplicate several functionalities at the application level.

FIA proposals employ traditional intra node and inter node inter service/process communication. The exception is RINA, which considers IPC as the main design abstraction—redesigning IPC from scratch. FIA initiatives need to better investigate and address IPC requirements, since IPC is fundamental for good performance. Among the open challenges are: zero copy, concurrency (when accessing kernel and network interface cards), kernel context switching, threads, and many other IPC issues. FIA proposals should incorporate state-of-the-art IPC research. More research is required to advance FIA proposals regarding intra node and inter node IPC.

The convergence of distributed systems and IoT is another topic related to FIA design. Our chapter revealed that it is not clear whether the proposed architectures can run embedded on IoT devices, as well as that the architectures have not been evaluated for IoT yet. Considering that IoT devices could be the majority of devices in future ICT, this fact is quite surprising. Some of them probably can run embedded when Linux is available on IoT devices. However, there are some concerns: the scalability of the proposed architectures for billions of nodes, the heterogeneity of IoT technologies, the support for time sensitive M2M communications, the self-organization of devices and software, the quest for name-oriented routing, the in-network cache of IoT data and information, the contextualization of IoT data, and the knowledge representation and exchange, among others. There is a great opportunity here to experimentally evaluate the fitness of these architectures regarding IoT scenarios. FIAs should support thousands of distributed architectures and diverse naming strategies, requiring scalable and secure name resolution approach.

Distributed systems and FI are both adopting the software-as-a-service paradigm. Named-service orchestration is an emerging approach that is supported by XIA, RINA, and NG. Ideally, FIA initiatives would support the entire service life-cycle, preferably adopting self-certifying names to identify services, as it is proposed on XIA's XIDs or NG's SCNs. NG considers service life-cycling as a fundamental requirement, which embraces the core protocol implementations.

Naming has an important role in linking human and machine languages. Current DSs employ port names to establish collaboration among distributed processes. Operating systems enable users to name files, directories, and processes. They also attribute names to PIDs, as well as IPC resources, like SHMIDs or MBIDs. Network interfaces are also named (e.g., "eth0" on Linux). Certainly, new Internet architectures will support thousands of DSs, and for this reason, their naming strategies must be comprehensive, secure, and flexible. They need to support natural language names (multilingual), hierarchical names (follow some naming structure), or flat names (self-certifying or public key) for all kinds of existences. XIA and NG can accommodate these requirements. However, XIDs are restricted to XIA principals. NG allows any kind of name to be used as a principal and proposes a name-binding service to represent relationships among named existences. This is a unique feature of NG. All traditional networking ingredients (like forwarding, routing, retransmission, multiple access, and management) can be name-based, and all existences can be a principal.

Also related to naming, are the identification and routing requirements. To identify means to determine unequivocally the communication target. To locate means to determine unequivocally the position of the communication target. NetInf, XIA, RINA, and NG enable name-based routing (i.e., the communication target can be identified by a name that is mapped to a locator for delivery purposes). OConS and CloNe do not employ the name-based approach inside SAIL. However, identification and path definition permeate all architectural levels. For example, a DS could require the identification of a certain process inside a kernel or the definition of a path among process for certain content. Therefore, to support these requirements, it is desirable that FIA approaches can generalize naming, identification, and location for all existences (e.g., content, services, shared memory, paths, hosts, networks, domains, etc.). Additionally, to avoid unnecessary duplication at the application level, architectures should provide generic mapping systems at the FIA core, which can be transparently accessed by distributed systems. This is exactly the NG approach, implemented by PSS, GIRS, and HTS—a distributed pub/sub name resolution service.

Future distributed systems require consistency and continuity of service during any existence mobility. Identification decoupling from location for all levels can provide efficient and flexible mobility support. Host mobility inside a domain is supported by local rebinding of mappings in OConS, XIA, RINA, and NG. XIA and NG provide similar support for service mobility (i.e., rebinding of service ID to host ID). Content mobility (e.g., in-network caching) is also supported by XIA, NetInf, and NG. DSs can ask for the closest copy of some uniquely identified content in these architectures.

From the distributed systems point of view, communication and computing hardware resources need to be exposed to software for adequate orchestration. For example, some network interface card could be shared by several applications in a certain host. By exposing this hardware to software, new data flows can be admitted and the QoS required by each flow guaranteed. The same exposition is wanted for CPUs, disks, etc. Computing hardware (e.g., core in a multicore system) exposition is in the beginning stage and highly required to solve some of the current drawbacks on MPI, for example. FIA approaches are far from named-exposition of computing resources. SAIL supports a top down model for transport (OConS) and virtual machine/network (CloNe) resources reservation. RINA adopts a similar top down model for DIF resources reservation. XIA does not cover this issue. NG exposes substrate resources using a proxy/gateway model. This bottom-up model can be applied to ICT resources, including CPUs, etc.

Finally, there is the communication model issue. From a DDs perspective, the indirect communication (loosely coupled, opportunistic) model is an emerging approach. Many systems, such as financial trading, can be classified as information-dissemination systems wherein a large number of producers (or publishers) distribute information items of interest (events or information) to a similarly large number of consumers (or subscribers). Additionally, senders and receivers do not need to exist at the same time (time uncoupling). FIA initiatives employ a broad spectrum of communication models: from "receiver accepts all" up to "request/response" and "publish/subscribe." FIA proposals should provide support for a diversity of communication models while respecting adopted design principles.

4.5.5 Analysis

The four FIA initiatives compared in the previous subsection bring important benefits to future distributed systems: named-applications, new abstractions for sockets, mobility-friendly design, support for heterogeneity at transport and computer technologies, support to emergent process collaboration and communication models, and software-as-a-service, among other benefits. However, many issues are still to be addressed by the research community:

■ Can FIA solutions be embedded in IoT devices? Right now, it is possible to include some minimal code that follows the FIA's philosophy.

However, too little work has been done in this direction up to now. In the future, the things will probably have supercomputer power, so it is going to be easy to support any FIA. This is a future requirement for NG implementation. The next version will have a very light core that can run embedded in IoT devices.

■ Can FIA approaches fit to IoT scales? They should fit to IoT scales, since there is no Internet without things in the next decades. All future information architectures—DSs, FIAs, and OSs will probably face strong pressure in this direction.

■ Can developed protocols support real time M2M communication? The DDS approach for IoT claims to support this prerequisite. Therefore, it is possible to support real time M2M in clean slate architectures. An FIA approach that is more efficient than current stack is better aligned to this need. However, the real time aspect has been part of the wish list since the beginning of DSs and FIAs.

■ Can FIA support future DSs self-organization requirements? Not all FIAs will support future DSs self-organizing requirements. However, a clean slate convergent information architecture that is solidly build from lower levels can have a better chance of this, since self-organization is intrinsically an emergent phenomenon, a bottom-up collaboration based behavior.

■ Are FIA naming solutions adequate for DSs? Certainly, several of them are not. Especially those that are limited to one or two principals. To support DSs naming, the naming structure needs to be general enough like the one proposed at NG. In other words, several other principals are the underlaying distributed systems, operating systems, and other systems close to the hardware level.

■ Is FIA components' integration cohesive enough? Many of them are poorly integrated. We believe NG is better integrated than the other FIAs presented in this chapter. We considered an integrated approach to all existing life-cycles.

■ Is substrate resources' exposition adequate for DS requirements? Substrate resources' exposition is at an embryonic stage. DSs are already requiring this and providing solutions for it. However, what they need is typically under the scope of FIAs and FIAs do not cover this issue. NG can address this issue via proxy/gateway services. PGSs can expose hardware resources to higher level services, including protocol implementations.

■ How to avoid functionality duplication at DS level? This is exactly one of the main aims of NG and of this chapter. To explore the common issues among Internet, clouds, FIAs, and convergent information architectures to avoid unnecessary duplication by designing deeply integrated and synergistic architectures. NG's unique approach to information exchanging, storage, and processing, including PIaaS, has the potential to reduce functionality duplication. Also, in a contract-based architecture like NG, unnecessary services have no appeal to establish contracts, and would be naturally left behind.

4.6 Conclusion

This chapter put in evidence that the development of a new architecture for the Internet faces many challenges in the landscape of distributed systems' technologies. However, the research in DSs and new Internet architectures advances with less synergy than one could expect. Typically, the DSs research adopts the current Internet as the processes' communication stack. Thus, the FI research can be easily ossified, since it depends on contemporary DSs technologies, which in turn depends on the current Internet protocols. A vicious cycle is formed. Many FI initiatives adopt already established DS solutions, which in general are tightly bound to TCP/IP. Thus, it is very difficult to advance with clean-slate FI approaches without breaking this technological cycle.

A possible solution to this *status quo* is to develop a convergent information architecture that can explore the synergies between both fronts, eliminating unnecessary overlappings, and integrating adopted ingredients towards a unique design. Such an effort is feasible, since FIA and future DS design have several common challenges, which are recurrent in both domains. This is the aim of the NG architecture—to address simultaneously the recurring challenges in DSs and FIA. To this end, NG adopts a self-similar approach where the same design ingredients are employed at both levels. Considering all issues analyzed in this chapter, we contend that NG provides the best combination of ingredients to meet future DSs requirements, addressing them broadly, instead of adopting punctual solutions (that possibly require functionality duplication) or application level implementations. As its main scientific contribution, NG offers the PIaaS concept, where protocols are on-the-fly assembled and disassembled in order to compose a semantical communication between endpoints.

Finally, this chapter presented a theoretical view of FIA and DS synergies and challenges, then presenting the concepts of NG. Future work includes the conclusion of the first NG prototype and its performance, functionality, and scalability evaluations. In this way, besides a qualitative comparison with related work, we intend to demonstrate NG features in a quantitative fashion by both running and comparing them with other FIA approaches.

Acknowledgements

This work was partially supported by Finep, with resources from Funttel, Grant No. 01.14.0231.00, under the Radiocommunication Reference Center (Centro de Referência em Radiocomunicações- CRR) project of the National Institute of Telecommunications (Instituto Nacional de Telecomunicações—Inatel), Brazil. Prof. Rodrigo was partially funded by the FAPERGS (Fundação de Amparo a Pesquisa do Estado do Rio Grande do Sul) agency. Prof. Dhananjay was supported by Hankuk University of Foreign Studies research fund.

Bibliography

[1] A. S. Tanenbaum, *Operating Systems: Design and Implementation*, Prentice-Hall, Inc., Upper Saddle River, NJ, 1987.

[2] A. Zúquete, and C. Frade, A new location layer for the TCP/IP protocol stack, *SIGCOMM Comput. Commun. Rev.* 42 (2) (2012) 16–27. doi:10.1145/ 2185376.2185379.

[3] H. Lowe, Internet/osi application migration/portability, *StandardView* 2 (1) (1994) 46–49. doi:10.1145/224145.224152.

[4] R. T. Fielding, and R. N. Taylor, Principled design of the modern web architecture, *ACM Trans. Internet Technol.* 2 (2) (2002) 115–150. doi:10.1145/ 514183.514185.

[5] G. Carofiglio, and L. Muscariello, On the impact of TCP and per-flow scheduling on internet performance, *IEEE/ACM Trans. Netw.* 20 (2) (2012) 620–633. doi:10.1109/TNET.2011.2164553.

[6] B. E. Carpenter, IP addresses considered harmful, *SIGCOMM Comput. Commun. Rev.* 44 (2) (2014) 65–69. doi:10.1145/2602204.2602215.

[7] Y. Ren, T. Li, D. Yu, S. Jin, T. Robertazzi, B. L. Tierney, and E. Pouyoul, Protocols for wide-area data-intensive applications: Design and performance issues, in: Proceedings of the *International Conference on High Performance Computing, Networking, Storage and Analysis, SC '12*, IEEE Computer Society Press, Los Alamitos, CA, 2012, pp. 34:1–34:11. http:// dl.acm.org/citation.cfm?id=2388996.2389043.

[8] G. Coulouris, J. Dollimore, T. Kindberg, and G. Blair, *Distributed Systems: Concepts and Design*, 5th Edition, Addison-Wesley Publishing Company, Boston, MA, 2011.

[9] J. Al-Jaroodi, and N. Mohamed, Review: Service-oriented middleware: A survey, *J. Netw. Comput. Appl.* 35 (1) (2012) 211–220. doi:10.1016/j. jnca.2011.07.013.

[10] A. N. Toosi, R. N. Calheiros, and R. Buyya, Interconnected cloud computing environments: Challenges, taxonomy, and survey, *ACM Comput. Surv.* 47 (1) (2014) 7:1–7:47. doi:10.1145/2593512.

[11] G. Tyson, N. Sastry, R. Cuevas, I. Rimac, and A. Mauthe, A survey of mobility in information-centric networks, *Commun. ACM* 56 (12) (2013) 90–98. doi:10.1145/2500501.

[12] C. Dobre, and F. Xhafa, Parallel programming paradigms and frameworks in big data era, *Int. J. Parallel Program.* 42 (5) (2014) 710–738. doi:10.1007/s10766-013-0272-7.

[13] J. Landay, Balancing design and technology to tackle global grand challenges, in: Proceedings of the *12th Annual International Conference on Mobile Systems, Applications, and Services, MobiSys '14*, ACM, New York, 2014, pp. 1–1. doi:10.1145/2594368.2620048.

[14] T. Leighton, Improving performance on the internet, *Commun. ACM* 52 (2) (2009) 44–51. doi:10.1145/1461928.1461944.

[15] A. M. Alberti, A conceptual-driven survey on future internet requirements, technologies, and challenges, *J. Braz. Comput. Soc.* 19 (3) (2013) 291–311. doi:10.1007/s13173-013-0101-2.

[16] A. M. Alberti, A. Vaz, R. Brandão, and B. Martins, Internet of information and services (IOIS): A conceptual integrative architecture for the future internet, in: Proceedings of the *7th International Conference on Future Internet Technologies, CFI '12*, ACM, New York, 2012, pp. 45–45.

[17] C. Partridge, Helping a future internet architecture mature, *SIGCOMM Comput. Commun. Rev.* 44 (1) (2013) 50–52. doi:10.1145/2567561. 2567570.

[18] H. Iqbal, and T. Znati, Overcoming failures: Fault-tolerance and logical centralization in clean-slate network management, in: 2010 Proceedings IEEE INFOCOM San Diego, CA, 2010, pp. 1–5. doi:10.1109/INFOCOM. 2010.5462199.

[19] S. Paul, J. Pan, and R. Jain, Architectures for the future networks and the next generation internet: A survey, *Comput. Commun.* 34 (2011) 2–42.

[20] G. group, Geni design principles, *Computer* 39 (9) (2006) 102–105. doi:10. 1109/MC.2006.307.

[21] P. Stuckmann, and R. Zimmermann, European research on future internet design, *IEEE Wireless Commun.* 16 (5) (2009) 14–22. doi:10.1109/ MWC.2009.5300298.

[22] J. Rexford, and C. Dovrolis, Future internet architecture: Clean-slate versus evolutionary research, *Commun. ACM* 53 (9) (2010) 36–40. doi:10.1145/1810891.1810906.

[23] G. J. Chaitin, Algorithmic information theory, *IBM J. Res. Dev.* 21 (4) (1977) 350–359.

[24] V. Jacobson, CCN routing and forwarding, *Tech. Rep.*, Stanford NetSeminar (2011).

[25] G. Lin, D. Fu, J. Zhu, and G. Dasmalchi, Cloud computing: IT as a service, *IT Professional* 11 (2) (2009) 10 –13. doi:10.1109/MITP.2009.22.

[26] T. Aoyama, A new generation network: Beyond the internet and NGN, *IEEE Commun. Mag.* 47 (5) (2009). doi:10.1109/MCOM.2009.4939281.

[27] S. Paul, J. Pan, and R. Jain, Architectures for the future networks and the next generation internet: A survey, *Comput. Commun.* 34 (2011) 2–42.

[28] J. Pan, S. Paul, and R. Jain, A survey of the research on future internet architectures, *IEEE Commun. Mag.* 49 (7) (2011) 26–36. doi:10.1109/MCOM.2011.5936152.

[29] J. Rexford, and C. Dovrolis, Future internet architecture: Clean-slate versus evolutionary research, *Commun. ACM* 53 (9) (2010) 36–40. doi:10.1145/1810891.1810906.

[30] A. M. Alberti, Searching for synergies among future internet ingredients, in: G. Lee, D. Howard, D. Slezak, and Y. Hong (Eds.), *Convergence and Hybrid Information Technology,* Vol. 310 of Communications in Computer and Information Science, Springer, Berlin and Heidelberg, 2012, pp. 61–68.

[31] I. Standardization, Iso/iec 7498-1: 1994 information technology-open systems interconnection-basic reference model: The basic model, *International Standard ISOIEC* 74981 (1996) 59.

[32] A. S. Tanenbaum, *Computer Networks* 4th edition Andrew S. Tanenbaum, Prentice-Hall Englewood Cliffs, NJ, 1989.

[33] A. Montresor, Designing extreme distributed systems: Challenges and opportunities, in: Proceedings of the *8th International ACM SIGSOFT Conference on Quality of Software Architectures, QoSA '12,* ACM, New York, 2012, pp. 1–2. doi:10.1145/2304696.2304698.

[34] H. Jamil, Integrating large and distributed life sciences resources for systems biology research: Progress and new challenges, in: *Transactions on Large-scale Data- and Knowledge-centered Systems iii,* Springer-Verlag, Berlin and Heidelberg, 2011, pp. 208–237. http://dl.acm.org/citation.cfm?id=2028190.2028199.

[35] J. Cleland-Huang, and J. Guo, Towards more intelligent trace retrieval algorithms, in: Proceedings of the *3rd International Workshop on Realizing Artificial Intelligence Synergies in Software Engineering, RAISE 2014,* ACM, New York, 2014, pp. 1–6. doi:10.1145/2593801.2593802.

[36] C. Siaterlis, and B. Genge, Cyber-physical testbeds, *Commun. ACM* 57 (6) (2014) 64–73. doi:10.1145/2602575.

[37] S. K. Das, K. Kant, and N. Zhang, *Handbook on Securing Cyber-Physical Critical Infrastructure,* 1st Edition, Morgan Kaufmann Publishers Inc., San Francisco, CA, 2012.

[38] J. Shen, A. L. Varbanescu, P. Zou, Y. Lu, and H. Sips, Improving performance by matching imbalanced workloads with heterogeneous platforms, in: Proceedings of the *28th ACM International Conference on Supercomputing, ICS '14,* ACM, New York, 2014, pp. 241–250. doi:10.1145/2597652.2597675.

[39] R. Fernandes, K. Pingali, and P. Stodghill, Mobile mpi programs in computational grids, in: *PPoPP '06: Proceedings of the Eleventh ACM SIGPLAN Symposium on Principles and Practice of Parallel Programming* ACM Press, New York, 2006, pp. 22–31.

[40] IBM CORPORATION, Message queue telemetry transport (MQTT) (June 2014). http://www-03.ibm.com/software/products/en/wmq-telemetry.

[41] P. Saint-Andre, Extensible Messaging and Presence Protocol (XMPP): Core, RFC 3920 (Proposed Standard) (October 2004). http://www.ietf.org/rfc/rfc3920.txt.

[42] S. Vinoski, Advanced message queuing protocol, *IEEE Internet Comput.* 10 (6) (2006) 87–89. doi:http://doi.ieeecomputersociety.org/10.1109/MIC.2006.116.

[43] G. Pardo-Castellote, Omg data-distribution service: Architectural overview, in: Proceedings of the *2003 IEEE Conference on Military Communications - Volume I, MILCOM'03,* IEEE Computer Society, Washington, DC, 2003, pp. 242–247. http://dl.acm.org/citation.cfm?id=1950503.1950555.

[44] OpenMP Architecture Review Board, OpenMP application program interface version 3.0 (May 2008). http://www.openmp.org/mp-documents/spec30.pdf.

[45] MPI Forum, MPI: A Message-Passing Interface Standard. Version 2.2. http://www.mpi-forum.org (Dec. 2009) (September 4th 2009).

[46] R. L. Graham, T. S. Woodall, and J. M. Squyres, Open mpi: A flexible high performance MPI, in: R. Wyrzykowski, J. Dongarra, N. Meyer, and J. Wasniewski (Eds.), *PPAM*, Vol. 3911 of Lecture Notes in Computer Science, Springer, Berlin and Heidelberg, 2005, pp. 228–239.

[47] A. Bouteiller, F. Cappello, T. Herault, G. Krawezik, P. Lemarinier and F. Magniette, Mpich-v2: A fault tolerant mpi for volatile nodes based on pessimistic sender based message logging, in: Proceedings of the *2003 ACM/IEEE Conference on Supercomputing, SC '03,* ACM, New York, 2003, pp. 25. doi:10.1145/1048935.1050176.

[48] T. Hamada, N. Nakasato, Infiniband trade association, infiniband architecture specification, vol. 1, release 1.2.1, in: *International Conference on Field Programmable Logic and Applications,* Tampere, Finland, 2005, pp. 366–373. http://www.infinibandta.com.

[49] R. M. Metcalfe, and D. R. Boggs, Ethernet: Distributed packet switching for local computer networks, *Commun. ACM* 19 (7) (1976) 395–404. doi:10.1145/360248.360253.

[50] C. R. R. Maule, iwarp ethernet: Key to driving ethernet into high performance environments, in: *Proceedings of the 2006 ACM/IEEE Conference on Supercomputing, SC '06,* ACM, New York, 2006. doi:10.1145/1188455. 1188726.

[51] N. Boden, D. Cohen, R. Felderman, A. Kulawik, C. Seitz, J. Seizovic, and W.-K. Su, Myrinet: A gigabit-per-second local area network, *IEEE Micro* 15 (1) (1995) 29–36. doi:10.1109/40.342015.

[52] L. Buttyan, and J.-P. Hubaux, *Security and Cooperation in Wireless Networks: Thwarting Malicious and Selfish Behavior in the Age of Ubiquitous Computing,* Cambridge University Press, New York, 2007.

[53] J. Zhang, A survey on trust management for vanets, in: *2011 IEEE International Conference on Advanced Information Networking and Applications,* Biopolis, Singapore 2011, pp. 105–112. doi:10.1109/AINA.2011.86.

[54] M. Raya, P. Papadimitratos, V. D. Gligor, and J.-P. and Hubaux, On datacentric trust establishment in ephemeral ad hoc networks, in: *IEEE INFO-COM 2008 - The 27th Conference on Computer Communications,* Phoenix, AZ, 2008, pp. 1238–1246.

[55] M. M. Mejia, N. M. Peña, J. L. Muñoz, O. Esparza, and M. A. Alzate, Decade: Distributed emergent cooperation through adaptive evolution in mobile ad hoc networks, *Ad Hoc Networks* 10 (7) (2012) 1379–1398.

[56] B. Ahlgren, P. Aranda, P. Chemouil, S. Oueslati, L. Correia, H. Karl, M. Söllner, and A. Welin, Content, connectivity, and cloud: Ingredients for the network of the future, *IEEE Commun. Mag.* 49 (7) (2011) 62–70. doi:10.1109/MCOM.2011.5936156.

[57] D. Han, A. Anand, F. Dogar, B. Li, H. Lim, M. Machado, A. Mukundan, W. Wu, A. Akella, D. G. Andersen, J. W. Byers, S. Seshan, and P. Steenkiste, Xia: Efficient support for evolvable internetworking, in: *Proceedings of the 9th USENIX Conference on Networked Systems Design and Implementation, NSDI'12,* USENIX Association, Berkeley, CA, 2012, pp. 23–23.

[58] S. Vrijders, D. Staessens, D. Colle, F. Salvestrini, E. Grasa, M. Tarzan, and L. Bergesio, Prototyping the recursive internet architecture: The irati project approach, *IEEE Netw.* 28 (2) (2014) 20–25. doi:10.1109/MNET. 2014.6786609.

[59] C. Dannewitz, D. Kutscher, B. Ohlman, S. Farrell, and B. Ahlgren, H. Karl, Network of information (netinf)—An information-centric networking architecture, *Comput. Commun.* 36 (7) (2013) 721–735. doi:10.1016/j.comcom.2013.01.009.

[60] C. Dannewitz, M. Herlich, and H. Karl, OpenNetInf - Prototyping an information-centric network architecture, in: *37th Annual IEEE Conference on Local Computer Networks - Workshops,* Clearwater, FL, pp. 1061–1069. doi:10.1109/LCNW.2012.6424044.

[61] K. Varda, Protocol buffers: Google's data interchange format, *Tech. rep.,* Google (6 2008).

[62] R. Khare, and A. Rifkin, XML: A door to automated web applications, *IEEE Internet Comput.* 1 (4) (1997) 78–87. doi:10.1109/4236.612222.

[63] R. Vanbrabant, *Google Guice: Agile Lightweight Dependency Injection Framework* (Firstpress),192 pages, 1st ed., APress, 2008.

[64] SPARQL 1.1 Query Language, Tech. rep., W3C (2013). http://www. w3.org/TR/sparql11-query.

[65] D. Brickley, and R. Guha, RDF vocabulary description language 1.0: RDF Schema, *Tech. rep.* (2004). http://www.w3.org/TR/rdf-schema/.

[66] A. I. T. Rowstron, and P. Druschel, Pastry: Scalable, decentralized object location, and routing for large-scale peer-to-peer systems, in: *Proceedings of the IFIP/ACM International Conference on Distributed Systems Platforms Heidelberg, Middleware '01,* Springer-Verlag, London, 2001, pp. 329–350. http://dl.acm.org/citation.cfm?id=646591.697650.

[67] M. D'Ambrosio, C. Dannewitz, H. Karl, and V. Vercellone, MDHT: A hierarchical name resolution service for information-centric networks, *Proceedings of the ACM SIGCOMM Workshop on Information-Centric Networking, ICN '11,* ACM, New York, 2011, pp. 7–12. doi:10.1145/2018584. 2018587.

[68] J. Keeney, D. Roblek, D. Jones, D. Lewis, and D. O'Sullivan, Extending siena to support more expressive and flexible subscriptions, in: *Proceedings of the Second International Conference on Distributed Event-based Systems, DEBS '08, ACM,* New York, 2008, pp. 35–46. doi:10.1145/ 1385989. 1385995.

[69] P. Pietzuch, and J. Bacon, Hermes: A distributed event-based middleware architecture, in: *Proceedings 22nd International Conference on Distributed Computing Systems Workshops,* 2002 pp. 611–618. doi:10.1109/ ICDCSW.2002.1030837.

[70] R. Moskowitz, and P. Nikander, Host identity protocol (HIP) architecture, RFC 4423 (Informational) (May 2006). http://www.ietf.org/rfc/rfc4423.txt.

Chapter 5

Collective Intelligence in Networking

Ratneshwer

Jawaharlal Nehru University

Vandana Kushwaha

Banaras Hindu University

CONTENTS

5.1   Introduction

Collective intelligence (CI) is shared or group intelligence that emerges from the collaboration, collective efforts, and competition of many individuals, and appears in consensus decision making. The term appears in socio-biology, political science, and in context of mass peer review and crowd sourcing applications [1]. Although the expression may bring to mind ideas of group consciousness or supernatural phenomena, when technologists use this phrase they usually mean the combining of behavior, preference, or idea of a group of people to create novel insights [2]. (CI) can enhance business outcomes by improving how organizations access the untapped knowledge and experience of their networks to discover and share new ideas, augment skills and distribute workload, and improve forecasting effectiveness. Harnessing (CI) can play an important role in solving age-old problems, disaggregating and distributing work in new and innovative ways, and making better and more informed decisions about the future [3].

Networks, in today's scenarios, support a wide range of applications on them. A network system is a large distributed complex system, with different, often highly nonlinear, time-varying and chaotic behavior. There is an inherent fuzziness in the definition of the controls. Dynamic or static modelling of such a system for control is extremely complex [4]. The traditional solution should be supplemented with (CI). (CI) can effectively be used in the networking domain to enhance the quality of routing algorithms, manage the congestion problems, security, etc.

It is therefore very promising to identify opportunities for integration and inculcation of (CI) methodologies and technologies into networking. This requires judiciously considering the various activities in a computer network for this purpose. Finally, such an effort would aim at the remodeling of computer networks in the context of (CI).

The rest of the chapter is organized as follows. We start with the general discussion of (CI) in Section 5.2. In Section 5.2.1, a framework of (CI) is discussed. In Section 5.2.2, a classification of (CI) is given. Types of problems that can be solved by (CI) are mentioned in Section 5.2.3. In Section 5.2.4, benefits of (CI) are pointed out. Challenges in (CI) implementations are given in Section 5.2.5. In Section 5.2.6, a difference between (CI) and swarm intelligence is given in brief. In Section 5.3, the use of (CI) in computer network domain is explained. In the following subsections, some approaches to implement (CI) are explained. In Section 5.3.1, collaborative filtering approach (CF) and its application in computer networks is given. In Section 5.3.2, the searching and ranking approach and its use in computer networks is mentioned. In Section 5.3.3, the clustering approach and its application in computer networks is given. In Section 5.3.4, optimization problems and their solution in computer networks are mentioned. In Section 5.3.5, decision tree and its applications in computer networks are mentioned.

5.2   What Is Collective Intelligence?

The term "collective intelligence," or CI for short, was originally coined by French philosopher Pierre Levy in 1994 to describe the impact of Internet technologies on the cultural production and consumption of knowledge. Levy argued that because the Internet facilitates a rapid, open, and global exchange of data and ideas, over time the network should "mobilize and coordinate the intelligence, experience, skills, wisdom, and imagination of humanity" in new and unexpected ways [5].

Some other definitions of collective intelligence observed in the literature are mentioned in the next paragraphs.

CI is a phenomenon in sociology where a shared or group intelligence emerges from the collaboration and competition of many individuals. CI is also a form of networking enabled by the rise of communications technology, which has enabled interactivity and users generating their own content [6].

Smith (1994) defines it as "a group of human beings carrying out a task as if the group, itself, were a coherent, intelligent organism working with one mind, rather than a collection of independent agents" [7].

CI is commonly used for generating ideas; discussions; and solutions with networks of employees, customers, and other external parties. Many CI tools consist of virtual environments where participants can interact to discuss ideas and opinions or provide feedback on particular topics. Discussions can be ongoing or fixed for a specific length of time. Typically, participants can make suggestions and receive feedback on their suggestions from other participants, as well as rate and comment upon the suggestions of others. Community ratings and commentary on participants' suggestions can be analyzed to identify themes and comments that resonate most with the community. In addition to allowing organizations to gather ideas and feedback from a diverse, large group of people, CI approaches also help organizations to act on those insights by distributing work to people who are best placed to do it. This not only has the potential to enhance quality, but also to increase efficiency as tasks can be conducted in parallel by many people. CI can also be used to predict the outcomes of future events. Aggregating diverse perspectives, knowledge, and experience of employees and customers can improve the accuracy of predictions. This allows organizations to make more informed, evidence-based decisions that can complement traditional forecasting approaches [8].

There are many different kinds of CI [9], including

- The CI generated by high quality conversations among diverse people working together.

- The CI generated by independent consumers in a market.

- The CI of global information systems we reach through computers.

- The psycho-spiritual fields of CI we can reach through meditation and deep dialogue.

- The CI of whole societies who weave all of these into their cultures and into their political, governmental, and economic institutions.

Spectacular examples of collective online action, such as Wikipedia, the Linux operating system, or numerous other results from the open source movement, have shown the potential of digital tools and a distributed networked public to generate forms of CI [10].

CI systems may significantly vary in nature, from collaborative systems, like open source software development communities, to competitive ones, like problem solving companies that benefit from the competition among participating teams to identify solutions to various R&D problems. The advantage that CI systems earn user communities, together with the fact that they share a number of basic common features, provided the potential for designing a general methodology for their efficient modelling, development, and evaluation [11]. CI not only allows us to harness the knowledge and brainpower of other individuals but also makes use of advanced functional capabilities provided by contemporary software tools that can enhance human reasoning and analytical capabilities, all in a connected real time environment such as the web [12]. As a matter of fact, in the last 20 years, the collective behavior of insect societies related to operations such as foraging, labor division, nest building and maintenance, cemetery formation, etc., has provided the impetus for a growing body of scientific work, mostly in the fields of telecommunication, distributed systems, operation research, and robotics [13].

5.2.1 A framework for CI [14]

During the last few years, the web has evolved to become an amazing source of knowledge for many individual users and communities who share interests and experiences in many fields. The Web 2.0 wave has changed user habits; users have become more active in dealing with the web. Social networks have emerged, opening the way to social software that is becoming very popular. Users are involved in diverse social activities such as social tagging, discussion forums, online group games, bookmark sharing, image and video sharing, blogs, wikis, podcasting, etc. Such social activities initiated by users' interactions, collaboration, and competition generate CI – an important ingredient of Web 2.0 applications [2]. Web applications and services that exploit CI are valuable to users if they are reactive to changes that affect their environment, are context aware, and employ efficient resource management to access and retrieve web resources.

Berri [14] proposed a framework for CI that aims to provide users with different applications to become aware with knowledge for their daily tasks. The framework exhibits four layers; namely web services management layer, knowledge management layer, application layer and user layer.

5.2.1.1 Web services management layer

Web Services Management (WSM) is the forefront layer; it is dedicated to digging into the web in order to gather information. Specialized web agents are used to navigate into the web in order to gather multimedia information. This layer is similar to a web crawler; it is continuously gathering, filtering, and organizing data collected from the web.

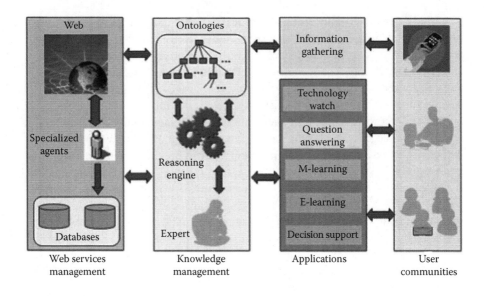

Figure 5.1 A framework for collective intelligence [14].

5.2.1.2 Knowledge management layer

Knowledge Management includes a reasoning engine that uses a set of rules to create and update specific purpose ontologies. Ontologies are used as a backbone structure to organize all concepts related to a given domain of interest. The reasoning engine retrieves semantic relationships from the upper layer WSM, and organizes the whole as a tree of concepts. Ontology creation and update is done from information sources and might be supervised by an expert who intervenes whenever an expert decision needs to be made to refine or update an existing ontology, or to describe a web multimedia resource. In order to be able to automate ontology creation, natural language techniques need to be added to the Web Information Extractor. Two main modules need to be implemented: the first one to extract domain concepts and the second to identify semantic relationships.

5.2.1.3 Application layer

The third framework layer includes applications destined to a variety of user communities. Various applications can be offered ranging from the basic question answering where the user expects an informative answer to his query, to e-learning necessitating the production of a courseware including the generation of learning objects, a recommended learning path, and the management of user learning sessions. Applications aim to produce information that is relevant, instructive, and adaptive. Relevancy is an important criterion in order to not submerge the user with useless information that is not related to his or her request.

The objective is to produce information that precisely fits the user's needs. This is why the user query is analyzed so that concepts and semantic relationships in the query are taken into account by the reasoning engine while retrieving the response.

5.2.1.4 User layer

The user layer manages all possible user communities that interact with the existing applications. They can be individual users or members of communities sharing knowledge with their peers in specific situations. Also, users can be consumers of knowledge by requesting information, but they can be active by creating content to be shared by other community users. Managing community users needs to be done carefully in order to preserve personal information confidentiality and privacy related to a contextual user's information, such as the present location. Also, special care should be given to intellectual property related to user generated resources.

5.2.2 Classification of CI [11]

CI systems need not exist only in the World Wide Web. Instead, any situation where large enough groups of people gather, act individually but also share some common community goals could potentially be – through the proper use of technology – transformed into a CI system.

CI systems may be divided into two main categories: active CI systems and passive CI systems.

5.2.2.1 Passive CI systems

In this type of CI system, individuals act as they normally would without the system's presence. Their behavior and actions, however, may present specific characteristics that can be used by the CI system to provide each one of them with

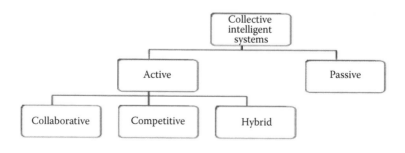

Figure 5.2 Classification of collective intelligence systems [11].

specific guidelines, hints, and coordination so that their shared goal will be more easily achieved. Passive CI systems can be used in almost any case where large groups of people already seem to exhibit collective-mind or swarm-resembling behavior, with each user performing individually but all users sharing a certain number of common goals.

An example of a passive CI system may be implemented in the field of vehicular network coordination as follows: Take the case of large city roads, where large numbers of vehicles move on a daily basis. Drivers perform a simple set of actions (e.g., follow the vehicle in front of them, brake and accelerate). In addition, they individually act in a variety of ways; for instance, some drivers may speed and accelerate suddenly, while others more smoothly; some drivers may prefer leaving a rather long distance between their vehicle and the leading vehicle, while others may leave a shorter distance. Imagine the case when a vehicle is forced to brake or significantly slow down its speed. Then a possible scenario is that vehicles following it will brake as well, reducing the distances among them and eventually ending up in a traffic congestion, which will not be resolved for quite a while; even after the first vehicle has gained its normal speed. The aforementioned scenario represents a typical swarm-resembling behavior and it possibly can be transformed into a passive CI system. That is, technology can be used –either in the form of fixed spots or in the form of an ad hoc communication among the vehicles– in order to warn, all or specific, following vehicles to slow down so that the first vehicles will have adequate time to move and thus avoid the traffic jam that is about to be created. Through this combination of the behavior of the crowd with technology, the collective capabilities and intelligence of the drivers can be facilitated to emerge.

5.2.2.2   Active CI systems

In this type of CI system, crowd behavior does not preexist but it is created and coordinated through specific system requests. This type of system can be further divided into the following categories:

- *Collaborative*: Individuals collaborate with one another in order to reach the community and individual targets.

- *Competitive*: In this type of CI, the system triggers user competition, so that the best solution may be reached.

- *Hybrid*: This last type of CI combines the collaborative and competitive types of systems through, for instance, the competition among groups of collaborating users.

An example of an active CI system of a collaborative nature is the popular online encyclopedia, namely Wikipedia. In this type of system, user behavior did not exist prior to system creation, but instead was created and triggered through it. In this system, decentralized users collaborate and build on the contributions of each other, in order to create encyclopedic articles.

Collaborative groups as a matter of course develop three basic types of information: tangible, intangible, and ephemeral. Tangible knowledge can be divided into target products that represent successful completion of the group's task and instrumental products that support the group's work on the target product but are not part of that product. The collection of target and instrumental products developed and maintained by a group during a project constitutes the group's artifact. Intangible knowledge does not take tangible form but, rather, remains within the heads of the members of the group. Some is shared to an approximation by all members of the group; other is private with respect to an individual or a subset of the group. Ephemeral products lie somewhere between tangible and intangible knowledge. This information is given physical form for brief periods of time, but unlike the instrumental products that are included within the artifacts, ephemeral products are destroyed or lost [7].

5.2.3   Types of problems that can be solved by CI [15]

Rey [15] has suggested seven types of problems or challenges that that can be suitable for open and participatory projects with good results:

- *Creativity*: CI is quite effective at generating ideas. The more people thinking, the more likely they will find a creative solution.

- *Bias assessment*: Activities that are highly susceptible to selection and assessment biases due to their inherent relativity or spurious interests. CI works well in data interpretation tasks subject to many different perspectives. Opening the analysis to a wide variety of points of view can help reduce the "expert bias" and achieve a more complete and balanced judgment.

- *Distributed Surveillance*: Activities in which the cost of failure is high. Any errors are best detected if more people are reviewing (Remember Linus's Law enunciated by Eric Raymond: "Given enough eyeballs, all bugs are shallow").

- *Prediction*: Tasks that involve predicting the future or estimating the probability of events, because in principle the large numbers help to lower certain kinds of biases. For example, in response to: "What product will be more successful? What technology is most suitable for...?" Here I would like to clarify that the CI-based prediction works best as a complement or in combination with expert assessments. A better solution could be that the collective evaluation team is composed of a significant number of "experts"; that is, people who understand the problem well.

■ *Passion*: Activities where enthusiasm and engagement make a difference. Working with highly committed people, with enthusiastic "pro-am" people who are involved only because they are motivated by the challenge, can translate into an influx of quality that is difficult to measure.

■ *Sense of community*: Activities that attract because of their strong collaborative nature and people pointing at them looking to enjoy socializing experiences. In this case what matters are the emotions and finding a sense of belonging. If the project moves through territories that awaken "socializing instincts" (share, talk, show, discuss, teach, share, etc.), it may be a good candidate for the use of CI.

■ *Multidisciplinary*: Challenges whose solution requires a complex and diverse mix of knowledge inputs. The more multidisciplinary is the problem, the better CI works because the necessary mix of know-how and skills will self-select without leaving out any viewpoint that can add value to the analysis.

5.2.4 Benefits of CI [16]

Applying CI to your application impacts it in the following manner:

■ *Higher retention rates*: The more users interact with the application, the stickier it gets for them, and the higher the probability that they will become repeat visitors.

■ *Greater opportunities to market to the user*: The greater the number of interactions, the greater the number of pages visited by the user, which increases the opportunities to market to or communicate with the user.

■ *Higher probability of a user completing a transaction and finding information of interest*: The more contextually relevant information that a user finds, the better the chances that he or she will have the information he or she needs to complete the transaction or final content of interest. This leads to higher click through and conversion rates for your advertisements.

■ *Boosting search engine ranking*: The more users participate and contribute content, the more content available in your application and indexed by search engines. This could boost your search engine ranking and make it easier for others to find your application.

5.2.5 Challenges in CI implementations [8]

The challenges for CI come in a variety of forms. These can be roughly divided into three groups: challenges in design, administration, and leadership.

5.2.5.1 Design challenges

The design choices made regarding the user interface of tools are extremely important because potential participants will only get involved if the user interface guides them in a straightforward and meaningful way. CI requires the design of both technical infrastructure and human-human interaction.

5.2.5.2 Administrative challenges

There are a number of administrative and operational hurdles to overcome in CI, and there has been a great deal of research to understand how active participation in online communities can be enhanced and maintained. Overcoming these challenges is crucial because human input is fundamental to CI–without enough fuel (and the right sort of fuel), the engine cannot run properly. The main administrative challenges are: lack of participation, lack of detail in participant input, anonymity, dealing with conflict and abuse, recruiting participants, motivating participants, and incentivizing participants.

5.2.5.3 Managerial and leadership challenges

CI in organizations poses some specific challenges for management and leadership because tapping into CI means taking a more open and transparent approach–and this requires significant cultural change. The main challenges are: loss of control, overcoming resistance, ethical treatment of crowd workers, intellectual property and security issues, lack of action in response, and creating specific roles.

5.2.6 Difference between CI and swarm intelligence (SI)

SI is a particular form of CI that relies on the capabilities of several *minimally intelligent but autonomous individuals* [17]. One of the possible forms of collective intelligence is the SI.

Alternatively, Fukuda and Ueyama [18] proposed the following definitions. Group intelligence can be divided into CI and swarm intelligence. CI arises from interaction among rational, intelligent individuals with negotiation capabilities while SI emerges from non-intelligent individuals. This taxonomy is clearly based on a different, more classical concept of intelligence, close to that proposed in [19].

Communication systems between individual insects in nature contribute to the formation of the "collective intelligence" of social insect colonies. Recently, the term "swarm intelligence" has been applied, denoting this "collective intelligence" [20].

CI has become a popular research topic over the past few years. However, the CI debate suffers from several problems such as that there is no unanimously agreed-upon definition of CI that clearly differentiates between CI and related terms such as SI and collective intelligence systems. Kornrumpf et al. [21] have argued that CI can be defined as the ability of sufficiently large groups of individuals to create an emergent solution for a specific class of problems or tasks.

5.3   CI in Computer Networks

CI overlaps with the subset of computer science that involves intelligent behavior by groups of people, computers, or both. For instance, "groups" of one person and one computer (human-computer interaction) are peripherally part of CI, but combining multiple people and computers to solve problems is central to CI (e.g., human computation, crowd-sourcing, social software, computer-supported cooperative work, groupware, collaboration technology) [22].

Computer network management is traditionally composed of stages executed in a logical and continuous sequence, having as a result the management itself. Network management also requires implementations that involve intelligence. Conditional data gathering or decentralized proactive detection are great examples of such utilization. CI can be used in all situations that involve decision-making in computer network management [23]. CI can effectively be used in networking domain to enhance the quality of routing algorithms, manage the congestion problems, security, etc. Collective intelligence may collect information at each router in the network to dynamically monitor the entire network to perform various management activities. The fact that nature has served as a major source of inspiration for the design of novel routing algorithms can be understood by noticing that these biological systems are characterized by the presence of a set of distributed, autonomous, minimalist units, that through local interactions self-organize to produce system level behaviors that show lifelong adaptation to changes and perturbations in the external environment [24].A proactive network has various points that are subject to reasoning and decision-making, each one with its characteristics within the process. There are situations of classification, of choices based on rules, or even conclusions based. For this reason, there exists the possibility of applying many paradigms of CI in network management, such as fuzzy logic, neural networks, genetic algorithms, tree-based decisions, or specialist systems. They may be combined or utilized individually, in one point or all points of decision-making [23].

Complex networks are frequently used for modelling interactions in real-world systems in diverse areas, such as sociology, biology, information spreading and exchanging, and many other different areas. One key topological feature of real-world complex networks is that nodes are arranged in tightly knit groups that are loosely connected to each other. Such groups are called communities. Nodes composing a community generally share common properties, are involved in a similar function, and have a similar role. Hence, unfolding the community structure of a network could give us much insight about the overall structure of a complex network [25]. In addition to allowing a network system to gather feedback from a diversity of routers, CI approaches also help the system to distribute network load to routers who are the best candidate for it. This not only helps to enhance quality, but increases efficiency as a task can be conducted effectively. Aggregating records from different routers, the network system can also predict the future behavior of the system, and some evidence based decisions can be made. Since network environments of the future will likely consist of both human and artificial intelligence (AI) agents, it is important to explore whether groups of humans exhibit the phenomenon of CI when working on the types of highly structured tasks where AI problem solving agents excel [26]. It may be a fruitful approach to understand CI from social interactions and use them in computer network areas.

There are various approaches to implement CI. We have mentioned these approaches briefly and discussed how they can be used in computer network domain. We limit our discussion to only computer networks and exclude other types of networks like electrical networks, etc.

5.3.1   Collaborative filtering

Collaborative filtering (CF) is a popular recommendation algorithm that bases its predictions and recommendations on the ratings or behavior of other users in the system. The fundamental assumption behind this method is that other users' opinions can be selected and aggregated in such a way as to provide a reasonable prediction of the active user's preference. This strategy is analogous to the common information retrieval method of producing relevance scores for each document in a corpus with respect to a particular query and presenting the top-scored items. Indeed, the recommended task can be viewed as an information retrieval problem in which the domain of items (the corpus) is queried with the user's preference profile [27]. CF is a widely used framework for rating prediction in recommender systems. Based on the assumption that users with similar tastes would rate items similarly, this framework first finds a group of users having similar interests. Ratings given by the users from that group are used to predict unknown ratings. User based CF algorithms assign weights to the users to capture similarities between them. The weighted average of similar users' ratings of the test item is output as prediction. Unlike content-based filtering, the key idea of CF is that users will prefer those items that people with similar interests prefer, or even that dissimilar people don't prefer.Amazon.com has been using CF for a decade to recommend products to their customers [28].

Next we explain the one approach of implementing CF—the Neighborhood-based approach.

5.3.1.1 Neighborhood-based approach [29]

Most CF systems apply the so called neighborhood-based technique. In the neighborhood-based approach a number of users are selected based on their similarity to the active user. A prediction for the active user is made by calculating a weighted average of the ratings of the selected users.

To illustrate how a CF system makes recommendations consider the example in the movie ratings table.

This shows the ratings of five movies by five people. A "+" indicates that the person liked the movie and a "−" indicates that the person did not like the movie. To predict if Ken would like the movie "Fargo," Ken's ratings are compared to the ratings of the others. In this case the ratings of Ken and Mike are identical, and because Mike liked Fargo, one might predict that Ken likes the movie as well.

	Amy	Jeff	Mike	Chris	Ken
The Piano	−	−	+		+
Pulp Fiction	−	+	+	−	+
Clueless	+		−	+	−
Cliffhanger	−	−	+	−	+
Fargo	−	+	+	−	?

Instead of just relying on the most similar person, a prediction is normally based on the weighted average of the recommendations of several people. The weight given to a person's ratings is determined by the correlation between that person and the person for whom to make a prediction. As a measure of correlation the Pearson correlation coefficient can be used. In this example a positive rating has the value 1 while a negative rating has the value −1, but in other cases a rating could also be a continuous number. The ratings of persons X and Y of the item k are written as X_k and Y_k, while $\bar{X}$ and $\bar{Y}$ are the mean values of their ratings. The correlation between X and Y is then given by:

$$r(X,Y) = \frac{\sum_k (X_k - \bar{X})(Y_k - \bar{Y})}{\sqrt{\sum_k (X_k - \bar{X})^2 \sum_k (Y_k - \bar{Y})^2}} \tag{5.1}$$

In this formula k is an element of all the items that both X and Y have rated. A prediction for the rating of person X of the item i based on the ratings of people who have rated item i is computed as follows:

$$p(X_i) = \frac{\sum_Y Y_i \cdot r(X,Y)}{n} \tag{5.2}$$

Where Y consists of all the n people who have rated item i. Note that a negative correlation can also be used as a weight. For example, because Amy and Jeff have a negative correlation, that Amy did not like "Fargo" could be used as an indication that Jeff will enjoy "Fargo".

The Pearson correlation coefficient is used by several CF systems. Group-Lens, a system that filters articles on Usenet, was the first to incorporate a neighborhood-based algorithm. In GroupLens a prediction is made by computing the weighted average of deviations from the neighbor's mean:

$$p(X_i) = \bar{X} + \frac{\sum\limits_{Y} (X_i - \bar{X}) \cdot r(X,Y)}{\sum\limits_{Y} r(X,Y)} \tag{5.3}$$

Note that if no one has rated item i the prediction is equal to the average of all the ratings person X has made. Ringo recommends music albums and artists a person might be interested in. Shardanand considers several collaborative algorithms and reports that a constrained Pearson r algorithm performs best for Ringo's information domain. The constrained Pearson measure is similar to the normal Pearson measure but uses the mean value of possible rating values (in this case the average is 4) instead of the mean values of the ratings of persons X and Y.

5.3.1.2 Collaborative filtering in computer networks domain

Recommender systems, in computer network domain, may help to build a model that provides a personalized and prioritized list of network nodes that could be significant for a particular network scenario. In other words recommender systems model the network behavior for the specific requirements. Nodes can be priorities according to the bandwidth consumption, data loss, network congestion, etc.

Some relevant applications of CF in computer network domain are mentioned in the next sections.

Adapting policy-based management of future networks using CF techniques [30]

Future networks constitute a complex and dynamic environment that network operators are called to orchestrate uniformly and efficiently. Among others, the increase in the number and heterogeneity of network infrastructure, the complexity of devices and protocols, and the explosion in traffic demands are only a "few" of the issues that network operators should take into consideration. Conventional network management schemes lack in automation, harmonization, and efficiency, preventing them from handling such chaotic environments.

Concentrating on ANM (Automatic Network Management) system design, research efforts are grouped into two sectors: those that propose distributed network management models and those that follow more centralized solutions [30].

In distributed approaches, several autonomic managers exist, sparsely placed in the network. The ANM system is responsible for defining how these managers should communicate and cooperate so as to converge to a global optimum. The proposed policy-based network management framework is made up of four basic components: (1) Human to Network Interface, (2) Policy Derivation and Management, (3) Conflict Resolution, and (4) ANE Enforcement. Together, those four modules enable closed-loop, scalable, end-to-end quality of service (QoS) control and resource management in converged future networks. Due to the fact that ANE elements' design and development are realized by several network and service providers, there is a strong need for defining a formal information model that manufacturers should conform to. Regarding the flow of information from ANE elements to the framework, a self-descriptive manifest of ANE instance is provided.

Initially, one can consider a set of ANEs that have been registered in the past and their manifests are stored in framework repository. Those ANEs are being held in groups, in accordance with how similar of behavior they present. The features used to evaluate the similarity between ANEs are the functionality family they belong to, the network technology they serve, the knowledge they produce, and the actions they enforce. For each group in the repository, a representative ANE is defined, for future calculations and field matching. Representative ANEs from all groups constitute a recommendation set (RS).During the registration of a new manifest, the extraction of the previously mentioned fields takes place. An iterative process then follows where the new ANE is compared to ANEs belonging to the RS. Similarity between couples of ANEs is calculated. In case ANEi presents similar behavior with one of the ANEs in the RS, then the considered ANE is inserted into the appropriate group, and its manifest is stored in framework's repository. Otherwise, a new entry to the RS takes place, along with the instantiation of the ontology with features extracted from the ANEi manifest. The performance with CF depends on the overhead from requests as well as on the probability that several ANEs exhibit similar behavior.

A distributed CF recommendation model for P2P networks [31]

With the rapid development of network technology and the continuous increase of resources, users need a more personalized service to help them find the information they need from enormous resources. The so-called personalized service considers the need and choice of individual users, applies different strategies for different users, and meets the users' personalized needs by providing pertinent content. The most prevailing and effective way is to use CF. CF is based on the fact that, if different users' rating of the information that has already been judged is similar, then their rating of other information that has not been judged should be similar. Thus, based on other users' evaluation and combining the target user's personal history record, the CF recommendation system provides the user with the information he or she may find most interesting.

Fuzzy trust recommendation based on CF for mobile ad-hoc networks [32]

Mobile ad-hoc networks (MANETs) are based on cooperative and trust characteristics of mobile nodes. Typically, nodes are both autonomous and self-organized without requiring a central administration or a fixed network infrastructure. Due to their distributed nature, MANETs are very vulnerable to various attacks. To enhance the security of MANETs, it is important to rate the trustworthiness of other nodes without central authorities to build up a trust environment. In MANETs, a trust relationship formed from direct interactions can be characterized as direct trust; a trust relationship or a potential trust relationship built from recommendations by a trusted node or a chain of trusted nodes, which create a trust path, is called indirect trust [33]. Regarding the aspect of recommendation, a node with high trust value does not correspond to the high or correct recommendation to other nodes. The trustworthiness of the recommendation of a node is different from that of the node itself, especially under some attack conditions, such as malicious collusion attacks. So, the hypothesis that nodes with the high trust value will give honest recommendations is questionable. The trust model in a MANET environment is hard to assess due to the uncertainties involved. The theory of fuzzy logic extends the ontology of mathematical research to be a composite that leverages quality and quantity, and which contains certain fuzziness. Luo et al. solve the issues associated with uncertainty in a MANET trust management by introducing fuzzy logic into the research of trust management by combining the CF. Mobile nodes need to be equipped with efficient facilities to calculate and evaluate trust and credibility values of other nodes in MANETs. The subjectivity and uncertainty contained in the individual notions and definitions of trustworthiness and credibility need a flexible and adjustable model. It needs to deal with cooperation risk measurement relying on the approximate estimation of a node's behavior instead of detailed and crisp data. Their trust model combines direct trust and trust recommendation information based on CF to allow nodes to represent and reason with uncertainty and imprecise information regarding other nodes' trustworthiness.

5.3.2 Searching and ranking

Ranking for different network components is among the most important collective intelligence algorithms. It is widely believed that Google's rapid rise from an academic project to the world's most popular search engine was based largely on the PageRank algorithm. Ranking with CI techniques has become increasingly popular and more important with the advent of new network technologies. Ranking individual elements within networks, including both nodes and links, allows for the identification of important subsets of elements for resource prioritization, such as finding the most authoritative web pages related to a search topic in the Internet, discovering the most influential people in a social network, evaluating the most cited scientific papers in a citation network, or identifying the most vulnerable components in infrastructure systems (e.g., transportation networks, power grids, and water systems). Quantification of the criticality of network components helps decision makers inform their management strategies. The ranking of network nodes or links addresses the question "Can important nodes or links in a network be meaningfully identified while keeping input data and computational resources low?" [30]

5.3.2.1 Ranking methods for networks [34]

Ranking objects in a network may refer to sorting the objects according to importance, popularity, influence, authority, relevance, similarity, and proximity, by utilizing link information in the network. For most of the ranking methods in networks, ranking scores are defined in a way that can be propagated in the network. Therefore, the rank score of an object is determined by other objects in the network, usually with stronger influence from closer objects and weaker influence from more remote ones.

Methods for ranking in networks can be categorized according to several aspects, such as global ranking vs. query-dependent ranking, based on whether the ranking result is dependent on a query; ranking in homogeneous information networks vs. ranking in heterogeneous information networks, based on the type of the underlying networks; importance-based ranking vs. proximity-based ranking, based on whether the semantic meaning of the ranking is importance related or similarity/proximity related; and unsupervised vs. supervised or semi-supervised, based on whether training is needed.

Centrality and prestige

In network science, various definitions and measures are proposed to evaluate the prominence or importance of a node in the network. Centrality and prestige are two concepts to quantify prominence of a node within a network, where centrality focuses on evaluating the involvement of a node no matter whether the prominence is due to the receiving or the transmission of the ties, whereas prestige focuses on evaluating a node according to the ties that the node is receiving [35].

In information network analysis, the most well-known ranking algorithm is PageRank [36], which has been successfully applied to the Web search problem. PageRank is a link analysis algorithm that assigns a numerical weight to each object in the information network, with the purpose of "measuring" its relative importance within the object set.

SimRank

SimRank [37] is used to calculate pair-wise similarity between objects in a network based on the link information. The intuition of the similarity model is based on the idea that "two objects are similar if they are related to similar objects." In other words, the similarity between objects can be propagated from pair to pair via links. SimRank can also be applied to bipartite networks, where similarity between one type enhances the quality of the other type alternatively. The time complexity of computing SimRank is high, as the similarity score between a pair of objects is dependent on the similarity between every other pair of objects.

Some significant efforts regarding searching and ranking methods in computer network domain are mentioned in the next section.

5.3.2.2 Searching techniques in peer-to-peer networks [38]

A peer-to-peer (P2P) network is a distributed system in which peers employ distributed resources to perform a critical function in a decentralized fashion. Nodes in a P2P network normally play equal roles; therefore, these nodes are also called peers. A typical P2P network often includes computers in unrelated administrative domains. These P2P participants join or leave the P2P system frequently; hence, P2P networks are dynamic in nature. P2P networks are overlay networks, where nodes are end systems in the Internet and maintain information about a set of other nodes (called neighbors) in the P2P layer. The desired features of searching algorithms in P2P systems include high-quality query results, minimal routing state maintained per node, high routing efficiency, load balance, resilience to node failures, and support of complex queries.

Searching in highly structured systems follows the well-defined neighboring links. For this reason, highly structured P2P systems provide guarantees on finding existing data and bounded data lookup efficiency in terms of the number of overlay hops; however, the strict network structure imposes high overhead for handling frequent node join-leave. Unstructured P2P systems are extremely resilient to node join-leave, because no special network structure needs to be maintained. Searching in unstructured networks is often based on flooding or its variation because there is no control over data storage. The searching strategies in unstructured P2P systems are either blind search or informed search. In a blind search such as iterative deepening [39], no node has information about the location of the desired data. In an informed search such as routing indices [40], each node keeps some metadata about the data location.

5.3.2.3 A new mutually reinforcing network node and link ranking algorithm [41]

Ranking individual elements within networks, including both nodes and links, allows for the identification of important subsets of elements for resource prioritization, such as finding the most authoritative web-pages related to a search topic in the Internet, discovering the most influential people in a social network, evaluating the most cited scientific papers in a citation network, or identifying the most vulnerable components in infrastructure systems (e.g., transportation networks, power grids, and water systems). Quantification of the criticality of network components helps decision makers inform their management strategies.

The ranking of network nodes or links addresses the question "Can important nodes or links in a network be meaningfully identified while keeping input data and computational resources low?" One of the simplest node importance measures in a network is the degree of a node $k(i)$, denoting the number of links connected to it. Traditionally, the adjacency matrix A is used to describe the connectivity patterns of nodes and links of a network.

5.3.2.4 Ranking of network algorithms using Analytical Hierarchy Process (AHP)

Emergence of different types of networks motivated the research community to design a new class of network algorithms that can perform efficiently in such diverse networks. As a result there are a number of algorithms proposed by the researchers and academicians in the last decade for different network domains. These algorithms are evaluated based on various performance issues like efficiency, friendliness, fairness, RTT fairness, convergence, etc. This is a tough problem to choose which algorithm will be appropriate for which network environment because every network environment may have their specific performance requirements.

AHP [42] is one of the ways to rank items based on different criteria. Xi et al. [43] have proposed an AHP model to assess computer network information security. They researched all kinds of factors influencing network security and constructed the evaluation indexes for computer network information security. Uddin et al. [44] have proposed an on-demand source routing protocol for MANET that works with six important QoS attributes by varying priority for different categories of traffic flow by incorporating Analytic Hierarchy Process in their proposal.

Congestion control is considered as an important issue for every network domain. Large numbers of approaches are available for network congestion control. Thus, there is a need of some model for proper selection of an optimal approach for congestion control in high speed networks. The AHP model can be used for modelling the selection of congestion control approach. A two phase AHP modeling approach [45] has been proposed in literature. The first phase modeling gives an idea about the best congestion control approach in each of three categories: loss based, delay based, and hybrid of congestion control algorithms for high speed networks. Further, the second phase modeling gives the best alternative among the three categories: loss based, delay based, and hybrid by using the results of the first phase. Such type of modelling is helpful in finding the relative ranking of considered approaches in both the phases. By using these rankings one can take the decision about the selection of an optimal approach for better control of congestion in high speed networks. Congestion control is also considered as an optimization problem as we have to achieve several conflicting goals while controlling congestion. Achieving efficiency, friendliness, and responsiveness simultaneously is one of the conflicting goals while controlling congestion. Thus, selection of an approach that performs optimally for every performance criterion is a difficult issue, and the AHP model can give an optimal solution to this issue.

5.3.3 Discovering groups using clustering in computer networks

CI can be utilized effectively in computer networks' challenging problems. Clustering may be the effective means to achieve this if we effectively include the concepts of CI with it. One can cluster network components/network data, retrieve the variations or solution patterns, and develop a method based on the observations. In computer networks, one can group network nodes into a number of clusters. The goal of such clustering is to make the network nodes in the same cluster share a high degree of similarity while being very dissimilar to the network nodes of other clusters. In this case CI-based clustering may be preferable to traditional clustering. Efforts should be made to aggregate the collective observations of different groups in order to infer newer information. With the advent of information technologies such as local area networks, real-time digital communications, and distributed software, researchers began exploring group problem solving in computerized network environments. Since network environments of the future will likely consist of both human and AI agents, it is important to explore whether groups of humans exhibit the phenomenon of CI when working on the types of highly-structured tasks where AI problem solving agents excel. A better understanding of human performance on these types of problem solving tasks is a first step towards the study of problem solving by groups composed of both human and non-human agents [46].

5.3.3.1 Objectives for network clustering

- *Load balancing:* Even distribution of sensors among the clusters is usually an objective for setups where CHs perform data processing or significant intra-cluster management duties [47]. Given the duties of CHs, it is intuitive to balance the load among them so that they can meet the expected performance goals [48]. Load balancing is a more pressing issue in WSNs where CHs are picked from the available sensors [49]. In such case, setting equal-sized clusters becomes crucial for extending the network lifetime since it prevents the exhaustion of the energy of a subset of CHs at high rate and prematurely making them dysfunctional.

- *Fault-tolerance:* In many applications, WSNs will be operational in harsh environments and thus nodes are usually exposed to increased risk of malfunction and physical damage. Tolerating the failure of CHs is usually necessary in such applications in order to avoid the loss of important sensors' data. The most intuitive way to recover from a CH failure is to re-cluster the network. However, reclustering is not only a resource burden on the nodes, it is often very disruptive to the ongoing operation. Therefore, contemporary fault-tolerance techniques would be more appropriate for that task.

■ *Increased connectivity and reduced delay:* Unless CHs have very long-haul communication capabilities (e.g., a satellite link) inter-CH connectivity is an important requirement in many applications. The connectivity goal can be limited to ensuring the availability of a path from every CH to the base-station [50]. When data latency is a concern, intra-cluster connectivity becomes a design objective or constraint.

■ *Minimal cluster count:* This objective is particularly common when CHs are specialized resource-rich nodes [51]. The network designer often likes to employ the fewest number of these nodes since they tend to be more expensive and vulnerable than sensors.

■ *Maximal network longevity:* Since sensor nodes are energy-constrained, the network's lifetime is a major concern; especially for applications of WSNs in harsh environments. When CHs are richer in resources than sensors, it is imperative to minimize the energy for intra-cluster communication [52]. If possible, CHs should be placed close to most of the sensors in its clusters [51,53]. On the other hand, when CHs are regular sensors, their lifetime can be extended by limiting their load.

Some significant approaches to apply clustering in computer network domain are discussed in the next section.

5.3.3.2 Clustering used in wireless sensor networks [54]

The past few years have witnessed increased interest in the potential use of wireless sensor networks (WSNs) in applications such as disaster management, combat field reconnaissance, border protection, and security surveillance. Sensors in these applications are expected to be remotely deployed in large numbers and to operate autonomously in unattended environments. Grouping sensor nodes into clusters has been widely pursued by the research community in order to achieve the network scalability objective. Clustering is done to relate similar nodes and saves necessary energy wasted in direct data transmission to the base station. Nodes in the network organize themselves into hierarchical tier-structures. Within a particular cluster, data aggregation and forwarding are performed at the cluster-head (CH) to reduce the amount of data transmitting to the base station. Cluster formation is usually based on the remaining energy of sensor nodes and the sensor's proximity to the CH [55]. Nodes other than the CH choose their CH right after deployment and transmit sensed information to the CH. The role of the CH, being itself a sensor node, is to forward these data and its own data to the base station after performing data aggregation.

Every cluster would have a leader, often referred to as the CH. Although many clustering algorithms have been proposed in the literature for ad-hoc networks, the objective was mainly to generate stable clusters in environments with mobile nodes. Many of such techniques care mostly about node reachability and route stability, without much concern about critical design goals of WSNs, such as network longevity and coverage. The cluster membership may be fixed or variable. CHs may form a second tier network or may just ship the data to interested parties (e.g., a base-station or a command center).

5.3.3.3 Adaptive bio-inspired clustered routing for MANET [56]

MANET is a mobile multi-hop wireless self-organized distributed ad-hoc network that does not require the basic internal construction. Routing in MANET is a challenging problem that draws researchers' vision, due to nodes mobility, dynamic topology, and lack of central point like base station or servers. Clustering of devices in MANET could reduce overhead, flooding, and collision in communication and make the network topology more stable. The ABC (Artificial Bee Colony) algorithm is a new meta-heuristic population based optimization technique inspired by the intelligent foraging behavior of honeybee swarms. Clustering provides scaling and eases in routing with efficient resource management in MANETs.

A cluster is a group of linked nodes, working together closely so that in many respects they form a single network, in which all nodes work together for the same purpose and belong to the same topological structure. In highly dynamic MANETs, where routing is a major problem for interrupted communication, clustering helps in maintaining a relatively stable effective topology [57,58]. As mobile nodes move around in the network, the structure of cluster changes according to the specified criterion in the algorithm. In a large network, complete information is maintained within a cluster using proactive routing algorithms; that is, intra-cluster routing, while the inter-clustering routing is achieved by using reactive routing algorithms. In a cluster structure a CH is allocated to each cluster for performing various task of inter- and intra-cluster maintenance. The performance of a cluster also depends upon the number of cluster member nodes in the cluster.

5.3.4 Optimization problems in computer networks

Optimization techniques are typically used in problems that have many possible solutions across many variables and that have outcomes that can change greatly depending on the combinations of these variables. These optimization techniques have a wide variety of applications: we use them in physics to study molecular dynamics, in biology to predict protein structures, and in computer science to determine the worst possible running time of an algorithm. Optimization finds the best solution to a problem by trying many different solutions and scoring them to determine their quality. Optimization is typically used in cases where there are too many possible solutions to try them all. The simplest but least effective method of searching for solutions is just trying a few thousand random guesses and seeing which one is best. More effective methods involve intelligently modifying the solutions in a way that is likely to improve them [2].

Network optimization has always been a core problem domain in computer science, applied mathematics, and many fields of engineering and management. The varied applications in these fields not only occur "naturally" on some transparent physical network, but also in situations that apparently are quite unrelated to networks. Types of network optimization are: (1) shortest paths, (2) maximum flows, (3) minimum cost flows, (4) assignment problems, (5) matching, (6) minimum spanning trees, (7) convex cost flows, (8) generalized flows, (9) multi-commodity flows, (10) the travelling salesman problem, and (11) network design [59].

It is an important point that optimization of limited network resources include not only an understanding of the dynamic nature of the computer network itself, but also an understanding of how humans interact with the network and exploit the network resources. The basic requirements of efficient transfer of network data depend on optimal network routing and scheduling. Different routing algorithms vary widely in terms of their purpose, the characteristics of network nodes, and the types of network. As the network size grows, the optimization problem grows dynamically in size. CI can effectively be used to understand optimized use of network resources. In order to make good decisions within the network environments, nodes need to coordinate their activities. Collective intelligence may provide a framework by selecting a group of network nodes with the specified objective. The framework may consist of the node selecting process and a system of reward/punishment based upon some utility function.

Some effective methods that involve intelligently modifying the solutions in a way that is likely to improve them are mentioned in the next section.

5.3.4.1   The cost function

The cost function is the key to solving any problem using optimization, and it's usually the most difficult thing to determine. The goal of any optimization algorithm is to find a set of inputs that minimizes the cost function, so the cost function has to return a value that represents how bad a solution is. There is no particular scale for badness; the only requirement is that the function returns larger values for worse solutions [2].

Some significant applications of cost function in computer network domain are mentioned in the next section.

Cost function based energy aware routing algorithms for WSNs [60]

Cost function based routing has been widely studied in WSNs for energy efficiency improvement and network lifetime elongation. WSNs are collections of low-cost battery-powered devices, called sensors, which have integrated sensing, computing, and wireless communication capabilities [60]. In WSNs, one of the main design challenges is to maximize network lifetime without scarifying network sensing performances (e.g., coverage and reliability).To maximize network lifetime, an energy-efficient routing algorithm should be used for data communications. The algorithm needs to have the following three main features: (1) minimum total energy usage, (2) balanced energy consumption, and (3) distributed characteristics. Cost function based routing has been studied extensively because of its distributed nature and good energy performance [61,62]. In such routing algorithms, a node currently having a packet to transmit decides locally which of its neighbors is the next hop based on a cost function. A well-designed cost function will lead to energy-efficient decisions and prolonged network lifetime. A. Liu et al. proposed a novel double cost function based routing (DCFR) algorithm, which is decentralized, adaptive, and outperforms existing cost function based solutions in terms of energy efficiency improvement and network lifetime elongation. Additionally, DCFR includes energy consumption rate in its cost function. The cost function has a rapidly increasing slope such that a small difference in energy consumption rate or available energy level can lead to a big difference in function values. Hence, DCFR has an excellent capability of balancing energy usage during routing. Each node chooses next hop to forward data according to the energy consumption rates as well as node remaining energy of its neighbors, so energy consumption is balanced in the entire network.

Sensor selection cost function to increase network lifetime with QoS support [63]

To maximize the network lifetime, one wants ideal use of the sensors, so that every sensor can contribute all its energy in supporting the application QoS. To achieve this goal, both energy constraints and QoS requirements need to be considered. First of all, consider the scenario where there are no energy constraints. In this case, the base station can randomly select one sensor set from all possible sensor sets to meet the application QoS. Hence, over time, when the base station has made numerous random selections from the possible sensor sets, each sensor set will support the application for approximately the same amount of time. Consequently, the more often a sensor appears in the possible sensor sets, the longer it tends to be used to support the application QoS. However, if we consider energy constraints, sensors cannot be alive forever. The lifetime of a sensor is determined by its initial energy and its power consumption, and hence the sensor may not be able to support the application in the same way as it did in the scenario of no energy constraints. Single-hop centralized wireless sensor networks are widely used for applications ranging from security and surveillance to medical monitoring. Often the goal of these networks is to provide satisfactory QoS to the application under different system states, but it is difficult to determine how the appropriate sensor sets should be selected over time, given the knowledge of all possible sensor sets to support the application QoS requirements. Yang and Heinzelman [63] proposed a novel cost function called Sensor Usage Index that is based on a sensor's relative ideal lifetime and can be used to select sensor sets so as to meet application QoS requirements for extended periods of time.

5.3.4.2   Hill climbing

In computer science, hill climbing is a mathematical optimization technique that belongs to the family of local search. It is an iterative algorithm that starts with an arbitrary solution to a problem, then attempts to find a better solution by incrementally changing a single element of the solution. If the change produces a better solution, an incremental change is made to the new solution, repeating until no further improvements can be found. For example, hill climbing can be applied to the travelling salesman problem. It is easy to find an initial solution that visits all the cities but will be very poor compared to the optimal solution. The algorithm starts with such a solution and makes small improvements to it, such as switching the order in which two cities are visited. Eventually, a much shorter route is likely to be obtained [64].

An application of hill climbing in computer network domain is mentioned in the next section.

Hill climbing approach for solving QoS multicast routing problem [65]

QoS multicast routing is essential for many network applications such as IPTV, Internet radio, multimedia broadcasting, and real-time telecommunication. Multicast routing involves transport of information from one single sender to multiple destinations. There are two requirements of multicast routing in many multimedia real time applications: one is optimized network cost and the other is bandwidth, bounded delay constraints. Mahmoud et al. proposed an efficient genetic algorithm based on clonal selection, and hill climbing is proposed to solve the least cost multicast routing problem with bandwidth and end-to-end delay constraints.

5.3.4.3   Genetic algorithms

Another set of techniques for optimization, also inspired by nature, is called genetic algorithms (GA). These work by initially creating a set of random solutions known as the population. At each step of the optimization, the cost function for the entire population is calculated to get a ranked list of solutions [2].

A GA is a search heuristic that mimics the process of natural selection. This heuristic (also sometimes called a meta-heuristic) is routinely used to generate useful solutions to optimization and search problems [66]. GAs belong to the larger class of evolutionary algorithms, which generate solutions to optimization problems using techniques inspired by natural evolution, such as inheritance, mutation, selection, and crossover. In a GA, a population of candidate solutions (called individuals, creatures, or phenotypes) to an optimization problem is evolved toward better solutions. Each candidate solution has a set of properties (its chromosomes or genotype) that can be mutated and altered; traditionally, solutions are represented in binary as strings of 0s and 1s, but other encodings are also possible [67].

The evolution usually starts from a population of randomly generated individuals and is an iterative process, with the population in each iteration called a generation. In each generation, the fitness of every individual in the population is evaluated; the fitness is usually the value of the objective function in the optimization problem being solved. The more fit individuals are stochastically selected from the current population, and each individual's genome is modified (recombined and possibly randomly mutated) to form a new generation. The new generation of candidate solutions is then used in the next iteration of the algorithm. Commonly, the algorithm terminates when either a maximum number of generations has been produced, or a satisfactory fitness level has been reached for the population [67].

Some significant applications of GA in Computer Network Domain are discussed in the next section.

A GA for designing distributed computer network topologies [68]

The topological design of distributed packet switched networks consists of finding a topology that minimizes the communication costs by taking into account a certain number of constraints such as the delay and the reliability. The information emitted by a source is forwarded to the destination node of the network according to the packet switching technique. In this way, messages are broken into blocks of a certain size called packets; the packets, when they contain the destination address, can follow different routes toward their destination. In the design of distributed packet switched networks, the aspects to be taken into account are: the topological configuration, the traffic, the capacity assignment, the routing schemes, and the flow control procedures. Topological design of distributed packet switched networks can be viewed as a search for topologies that minimize communication costs by taking into account delay and reliability constraints. This is a very hard optimization problem that is usually solved by means of heuristic approaches. Pierre and Legault [68] proposed a GA to deal with the topological design of distributed computer networks. Such an approach is inspired by biology; mainly Darwin's theory of evolution according to which the strongest species survive, and thus reproduce themselves, creating more outstanding offspring.

GA-based reliability optimization for computer network expansion [69]

The advent of low cost computing devices has led to explosive growth in computer networks. There are several benefits from this dispersal of computing across a network (e.g., resource sharing and improved reliability). One of the major advantages of computer networks over the centralized systems is their potential for improved system reliability. The reliability of a system depends not only on the reliability of its nodes and communication links, but also on how nodes are connected by communication links. Network expansion deals with an ever growing need for more computing across a network. In order to meet this demand, the size of the network is incrementally expanded according to the user requirements. In such an environment, it is critical that the new nodes and links are added in a prudent manner to maximize the performance and reliability characteristics of the expanded network. A generalized framework is developed for expanding existing networks. In most network-design problems, the reliability and availability measures are used as constraints but not as the objective functions for optimization. Such algorithms deal with the problem of overall network design and do not apply directly to incremental network-expansion.

A GA-Based Computer-Network Expansion Methodology is developed by Kumar, Pathak and Gupta to optimize a specified objective function (reliability measure) under a given set of network constraints. The objective of this work is to add new communication links and computer nodes to an existing network such that the cost factor(s) (representing the reliability measures of a network) are minimized/maximized and all the specified constraints are satisfied. The versatility of the genetic algorithm is illustrated by applying it to solve various network expansion problems (optimize diameter, average distance, and computer network reliability for network expansion).

5.3.5 Decision trees

You can understand the reasoning process of a decision tree just by looking at it, and you can even convert it to a simple series of if-then statements [2]. Decision making can be broken into two tasks: the generation of potential solutions and the evaluation of them. Each of those tasks can be negatively influenced by numerous human biases. Those biases, though, can be mitigated through the use of CI approaches: outreach, additive aggregation, and self-organization. In the field of operations research, solving a problem entails two high-level tasks: (1) generating solutions, which include framing the problem and establishing a set of working assumptions about it and (2) evaluating the different alternatives generated in the first step. Each of the tasks is subject to various biases. When generating solutions, for example, we tend to seek information that confirms our assumptions and maintain those beliefs even in the face of contrary evidence. With respect to the evaluation of solutions, we tend to see patterns where none exist (pattern obsession) and to be unduly influenced by how a solution is presented (framing). Those common traps represent just a few from among a much larger number of ways in which our basic human nature can lead us astray when we are making important decisions. Collective intelligence can help mitigate the effects of those biases. For instance, it can provide a diversity of viewpoints and input that can deter self-serving bias and belief perseverance. Diversity can also help combat pattern obsession and negative framing effects. Because of these benefits, decision trees combined with CI may work as an effective solution [70].

Some significant applications of decision tree in Computer Network Domain are discussed in the next section.

5.3.5.1 *Decision tree based algorithm for intrusion detection*

An Intrusion Detection System (IDS) is a defense measure that supervises activities of the computer network and reports the malicious activities to the network administrator. Intruders make many attempts to gain access to the network and try to harm the organization's data. Thus the security is the most important aspect for any type of organization. Due to these reasons, intrusion detection has been an important research issue.

Intrusion detection can be considered as a classification problem where each connection or user is identified either as one of the attack types or normal based on some existing data. Decision trees can solve this classification problem of intrusion detection as they learn the model from the data set and can classify the new data item into one of the classes specified in the data set. Decision trees can be used as misuse intrusion detection as they can learn a model based on the training data and can predict the future data as one of the attack types or normal based on the learned model. IDSs are categorized as signature based and anomaly based. Signature or misuse based IDS uses various techniques to locate the similarity among system behavior and previously known attacks stored in the signature database. Anomaly based IDS detects activities in a network that deviate from normal behaviors stored in a system profiles database. There are various classifiers that are applicable to misuse based detection. Some are tree based such as decision tree [71] and random forest [72].

Decision trees can analyze data and identify significant characteristics in the network that indicate malicious activities. They can add value to many real-time security systems by analyzing a large set of intrusion detection data. They can recognize trends and patterns that support further investigation, the development of attack signatures, and other activities of monitoring. The main advantage of using decision trees instead of other classification techniques is that they provide a rich set of rules that are easy to understand and can be effortlessly integrated with real-time technologies [73].

Benefits of decision tree analysis to cybersecurity

Decision trees can help perform the following functions for an organization [74]:

■ Supplement honeypot analysis by learning from adversary trends and creating rules to detect malicious activity.

■ Supplement penetration testing efforts by learning from the pen tester's actions and creating rules to detect their tactics, techniques, and procedures.

■ Identify and highlight malicious traffic.

■ Prioritize alerting by identifying and tagging low priority alerts.

■ Support IDS signature development by providing a set of rules to identify malicious activity.

After decision trees are built, they have the potential to reduce the amount of data required for analysis, help identify anomalous malicious activity, and provide analytic insight into the differences between malicious and benign network traffic. In addition, as the amount of data handled by IT security experts increases, flagging the priority of malicious activity in an automated fashion becomes more and more beneficial for the mission.

5.3.5.2   Network anomaly detection by cascading K-means clustering and C4.5 Decision Tree algorithm [75]

Intrusions pose a serious security risk in a network environment. A network intrusion detection system aims to identify attacks or malicious activity in a network with a high detection rate while maintaining a low false alarm rate. Anomaly detection systems monitor the behavior of a system and flag significant deviations from the normal activity as anomalies.

A supervised anomaly detection method has been proposed [75], called "K-Means + C4.5" developed by cascading two machine learning algorithms: (1) the k-Means clustering and (2) the C4.5decision tree. In the first stage, k-Means clustering is performed on training instances to obtain k disjoint clusters. Each k-Means cluster represents a region of similar instances, "similar" in terms of Euclidean distances between the instances and their cluster centroids. K-Means method is cascaded with the C4.5 by building decision trees using the instances in each k-Means cluster. Cascading the k-Means clustering method with C4.5 decision tree learning alleviates two problems in k-Means clustering: (1) the forced assignment problem and (2) the class dominance problem. Cascading the decisions from the k-Means and C4.5 methods involves two phases: (1) the selection phase, and (2) the classification phase. In the selection phase, the closest cluster (i.e., the nearest neighbor cluster to the test instance) is selected.

In the selected cluster the decision tree corresponding to that cluster is generated. In the classification phase, the test instance is classified into normal or anomaly using the decision tree result, and then it is included in the cluster with the classified label as normal or anomaly.

5.4 Conclusion

While CI is considered as a powerful approach to enhance the decision capability, very few attempts have been made to utilize its potential in computer network domain. This chapter is intended to provide the general overview of CI with a discussion of how it can be effectively applied in different areas of computer networks. More studies are required to inculcate the idea of CI in various areas of computer networks. It is therefore very promising to identify opportunities for integration and inculcation of collective intelligence methodologies and technologies into networking. This would require judiciously considering the various activities in a computer network for this purpose. Finally, such an effort would aim at remodeling of computer networks in context of CI.

Bibliography

[1] Nguyen, N. T. ed., *Transactions on Computational Collective Intelligence III*. Springer, Berlin. p. 63. ISBN: 978-3-642-19967-7.

[2] Segaran, T., *Programming Collective Intelligence: Building Smart Web 2.0 Applications*. O'Reilly Publication, Sebastopol, CA, 2007, 1–2.

[3] Web Article, Collective intelligence—Access the untapped knowledge of your networks. http://www.935.ibm.com/services/us/gbs/ thoughtleadership/ ibv-collective-intelligence.html (Accessed 04 September 2015).

[4] Pedrycz, W. and Vasilakos, A., *Computational Intelligence in Telecommunication Networks*. CRC Press, Boca Raton, FL, 2000, 1–2.

[5] Levy, P., *Collective Intelligence: Mankind's Emerging World in Cyberspace, trans.* Bononno, R., Perseus Books, Cambridge, MA, 1997, xxiv.

[6] Collective intelligence, Dictionary.com. http://www.dictionary.com/ browse/ collective-intelligence (Accessed 10 April 2016).

[7] Smith, J. B., *Collective Intelligence in Computer-Based Collaboration*. Lawrence Erlbaum, Hillsdale, NJ, 1994, 1–10.

[8] Silverman, M. and Picton, M., Collective intelligence in organisations: Uses and challenges, Executive Summary, Silverman Research,

London. http://www.silvermanresearch.com/wp-content/uploads/2015/05/ Collective-Intelligence-in-Organisations-Full-Report.pdf (Accessed 10 March 2016).

[9] Collective intelligence, The Co-Intelligence Institute. http://www.co-intelligence.org/Collective_Intelligence.html (Accessed 10 February 2016).

[10] Broadbent, S. and Gallott, M., Collective intelligence: How does it emerge? https://www.nesta.org.uk/sites/default/files/collective_intelligence.pdf (Accessed 10 February 2016).

[11] Lykourentzou, I., Vergados, D. J., Kapetanios, E., Loumos, V., Collective intelligence systems: Classification and modeling, *Journal of Emerging Technologies in Web Intelligence*, Vol. 3, No. 3 (2011): 217–226.

[12] DiMaio, P., Making sense of collective intelligence, Cutter Consortium Executive Report, Business Intelligence, Vol. 8, No. 9, 2008.

[13] Farooq, M. and Di Caro, G. A., Routing protocols for next-generation networks inspired by collective behaviours of insect societies: An overview. In: *Swarm Intelligence, Natural Computing*, pp. 101–160, 2008.

[14] Berri, J., Towards a framework for collective intelligence, In: *Proceedings of ICDIM'10 (International Conference on Digital Knowledge Management)*, Thunder Bay, ON, pp. 454–459, July 5–8, 2010.

[15] Rey, A., Types of problems that can be solved by collective intelligence. The Collective Intelligence Blog. http://collectiveintelligenceblog. com/types-problems-can-solved-collective-intelligence/ (Accessed 10 February 2016).

[16] Gilbreath, P., Web 2.0 Collective intelligence—How to use collective intelligence techniques in your web application. http://www.slideshare.net/ paulgilbreath/web-20-collective-intelligence-how-to-use-collective-intelligence-techniques-in-your-web-application (Accessed 15 February 2015).

[17] Parh, D. R., Pothal, J. K., Singh, M. K., Navigation of multiple mobile robots using swarm intelligence. In: *World Congress on Nature and Biologically Inspired Computing (NaBIC)*, pp. 1145–1149, 2009.

[18] Fukuda T. and Ueyama T., *Cellular Robotics and Micro Robotic Systems*. World Scientific Series in Robotics and Automated Systems, Vol. 10, World Scientific, Singapore, 1994.

[19] Minsky M., *The Society of the Mind*. Simon and Schuster, New York, 1986.

[20] Lucic, P. and Teodorovic, D., Transportation modeling: An artificial life approach. In: *Proceedings of the 14th IEEE International Conference on Tools with Artificial Intelligence ICTAI*, Washington, DC, pp. 216–223, 2002.

[21] Kornrumpf, A. and Bauml, U., From collective intelligence to collective intelligence sytems: Definitions and a semi-structured model, *International Journal of Cooperative Information Systems* Vol. 22, No. 3 (2013): 1340002.

[22] Malone, T. W. and Bernstein, M. S., Introduction: Chapter 1. In: *The Collective Intelligence Handbook*, MIT Press, 2015.

[23] Xavier, E., Koch, F., Westphall, C., Automations in computer network management utilizing computational intelligence. In: *3rd Latin American Network Operations and Management Symposium, LANOMS 2003*, Iguassu Falls, Brazil, 4–6 September, 2003.

[24] Farooq, M. and Di Caro, G. A., Routing protocols for next generation networks inspired by collective behaviors of insect societies: An overview. In: Blum C., Merkle D. (Eds), *Swarm Intelligence: Introduction and Applications*, Natural Computing Series, Springer, Berlin, 2008.

[25] Kanawati, R. YASCA: An ensemble-based approach for community detection in complex networks. *In: Proceedings of the 20th International Computing and Combinatorics Conference (COCOON'14)*, Atlanta, GA, pp. 657–666, 4–6 August, 2014.

[26] Wolpert, D., Tumer, K., Frank, J., Using collective intelligence to route Internet traffic. In: *Proceedings of the 1998 Conference on Advances in Neural Information Processing Systems II*, pp. 952–958, July 1999.

[27] Ekstrand, M. D., Riedl, J. T., Konstan, J. A., Collaborative filtering recommender systems, *Foundations and Trends in Human–Computer Interaction* Vol. 4, No. 2 (2010): 81–173.

[28] Bakker, A., Ogston, E., Steen, M., Collaborative filtering using random neighbours in peer-to-peer networks. In: *Proceedings of Complex Networks in Information and Knowledge Management (CNIKM), CNIKM 2009*, Hong Kong, China, 2–6 November, 2009.

[29] Collaborative filtering. http://recommender-systems.org/collaborative-filtering/ (Accessed 12 March 2016).

[30] Arapoglou, R., Rodis, I. D., Magdalinos, P., Alonistioti, N., Adapting policy-based management of future networks using collaborative filtering techniques. In: *2014 IEEE 19th International Workshop on Computer Aided Modeling and Design of Communication Links and Networks (CAMAD)*, Athens, pp. 224–228, 1–3 December 2014.

[31] Wang, J., Peng, J., Cao, X., A distributed collaborative filtering recommendation model for P2P networks. In: *Collaborative Computing: Networking, Applications and Work sharing*, pp. 1–10, 2009.

[32] Luo, J., Liu, X., Zhang, Y., Ye D., Xu Z., Fuzzy trust recommendation based on collaborative filtering for mobile ad-hoc networks. In: *Proceedings of the 33rd IEEE Conference on Local Computer Networks (LCN 2008)*, October 2008, pp. 305–311.

[33] Wu, Z. and Weaver, A. C., Application of fuzzy logic in federated trust management for pervasive computing. In: *COMPSAC '06: Proceedings of the 30th Annual International Computer Software and Applications Conference (COMPSAC'06)*, IEEE Computer Society, Washington, DC, pp. 215–222, 2006.

[34] Sun, Y. and Han, J., Ranking methods for networks. In: Alhajj, R. and Rokne, J. (Eds), *Encyclopaedia of Social Network Analysis and Mining*, pp. 1488–1497, October 2014.

[35] Wasserman, S. and Faust, K., *Social Network Analysis: Methods and Applications*. Cambridge University Press, 1994.

[36] Brin, S. and Page, L., The anatomy of a large-scale hypertextual web search engine, *Computer Networks* Vol. 30, No. 1–7 (1998): 107–117.

[37] Jeh, G. and Widom, J., Simrank: A measure of structural-context similarity. In: *Proceedings of the Eighth ACM SIGKDD International Conference on Knowledge Discovery and Data Mining, KDD'02*, pp. 538–543, doi:10.1145/775047.775126, 2002.

[38] Li, X. and Wu. J., Searching techniques in peer-to-peer networks. In: Wu, J. (Ed.), *Handbook of Theoretical and Algorithmic Aspects of Ad-Hoc, Sensor, and Peer-to-Peer Networks*, Auerbach Publications, New York, 2006.

[39] Yang, B. and Garcia-Molina, H., Improving search in peer-to-peer network's. In: *Proceedings of the 22nd IEEE International Conference on Distributed Computing (IEEE ICDCS'02)*, 2002.

[40] Crespo, A. and Garcia-Molina, H., Routing indices for peer-to-peer systems. In: *Proceedings of the 22nd International Conference on Distributed Computing (IEEE ICDCS'02)*, 2002.

[41] Wang, Z., Dueas Osorio, L., Padgett, J. E., A new mutually rein-forcing network node and link ranking algorithm. Scientific Reports. www.nature.com/scientificreports, 2014.

[42] Saaty, T. L., *The Analytic Hierarchy Process*. McGraw Hill, New York, 1980.

[43] Xi, Z., Chen, H., Wang, X., Sheng, J., Fan, Y., Evaluation model for com-puter network information security based on analytic hierarchy process. In: *Proceedings of 3rd International Symposium on Intelligent Information Technology Application*, pp. 186–189, 2009.

[44] Uddin, A. H. M. M., Monir, M. I., Iqbal, S. M. A., A QoS aware route selection mechanism using analytic hierarchy process for mobile ad hoc network. Paper published in *12th International Conference on Computer and Information Technology (ICCIT 2009)*, IEEE, Independent University, Dhaka, Bangladesh, 21–23 December, 2009.

[45] Kushwaha, V. and Ratneshwer, Ranking of source based congestion control approaches for high-speed networks using AHP, *International Journal of Communication Networks and Distributed Systems (IJCNDS)-Communicated*, Vol. 17, No. 4 (2016): 387–411.

[46] Steinbock, D., Kaplan, C., Rodriguez, M. A., Diaz, J., Der, N., Garcia, S., Collective intelligence quantified for computer-mediated group problem solving. Technical report, University of California at Santa Cruz, 2004.

[47] Gupta, G. and Younis, M., Load-balanced clustering in wireless sensor net-works. In: *Proceedings of the International Conference on Communication (ICC 2003)*, Anchorage, AK, May 2003.

[48] Younis, M., Akkaya, K., Kunjithapatham, A., Optimization of task alloca-tion in a cluster–based sensor network. In: *Proceedings of the 8th IEEE Symposium on Computers and Communications (ISCC'2003)*, Antalya, Turkey, June 2003.

[49] Younis, O. and Fahmy, S., HEED: A hybrid, energy-efficient, distributed clustering approach for ad hoc sensor networks, *IEEE Transactions on Mobile Computing*, Vol. 3, No. 4 (2004): 366–379.

[50] Bandyopadhyay, S. and Coyle, E., An energy efficient hierarchical clus-tering algorithm for wireless sensor networks. In: *Proceedings of the 22nd Annual Joint Conference of the IEEE Computer and Communications Soci-eties (INFOCOM 2003)*, San Francisco, CA, April 2003.

[51] IlkerOyman, E. and Ersoy, C., Multiple sink network design problem in large scale wireless sensor networks. In: *Proceedings of the IEEE International Conference on Communications (ICC 2004)*, Paris, France, June 2004.

[52] Younis, M., Youssef, M., Arisha, K., Energy-aware management in cluster-based sensor networks, *Computer Networks* Vol. 43, No. 5 (2003): 649–668.

[53] Hou, Y. T., Shi, Y., Sherali, H. D., On energy provisioning and relay node placement for wireless sensor networks, *IEEE Transactions on Wireless Communications*, Vol. 4, No. 5 (2005): 2579–2590.

[54] Mamalis, B., Gavalas, D., Konstantopoulos, C., Pantziou, G., Clustering in wireless sensor networks. In: Zhang, Y., Yang, L. T., Chen, J. (Eds), *RFID and Sensor Networks: Architectures, Protocols, Security and Integrations*, CRC Press, Boca Raton, FL, pp. 324–353, 2009.

[55] Lin, C. R. and Gerla, M., Adaptive clustering for mobile wireless networks, *IEEE Journal on Selected Areas in Communications*, Vol. 15, No. 7 (1997): 1265–1275.

[56] Santhiya, K. G. and Arumugam, N., A novel adaptive bio-inspired clustered routing for MANET, *Procedia Engineering*, Vol. 30 (2012): 711–717.

[57] McDonald, A. B. and Znati, T., A mobility based framework for adaptive clustering in wireless ad-hoc networks, *IEEE Journal on Selected Areas in communications*, Vol. 17, No. 8 (1999): 1466–1487.

[58] Dagdeviren, O., Erciyes, K., Cokuslu, D., Merging clustering algorithms in MANET. In: Chakraborty, G. (Ed.) *Distributed Computing and Internet Technology. ICDCIT 2005*. Lecture Notes in Computer Science, Vol. 3816. Springer, Berlin and Heidelberg.

[59] Ahuja, R. K., Magnanti, T. L., Orlin, J. B., Reddy, M. R., Applications of network optimization. In: Ball, M. O., Magnanti, T. L., Monma, C. L., Nemhauser, G. L. (Eds), *Handbooks in Operational Research and Management Science*, Volume 7: Network Models, North-Holland, Amsterdam, pp. 1–84, 1995.

[60] Liu, A., Ren, J., Li, X., Chen, Z., Shen, X., Design principles and improvement of cost function based energy aware routing algorithms for wireless sensor networks, *Computer Networks: The International Journal of Computer and Telecommunications Networking*, Vol. 56, No. 7 (2012): 1951–1967.

[61] Ok, C. S., Lee, S., Mitra, P., Kumara, S., Distributed energy balanced routing for wireless sensor networks, *Computers and Industrial Engineering* Vol. 57, No. 1 (2009): 125–135.

[62] Rahme, J., Viana, A. C., Al Agha, K., Looking for network functionalities extension by avoiding energy-compromised hotspots in wireless sensor networks, *Annals of Telecommunications*, Vol. 63 (2008): 487–500.

[63] Yang, O. and Heinzelman, W., Sensor selection cost function to increase network lifetime with QoS support. In: *ACM International Conference on Modeling, Analysis and Simulation of Wireless and Mobile Systems (MSWiM)*, Vancouver, BC, 27–31 October 2008.

[64] Hill climbing, Wikipedia, last modified 31 January, 2016, https://en.wikipedia.org/wiki/Hill_climbing.

[65] Mahmoud, T. M., Nashar, A. I. E., Eman, M., An efficient genetic algorithm based clonal selection and hill climbing for solving QoS multicast routing problem, *IJCSI International Journal of Computer Science Issues*, Vol. 11, No. 3 (2014): 83.

[66] Mitchell, M., *An Introduction to Genetic Algorithms*. MIT Press, Cambridge, MA. ISBN 9780585030944, 1996.

[67] Whitley, D., A genetic algorithm tutorial. *Statistics and Computing* Vol. 4, No. 2 (1994): 65–85. doi:10.1007/BF00175354.

[68] Pierre, S. and Legault, G., A genetic algorithm for designing distributed computer network topologies, *IEEE Transactions on Systems, Man, Cybernetics Part B*, Vol. 28 (1998): 249–258.

[69] Kumar, A., Pathak, R. M., Gupta, Y. P., Genetic-algorithm-based reliability optimization for computer network expansion, *IEEE Transactions on Reliability*, Vol. 44, No. 1 (2002): 63–72.

[70] Bonabeau, E., Decisions 2.0: The power of collective intelligence, *MIT Sloan Management Review*, Vol. 50, No. 2 (2009): 45–52.

[71] Aggarwal, P. and Sharma, S. K., An empirical comparison of classifiers to analyze intrusion detection. In: *Proceedings of Fifth International Conference an Advanced Computing and Communication Technologies*, 2015.

[72] Ho, T. K., Random decision forests. In: *Proceedings of the 3rd International Conference on Document Analysis and Recognition*, Montreal, QC, pp. 278–282, 14–16 August 1995.

[73] Markey, J., Using decision tree analysis for intrusion detection: A how-to guide, SANS Institute InfoSec Reading Room, June, 2011.

[74] Kumar, M., Hanumanthappa, M., Kumar, T. V. S., Intrusion detection system using decision tree algorithm. In: *Proceedings of IEEE 14th International Conference on Communication Technology ICCT*, pp. 629–634, 2012.

[75] Muniyandi, A. P., Rajeswari, R., Rajaram, R., Network anomaly detection by cascading KMeans clustering and C4.5 decision tree algorithm, *Procedia Engineering*, Vol. 30 (2012): 174–182.

Chapter 6

QoS Route Search for Mobile Ad Hoc Network Using Genetic Algorithm

J. Abdullah

Universiti Tun Hussein Onn Malaysia

CONTENTS

6.1 Introduction

Wireless mobile devices are very popular and provide users access to information and communication anywhere and anytime. The smartphone, cellular access system, Worldwide Interoperability for Microwave Access (WIMAX), Fourth Generation (4G) and Long Term Evolution (LTE) are all merging together providing services that never existed before. The paradigm of telco managed networks will be here forever, for the simple reason that it is the best choice for everyone to be connected. MANETs, on the other hand, offer a wireless network with a different paradigm. It is an infrastructureless network with multihop capabilities. It may be connected or not connected to the Internet infrastructure. MANET (Conti and Giordano 2014) consists of a group of autonomous mobile nodes with its own network management system. It becomes very useful in certain applications, such as infantry communications, crowd management, and communications during large-scale disaster and vehicular networking. Diverse multimedia applications such as voice, video, and data will certainly increase the popularity of MANET (Ghosekar et al. 2010). For quality data delivery, it is imperative that MANET provides QoS routing support, managing connectivity, node density, bandwidth, and delay constraints. QoS routing in MANET is a challenging effort, since the topology dynamics, the shared medium and node mobility will affect the QoS constraints. Over the years, researchers have proposed a number of mechanisms for QoS routing protocols (Abdullah 2011; Karimi 2011). On-demand protocols performed reasonably well compared to table-driven protocols (MANET IETF 2017). Among the on-demand QoS routing protocols proposed (Li and Nahrstedt 1998; Mohapatra et al. 2003), those with node mobility scenarios and distributed Medium Access Control MAC protocols, such as IEEE 802.11 Distributed Coordinated Function (DCF) (Perkins and Bhagwat 1994), are utilized effectively. The design problem is to choose a set of source–destination routes for a given set of nodes, either maximizing reliability given a cost constraint or minimizing cost given a minimum network reliability constraint. This kind of design problem is NP-hard (Lee et al. 2000), and additionally, the calculation of all node reliability is also NP-hard.

QoS routing support for MANET (Chen and Nahrstedt1999) is needed for multimedia applications, such as collaborative computing requiring data sharing mode. In order to support QoS routing, the link state information such as delay, bandwidth, cost, lost rate, connectivity, and error rate should be managed efficiently. Link state management of MANETs is difficult due to the variation in quality of the wireless links. Hence, implementing QoS functionality in a dynamic environment is really a challenging effort. Most mechanisms proposed to achieve QoS in MANETs are based on mapping wireless networks to a wireline paradigm of nodes and links.

The nature of MANETs lacking central authority, shared medium, low bandwidth, and mobile nodes makes it a challenge to offer connections with sufficient quality for realtime applications such as voice and video. A potentially promising approach is to establish multiple paths between source and destination. Then, use routing schemes that take advantage of multiple paths to improve the performance of MANETs in realtime applications. Recent studies have focused on two aspects, developing multipath routing protocols for MANETs and developing schemes for multipath topologies (Barolli et al. 2011; Liu and Huang 2009). While the first group of studies mainly focuses on the efficiency of the protocols proposed, the second focuses on the schemes themselves and not on the characteristics of the network that has to support them.

When dealing with wireless communication, the critical resources are the electromagnetic spectrum in a space, where contention occurs due to multiple channel access paradigms. A node state is utilized where most of the important node parameters or status are saved and can be retrieved. All the related QoS route information is stored as node state. A node is a basic hardware and software unit in the network that has the capability to forward packets based on its local routing table. A host is another basic unit in the network that may attach to or act as a router. It can be either the source or the sink of a data flow in the network. Two nodes within the signal transmission range are neighbors to each other. A route is a sequence of nodes connecting two end hosts.

6.2 Multiple Routes of MANET and Its Reliability

There are two types of multiple routes between a source and a destination, namely node-disjoint and link-disjoint multiple routes. Node-disjoint routes do not have any nodes in common, except the source and destination. Figure 6.1a shows a typical non-disjoint network that can utilize the Non-Disjoint Multiple Routes Discovery (NDMRD) protocol for route discovery (Abdullah 2015). Figure 6.1b shows node-disjoint multiple routes with two initial branches. Node-disjoint routes are shown in Figure 6.1c. Nodes labeled S and T are source and destination nodes, respectively. The advantages of node-disjoint multiple routes are that they may fail independently of each other. Breakage on one route can be corrected by resuming data sessions through the other routes. Figure 6.1d shows link-disjoint multiple routes between S and T, formed with two segments. In Figure 6.1e, node-disjoint multiple routes to the destination are available on every node at every route. Figure 6.1f shows a fail-safe with multiple routes. Consider the networks in Figure 6.1a and b only. The network (a) shows rich non-disjoint connectivity, and network (b) shows disjointed $S - T$ connectivity. If we consider each node to have mobile probability p, then using reliability calculations we can determine the $S - T$ reliability of the two networks. Using the method of inclusion and exclusion on minimum paths shown by (Mosko and Garcia-Luna-Aceves 2005), the reliability polynomials are given as in the following equations:

$$R_{\text{disj}} = 2p^4 - p^8 \tag{6.1}$$

$$R_{\text{non-disj}} = 2p^4 - p^8 + (6p^4 - 12o^6 - 8p^7 + 15p^8 + 12p^9 - 20p^{10} + 8p^{11} - p^{12}) \tag{6.2}$$

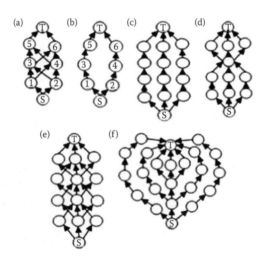

Figure 6.1 Different types of multiple routes.

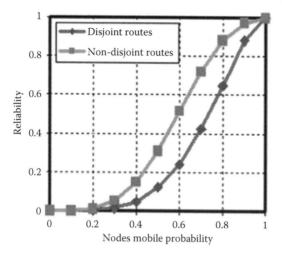

Figure 6.2 Routes reliability.

The non-disjoint routes in Figure 6.1a have eight minimum paths $\{S\text{-}1\text{-}3\text{-}5\text{-}T\}$, $\{S\text{-}1\text{-}4\text{-}6\text{-}T\}$, $\{S\text{-}1\text{-}3\text{-}6\text{-}T\}$, $\{S\text{-}1\text{-}4\text{-}5\text{-}T\}$, $\{S\text{-}2\text{-}4\text{-}6\text{-}T\}$, $\{S\text{-}2\text{-}3\text{-}5\text{-}T\}$, $\{S\text{-}2\text{-}3\text{-}6\text{-}T\}$, $\{S\text{-}2\text{-}4\text{-}5\text{-}T\}$. The disjoint routes in Figure 6.1b have two minimum paths: $\{S\text{-}1\text{-}3\text{-}5\text{-}T\}$, $\{S\text{-}2\text{-}4\text{-}6\text{-}T\}$. Figure 6.2 shows the network reliability for the disjoint and non-disjoint route configurations. As expected, the non-disjoint configuration has a significantly higher reliability. Nasipuri et al. (2001) use a non-disjoint multipath approach in their routing protocol, where it is assumed that the primary path be k hops. Each node along the primary path has an alternate disjoint route to T, so there are $k + 1$ minimum paths. In their first protocol, it has only two minimum paths, where they noted that the rate of path discovery decreases as the path length increases. Another minimum path is added because of each extra hop along the primary path. The method by Mosko and Garcia-Luna-Aceves (2005) used in the computation of time for node connectivity is adopted. Accordingly, the cumulative distribution function (CDF) for link operation is given as $F(t) = 1 - e^{(-\lambda t)}$. For a series of k links, the CDF is $F_s(t) = 1 - e^{(-k\lambda t)}$ (Nasipuri et al. 2001). For a set of m parallel paths, each has a CDF of $F_s(t)$, $F_p(t) = (F_s(t))^m$. Using these results, the CDF for the disjoint network in Figure 6.1b is $F_{\text{disj}}(t) = (1 - e^{(-4\lambda t)})^2$. Assuming this relationship, the expected value is $E[X] = \int_0^\infty 1 - F(x)\,dx$ and the mean lifetime of the disjoint routes is given in the following equation:

$$E_{\text{disj}}[X] = \int_0^\infty 2e^{-4\lambda t} - e^{-8\lambda t}\,dt = \frac{3}{8\lambda} \qquad (6.3)$$

To analyze the non-disjoint routes, we use the equation from the work of Mosko and Garcia-Luna-Aceves (2005) for reliability, shown as the following equation:

$$R = \sum_{j=1}^{h}(-1)^{j+1} \cdot \sum_{I \subseteq (1,2,\ldots h),\ |I|=j} \text{Prob}[E_I] \qquad (6.4)$$

where E_I is the event that all paths P_i with $i \in I$ operate no longer than time t. Let n be the number of distinct links in E_I, then Prob $[E_I] = 1 - e^{(-n\lambda t)}$. It is required that all paths with n distinct links operate no longer than time t and are exactly the same as a series path of n links. This will yield an equation almost identical to Equation 6.3, except that each term ap^b will be replaced by $(-ae^{(-b\lambda t)})$. Then, non-disjoint formulations are shown as in the following equations:

$$F_{\text{non-disj}}(t) = 1 - 8e^{-4\lambda t} + 12e^{-6\lambda t} + 8e^{-7\lambda t} + 14e^{-8\lambda t}$$

$$- 12e^{-9\lambda t} + 20e^{-10\lambda t} - 8e^{-11\lambda t} + e^{-12\lambda t} \qquad (6.5)$$

$$E_{\text{non-disj}}[X] = \int_{0}^{\infty} 1 - F_{\text{non-disj}}(t)\mathrm{d}t = \frac{44}{77\lambda} \qquad (6.6)$$

Comparing the two equations, we find that the non-disjoint routes last, on average, 1.52 times longer than the disjointed routes. Repeating the same calculations for 3-hop, 5-hop, and 6-hop routes, we get the comparisons of ratios. The ratios of node connectivity time of non-disjoint to disjoint networks for 3, 4, 5, and 6-hops are given as 1.28, 1.52, 1.79, and 2.01, respectively. While it is difficult to generalize the mean $S - T$ connectivity lifetime to an arbitrary network, the trend is obviously favoring the node non-disjoint network configuration.

6.3 Genetic Algorithms as Nature-Inspired Solution in Multi-Objective Optimizations

Genetic algorithms (GAs) are nature-inspired solutions that provide a guided random search and optimization technique, based on the principles of evolution, that is, the survival of the fittest and inheritance. It uses probabilistic transition rules and a payoff function to guide the search. Using the GA technique, it overcomes those limitations to provide a near-optimal solution in a few iterations. The steps involved when dealing with GA can be briefly summarized as follows: (i) generation of a population of chromosomes; (ii) decoding each chromosome to evaluate its fitness; (iii) performing the selection, crossover, and mutation operations; and (iv) then repeating steps (ii) and (iii) until a stopping criterion is satisfied. In order to solve any optimization problem, GA starts with chromosomal representation of the parameter of concern. A set of such chromosomes is termed a population. The fitness function is chosen in such a way that the good points in the search space possess high fitness values. Intuitively, GA mimics the natural inspiring evolution process through its genetic operations such as selection, crossover, and mutation operations. The key idea behind the crossover is to exchange information between two randomly selected parent strings to give birth to the offspring for the next generation (Younes 2011).

The successful application of GA is due to fact that GA is computationally simple and provides powerful parallel search ability. It is immune to the problem's lack of locality, because its search takes place simultaneously throughout the solution space. Functionally, GA emulates the criteria of natural selection in a search procedure. Organisms in nature have certain characteristics that affect their ability to survive and reproduce. The characteristics are represented by a chromosome of the organism. In a reproduction process, the offspring's chromosomes consist of a combination of the chromosomal information from each parent. Natural selection then operates, such that fitter individuals will have the opportunity to mate most of the time. In turn, this leads to the expectation that the offspring stand a good chance of being similarly fit. Mutations occasionally can occur, enhancing the individual's characteristics and improving its progress to the next generation. Conversely, the chromosomal changes may have no effect at all.

GA had been used in combinatorial optimization approaches for reliable design and successful search algorithm. Coley (1999) outlines wide-ranging practical areas in the field of engineering in which GA has been effectively applied, such as image processing, Very Large Scale Integration (VLSI) circuit routing, water networks, control, and communication networks. Gen and Cheng (2000) produced a detailed study of various GA-based industrial engineering applications, such as scheduling, spanning tree, transportation, reliability, optimization, network design, and network routing. In communication networks, GA was utilized to optimize network operations. R. Elbaum and Sidi (1996) used GA in designing Local Area Network (LAN) with objectives to minimize network delay when addressing the issue of clustering and routing. Mao et al. (2005) use a GA to formulate effectively a routing problem as applied to multiple description video in wireless ad hoc networks. Hwang et al. (2000) proposed a multicast routing algorithm based on GA in a real computer network. It produced an extension to the algorithm addressing the issue of QoS constraints. Researchers have applied GA to the shortest path routing problem (Shimamoto et al. 1993), the multicasting routing problem (Lu and Zhu 2013), the dynamic channel allocation problem (Wong and Wassell 2002), and also the dynamic routing problem (Mehboob et al. 2016). Munetomo et al. (1998) proposed GA for wireline and wireless networks, using variable-length chromosomes (Lee and Antonsson 2000) to encode a feasible solution. In contrast, Inagaki et al. (1999) proposed GA that employs fixed-length chromosomes.

6.4 QoS Route Selection Strategies

6.4.1 Complexity of QoS route selection

The route selection algorithm has a degree of computational complexity depending on the rule of metric composition. Since applications generate traffic with very diverse requirements in terms of QoS, the route selection algorithm must select routes that satisfy a set of constraints. The value of metric along a route, based on its value in each hop, depends on the nature of the metric. There are additive, multiplicative, and concave metrics. The rule for additive metrics composition is that its value of the metric over a path is the sum of the values of each hop. Delay and number of hops are examples of additive metrics. With a multiplicative metric, the value of a metric over a path is the product of its values in each hop, as is the case with packet losses. The value of a concave metric over a path corresponds to the minimum value observed in all hops of that path. Bandwidth is a common example of a concave metric.

6.4.2 The ordering of QoS metrics

Metric ordering requires the identification of the metric that has higher priority and the computation of the best paths according to this metric. The second metric is used in case of a tie to decide which are the best paths. This is the case of shortest-widest path and widest-shortest path algorithms. Shortest-widest path algorithms first find paths with maximum available bandwidth. Next, if there are paths with the same amount of available bandwidth, the path that has the shortest number of hops is selected. Widest-shortest path algorithms select the path that has higher bandwidth availability from the shortest paths with an equal number of hops. The Dijkstra algorithm can also be used to compute widest-shortest paths (Ma and Steenkiste 1997).

6.4.3 The sequential filtering

In the sequential filtering heuristics (Ma and Steenkiste 1997), links that cannot handle the required bandwidth are excluded from the network. The route selection algorithm, such as the shortest path, is computed. The threshold that determines the exclusion of a link depends on the instant of application of the route selection algorithm. When paths are computed on demand, the desired value of bandwidth and delay can be expressed on request. For path pre-computation, it is necessary to compute and store several pre-computed routes that satisfy the defined range of bandwidth values.

6.4.4 The scheduling disciplines of metrics

The complexity of path selection algorithms can be avoided using the relationship among QoS parameters determined by the nature of scheduling disciplines.

If it uses a weighted fair queuing scheduling mechanism, it is possible to use the relations between bandwidth, delay, and jitter to find a path, in polynomial time, subject to constraints of bandwidth, delay, and jitter (Ma and Steenkiste 1997).

6.4.5 Admission control techniques

For some QoS models, the admission of new flows may be subject to a mechanism of admission control. They interact with routing mechanism in which the topological network information can be used for admission control decisions. Admission control and QoS routing are tightly connected with resource reservation as in an In-band Signalling in Ad Hoc Network (Lee et al. 2000) and Stateless Wireless Ad Hoc Networks (SWAN) (Ahn et al. 2002) QoS model.

6.4.6 The QOSRGA protocol, using GA for route selection

1. QoS routing functionality
   QoS routing is a key MANET function for the transmission and distribution of multimedia services with the following objectives: (i) finding routes that satisfy QoS constraints and (ii) making efficient use of limited resources. The complexity involved in the networks may require the consideration of multiple objectives for the routing decision. In this chapter GA based QoS routing for MANET as a multiple objective optimization is presented.

2. Why GA?
   GA possesses an intrinsic parallel architecture that is suitable as a parallel computation. It addresses multiple objective problems such as QoS route selection. Additionally, GA is capable of handling problems with multiple constraints. It can solve multimodal, non-differentiable, and non-continuous problems. Within a function, there can be different local maximums or minimums, and GA is able to find the best of them. Three factors contribute to its ability to locate a global maximum or minimum: (i) searching from a population of potential solutions, (ii) using fitness information as the criterion used to determine which variables are the best; and (iii) randomized operators instead of deterministic ones in order to find the best solutions. Computational time for the convergence of GA operation might be long, but with extremely fast hardware such as the Field Programmable Gate Array (FPGA), it can support services such as voice, video, and teleconferencing (Bhandari et al. 2012).

6.5 Modeling the Mobile Ad Hoc Network

6.5.1 Mobile ad hoc network topology

The topology of mobile ad hoc networks is modeled as a graph $G = (E, Q\{nci, B_{\text{AVA}}, D_{\text{E2E}}, D_{\text{MAC}}\})$ where E is the set of all mobile nodes, and Q is the set of QoS parameters that determine the viable QoS connectivity between the nodes. Each mobile node $i \in E$ has a unique identity and moves arbitrarily. A radius R defines a coverage area within which every node can communicate with each other directly. Neighbors of node i are defined as a set of nodes $i \# j$, which are within radius R and reachable directly from the node i. Every pair of neighbors can communicate with each other in both directions. Figure 6.3 shows a typical mobile ad hoc network. Hence, there exists a connectivity between neighbors i and j with the index of nci. The variable nci is the node connectivity index, which indicates how reliable node pair connectivity is (Abdullah and Parish 2007). This connectivity constraint may appear and disappear in the nci matrix due to node mobility. A route P from source s to destination t is defined as a sequence of intermediate nodes, such that $P(s, t) = \{s, \ldots i, j, k, l, \ldots, t\}$ without loop. The connectivity constraint, $nci_{(i,j)}$ associated with a node pair is transmission cost. It is specified by the connectivity matrix $C = [nci_{(i,j)}]$, where $nci_{(i,j)}$ represents the node pair connectivity index. It is described as follows:

$$C = \begin{bmatrix} nci_{0,0} & \cdots & nci_{0,k-1} \\ \vdots & \ddots & \vdots \\ nci_{k-1,0} & \cdots & nci_{k-1,k-1} \end{bmatrix} \tag{6.7}$$

The connectivity matrix is built at the source upon receiving the route reply (RREP) packets from the destination after a certain period of time, given as route accumulation latency (RAL) of $nci_{(i,j)}$. The values continue to change as

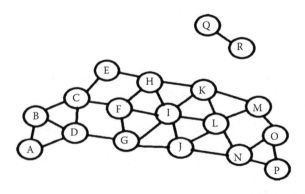

Figure 6.3 Mobile ad hoc network.

the topology changes. The protocol also removes the staled values to ensure that the contents remain current. A connection indicator $L_{i,j}$ mapped the connected nodes forming a chromosome. L_{ij} provides the information on whether the link from node i to node j is included in the routing path. It is defined as follows:

$$L_{i,j} = \begin{cases} 1 & \text{if there exists connectivity}(i, \ j) \\ 0 & \text{if otherwise} \end{cases} \tag{6.8}$$

The diagonal elements of L must always be zero. Another formulation in describing the MANET topology is node sequence in the routes, such that

$$N_k = \begin{cases} 1 & \text{if node} N_k \in \text{route} \\ 0 & \text{if otherwise} \end{cases} \tag{6.9}$$

Using the previous definitions, MANET QoS routing can be formulated as a combinatorial optimization problem minimizing the objective function. The sum of *nci* of the selected route should be minimized, since this would be the most preferred route due to the higher probability of being connected longer with next hop neighbors. Then, the formulation statement is to minimize the sum of node connectivity index of the route, shown as

$$C_{\text{sum}(S,T)} = \sum_{i=S}^{T} \sum_{\substack{j=S \\ j \neq i}}^{T} C_{ij} \cdot L_{ij} \tag{6.10}$$

The sum of *nci* of the route $P(s, t)$ constitutes the "cost" of the packet transmission process. In this approach, the *cost* of transmission is due to the lifetime of the node pair connection. The longer the connectivity lifetime, the lower the *cost* of the route is. The node pair connectivity index indicates the estimated length of time a given node pair is connected. The most important features of *nci* are the velocity and position of a node with respect to the other neighbor node. A node has longer connectivity time if its *nci* is smaller than the other node pair. Detailed descriptions and analysis of node pair connectivity index, *nci* can be found in the work of Abdullah and Parish (2007).

6.5.2 Constraints

The operation of GA will minimize the sum of node connectivity index of the route, $C_{\text{sum}(S,T)}$, subject to various constraints.

1. There must be no looping.
   This constraint ensures that the computed result is indeed an existing path and without loop between a source, S, and a designated destination, T,

such that

$$\sum_{\substack{j=S \\ j \neq i}}^{T} L_{i,j} - \sum_{\substack{j=S \\ j \neq i}}^{T} L_{j,i} = \begin{cases} 1 & \text{if} & i = S \\ -1 & \text{if} & i = T \\ 0 & \text{otherwise.} \end{cases} \qquad (6.11)$$

2. Available node bandwidth must be greater than request bandwidth.
   This constraint ensures that the node bandwidth can manage the requested
   bandwidth such that,

$$B_{AVA,i} \geq B_{REQ} \qquad (6.12)$$

and for the whole route,

$$B_{REQ} \leq \min(B_S, \ldots, B_i, B_j, \ldots, B_T) \qquad (6.13)$$

where B_{REQ} is the bandwidth requested by the source. The node bandwidth must be greater than the requested bandwidth. Generally, for a QoS
operation to be effective, the bandwidth available for the particular node
must be considered. Since the shared medium is being utilized by Car-
rier Sense Multiple Access/Collision Avoidance (CSMA/CA) at the link
layer of the MANET, the problem of medium contention among the nodes
within the transmission range must be taken into account. Hence, it is
necessary to estimate the instantaneous bandwidth $B_{AVA,i}$ and consumed
bandwidth $B_{CON,i}$, at the node concerned.

3. Total delay must be minimized.
   The constraints in terms of link delay and node delay must be considered.

$$D_w \geq \left\{ \sum_{i=1}^{m} \sum_{\substack{j=1 \\ j \neq 1}}^{|S \to T|} D_{i,j} \cdot L_{i,j} + \sum_{i=1}^{|S \to T|} D_j \cdot N_j \right\} \qquad (6.14)$$

If several routes exist, then the total delay for a route to be selected is the
one that is the shortest.

6.6 The Implementation of Genetic Algorithm for Route Selection

6.6.1 Encoding

The chromosome consists of sequences of positive integers, which represent the
identity of nodes through which a route passes (Figure 6.4). Each locus of the

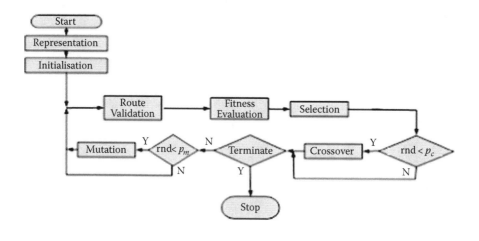

Figure 6.4 Procedure for genetic algorithm.

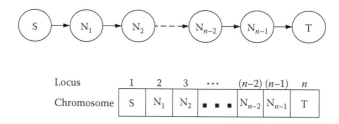

Figure 6.5 Chromosome representation of route from *S* to *T*.

chromosome represents a position of a node in a route. The gene of the first and the last locus is always reserved for the source node, *S*, and destination node, *T*. The length of the chromosome is variable, but it should not exceed the maximum length, which is equal to the total number of nodes in the network (Lee and Antonsson 2000). A chromosome that represents the route encodes the problem by listing up node identity from *S* to *T*, based on node information monitored from the network. The information is obtained and managed in real time by the NDMRD protocol (Abdullah 2015). The chromosome is essentially a list of nodes along the constructed path, *S* to *T*. In Figure 6.5, *n* represents the total number of nodes forming a path. The gene of the first locus encodes the source node, and the gene of the second locus is selected from the nodes connected with the source node. A chosen node is always checked to prevent it from being selected twice in the same chromosome, thereby avoiding loops in the route. This process continues until the destination node is reached. It is noted that an encoding is possible only if each step of a route passes through a physical link in the network.

Table 6.1 **An example of search space for different network sizes**

No of nodes	7	10	15	20	40
Links, $L = n(n-1)/2$	21	45	105	190	780
Search space, 2^L	2.10×10^6	3.51×10^{13}	4.05×10^{31}	1.56×10^{57}	6.4×10^{234}

6.6.2 Limited population initialization

GA process typically starts with a large number of initial populations. A large population results in better chances of finding good solutions. Generally, the initial populations are obtained by generating the chromosomes randomly. In QOSRGA, the initial populations are gathered as a result of NDMRD protocol. In a MANET with 5 nodes, the possible number of solutions was calculated as 10 according to the formula $n(n-1)/2$ (Gen and Cheng 2000). One approach is to generate the initial solutions randomly, and then remove the invalid solutions before they are fed to the GA module. Furthermore, the infeasible solutions can only be eliminated after the connectivity matrix is obtained by the NDMRD algorithm. The number of possible solutions increases enormously as the network gets bigger, as shown in Table 6.1. Clearly, a set of useful solutions is extracted before being processed by the GA module. This set of solutions has the characteristics of non-disjoint multiple routes, whereby each chromosome starts with the source node and ends with the target node. No looping is possible as it was done by the multiple routes discovery procedure, and each intermediate node must pass the two-node connectivity test.

6.6.3 Objective function

Fitness calculation is most crucial in the GA operation, whereby the best route can be identified. In this case the least value of fitness constitutes the lowest cost and the one that is to be chosen. The fitness value of routes is based on various QoS parameters: bandwidth, node delay, end-to-end delay, and the *nci*. Clearly it can be classified as a multi-objective optimization problem. According to Gen and Cheng (2000), each objective function can be assigned a weight, and then the weighted objectives are combined into a single objective function. For this MANET QoS routing protocol, the weighted-sum approach can be represented as follows. The fitness function operates to minimize the weighted-sum F, which is given as

$$F = \alpha \cdot F_1 + \beta \cdot F_2 + \gamma \cdot F_3 \qquad (6.15)$$

where F_1, F_2, and F_3 are the objective functions that describe *nci*, delay, and bandwidth, respectively. F_1, F_2, and F_3 are given as follows:

$$F_1 = \sum^{|s \to t|} C_{ij} \cdot L_{ij} \tag{6.16}$$

$$F_2 = \left(\sum_{j=1}^{S \to t} D_{ij} \cdot L_{ij} + \sum^{s \to t} d_j \cdot N_j \right) \tag{6.17}$$

$$F_3 = \begin{cases} 1/B_i & \text{if} \quad B_i - B_{QOS} > 0 \\ 1000 & \text{if} \quad B_i - B_{QOS} \leq 0 \end{cases} \tag{6.18}$$

The weights α, β, and γ are interpreted as the relative emphasis of one objective as compared to the others. The values of α, β, and γ are then chosen to increase the selection pressure on any of the three objective functions. The fitness function (Gen and Cheng 2000) measures the quality and the performance of a specific node state. It includes and correctly represents all or at least the most important parameters that affect QoS Routing. Having described these parameters—which are the bandwidth, *nci*, medium access delay, and end-to-end delay—the next issue is the decision on the importance of each parameter for QoS Routing as a whole. The significance of each parameter is defined by setting appropriate weighting coefficients to α, β, and γ in the fitness function that will be minimized by the GA operations. The values of these weighting coefficients were determined based on their equal importance towards the overall QoS Routing performance; hence, α, β, and γ are equal to 10^{-3}, 10^{-4}, and 10^{-3}, respectively. Concerning the function that involved bandwidth, we need to find the minimum bandwidth among the nodes and compare this with the demand bandwidth, B_{QOS}. If the minimum bandwidth is less than the B_{QOS}, the fitness is set to a high value so that in the selection process it will be eliminated. By doing so, all the nodes where the bandwidth is limited will have been eliminated simultaneously, the total delay being more than the typical delay and when the node pair connectivity index is high.

6.6.4 Mobile nodes crossover

Mobile nodes crossover examines the current solutions in order to find better ones. Physically, the crossover operation in the QoS routing problem plays the role of exchanging each partial route of two chosen chromosomes in such a manner that the offspring produced by the crossover represent only one route (Figure 6.6). This dictates selection of one-point crossover as a good candidate scheme for the proposed GA. One partial route connects the source node to an intermediate node, and the other partial route connects the intermediate node to the destination node. The crossover between two dominant parents chosen by the selection gives a higher probability of producing offspring having dominant

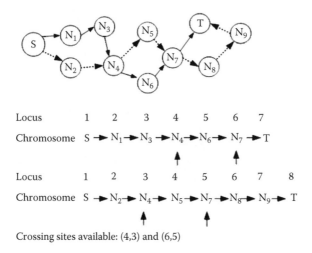

Locus   1   2   3   4   5   6   7

Chromosome   S → N_1 → N_3 → N_4 → N_6 → N_7 → T

Locus   1   2   3   4   5   6   7   8

Chromosome   S → N_2 → N_4 → N_5 → N_7 → N_8 → N_9 → T

Crossing sites available: (4,3) and (6,5)

Figure 6.6 Crossover operation.

traits. But the mechanism of the crossover is not the same as that of the conventional one-point crossover. In the proposed scheme, the two chromosomes chosen for crossover should have at least one common gene, except for source and destination nodes. It is not a requirement that they be located at the same locus. That is, the crossover is independent of the node position in routing paths. It shows a set of pairs of nodes that are commonly included in the two chosen chromosomes but without first determining positional consistency. Such pairs are called potential cross sites. Then, one pair (4,3) is randomly chosen and the locus of each node becomes a crossing site for each chromosome. The crossing sites of two chromosomes may be different from each other. Each partial route is exchanged and assembled, eventually leading to two new routes. It is possible that loops are formed during crossover. A simple restoration procedure is designed for this so that it can improve the rate of convergence and the quality of solution.

For the system of routes that can perform the crossover efficiently, it is done by first defining the adaptive back-off selection probability as shown by the following equation:

$$p = \frac{|N| - |M_{\text{weight\_}k}|}{|N|} \tag{6.19}$$

where $|N|$ is the number of individuals, and $|M_{\text{weight\_}k}|$ equals k. Adaptive back-off selection is an ideal way to implement family competition, which avoids two potentially negative effects: (i) loss of population diversity and (ii) trapping at a local optimal. Hence, when one individual is chosen randomly, the other one that participates in crossover could be selected by the mathematical expression in the following equation:

$$P\{T_S(\bar{X}) = X_j\} = \begin{cases} \frac{f(X_j)}{\sum_{k=1}^{n} f(X_k)} \cdot p & X_j \in M_{\text{weight}\_k} \\ \frac{f(X_j)}{\sum_{k=1}^{n} f(X_k)} \cdot (1-p) & X_j \in M_{\text{weight}\_k} \end{cases} \quad (6.20)$$

where $X(n)$ is the nth generation population, X stands for the current population, and X_i is the individual in X. To speed up convergence of QOSRGA, greedy algorithm is imported. Greedy, a useful and powerful means in many optimization problems, converges very quickly but it is liable to trap at a local optimal. That is the reason why we employed family competition; then the crossover operation can be described as follows:

Step 1. Select N individuals independently from the group $X(n)$ so as to get the population of $X(n) = (X_1, X_2, \ldots, X_N)$.

Step 2. Select two individuals for family competition: $X_i(n) = (v_1, v_2, v_3, \ldots, v_n)$ and $X_j(n) = (v_1, v_2, v_3, \ldots, v_n)$.

Step 3. Let v_1 be the first gene of $X_i(n+1)$; find the next gene of v_1. Evaluating their Pareto Relationship, choose one (such as v_2) that dominates the other as the second gene of $X_i(n+1)$ and set the corresponding unit of visiting vector to 1.

Step 4. Find the position v_2 in $X_i(n)$ and $X_j(n)$; compare their tail gene to confirm which is better, then choose it as the next gene of $X_i(n+1)$ and set the corresponding unit of visiting vector to 1.

Step 5. Repeat Steps 1–4, till $X_i(n+1)$ is formatted.

$X_j(n+1)$ can be generated similarly, just with the difference of getting the next gene from the reversed direction. Then crossover can be described as in the following equation:

$$P\{T_c(X_i, X_j)\} = \begin{cases} (l_1 + l_2) p_c & X_i \neq X_j \\ 1 - p_c & X_i = X_j \end{cases} \quad (6.21)$$

where l_1 and l_2 are the lengths of chromosomes. Following the theoretical aspect of chromosomal crossover, the following procedure represents the implementation of crossover as one of GA operator.

Step 1. Input a matrix that consists of the route array and the number of nodes in the environment.

$$ROUTE\_ARRAY = \begin{bmatrix} n_{0,0} & n_{0,1} & n_{0,2} & \cdots & n_{0,k-1} \\ n_{1,0} & \cdots & \cdots & \cdots & \cdots \\ n_{2,0} & \cdots & \cdots & \cdots & \cdots \\ \cdots & \cdots & \cdots & \cdots & \cdots \\ n_{m-1,0} & \cdots & \cdots & \cdots & n_{m-1,k-1} \end{bmatrix} \quad (6.22)$$

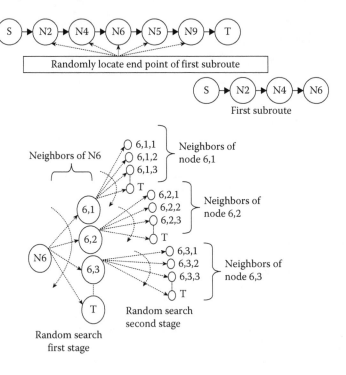

Figure 6.7 Schemes for multiple routes mutation.

Step 2. Test for crossover rate. Initialize the random number generator and the new route array. The population size must be positive and even.

Step 3. Consider a pair of chromosomes denoted as parents, V_1 and V_2, starting from the last chromosome within the population. The length might not be the same as originally stated.

Step 4. Locate the potential pair of crossing sites by searching for nodes common to both chromosomes.

Step 5. If more than one pair of crossing sites exist, apply a random number to establish one particular pair of crossing sites, i.

Step 6. Perform crossover of V_1 and V_2 by exchanging all nodes after the crossing site i. Two offspring, V_1' and V_2', were produced (Figure 6.7).

6.6.5 Route mutation

Mutation is used to change randomly the value of a number of the genes within the candidate chromosomes. It generates an alternative chromosome from a selected chromosome. It can thus be seen as an operator charged with maintaining the genetic diversity of the population, thereby keeping away from local optima. Mutation may also induce a subtle bias in which it generates an alternative partial route from the mutation node to the destination node. However, this small bias does not affect the performance of the algorithm. This is explained as follows: (i) mutation leads to an infinitesimal increase in the probability of inducing the bias and (ii) selection and crossover strongly influence the way this bias operates. Indeed, by the process of mutation, harmful effects may vanish altogether. The single point mutation is shown in the following equation:

$$P\left\{T_{\mathrm{m}}\left(\vec{X}\right) = Y\right\} = p_{\mathrm{m}}^{\mathrm{d}(X,Y)}(1 - p_{\mathrm{m}})^{l - \mathrm{d}(X,Y)} \tag{6.23}$$

where d(X, Y) denotes the number of gene pairs that the corresponding gene in X and Y is different from each other.

The general procedure for the mutation process is outlined in the following steps.

Step 1. Input two matrices: (i) the population matrix (Equation 6.24) and (ii) the connectivity matrix (Equation 6.25). The population matrix consists of a collection of chromosomes that result from a previous generation where each chromosome represents a QoS route from source to destination.

$$POP\_MATRIX = \begin{bmatrix} n_{0,0} & n_{0,1} & n_{0,2} & \cdots & n_{0,k-1} \\ n_{1,0} & \cdots & \cdots & \cdots & \cdots \\ n_{2,0} & \cdots & \cdots & \cdots & \cdots \\ \cdots & \cdots & \cdots & \cdots & \cdots \\ n_{m-1,0} & \cdots & \cdots & \cdots & n_{m-1,k-1} \end{bmatrix} \tag{6.24}$$

$$CONNECTIVITY\ MATRIX, L_{i,j} = \begin{bmatrix} l_{1,1} & l_{1,2} & l_{1,3} & \cdots & l_{1,n} \\ l_{2,1} & \cdots & \cdots & \cdots & \cdots \\ l_{3,1} & \cdots & \cdots & \cdots & \cdots \\ \cdots & \cdots & \cdots & \cdots & \cdots \\ l_{n,1} & \cdots & \cdots & \cdots & l_{n,n} \end{bmatrix} \tag{6.25}$$

Step 2. Randomly select a parent chromosome V, from the *POP_MATRIX*. It is selected with the probability P_m.

Step 3. Randomly select a mutation node i from V.

Step 4. Generate the first subroute r_1 from source node, S, to node i by deleting a set of nodes in the upline nodes after the mutation node.

Step 5. Generate a second subroute r_2 from i to the destination node T. It is done as follows:

Step 5-1. Determine node degrees of i, deg (i), neighbors of i. If deg$(i) = 1$ and $\{\deg(i)\} = T$, then terminate the search, since the second subroute consists of T. If deg$(i) = 1$ and $\{\deg(i)\}$ # T, then terminate the mutation process. If deg$(i) > 1$, go to Step 5-2.

Step 5-2. Select node $\{1, 2, 3,..., \deg(i)\}$. If deg$(1) = 1$ and $\{\deg(1)\} = T$, then second subroute is generated. Proceed with 2 and so on. If deg$(1) = 1$ and $\{\deg(1)\}$#T, proceed with 2 and so on. If deg$(1) > 1$, go to Step 5-3.

Step 5-3. Select node $\{1, 2, 3,..., \deg(1)\}$. If deg$(1) = 1$ and $\{\deg(1)\} = T$, then second subroute is generated. Proceed with 2 and so on. If deg$(1) = 1$ and $\{\deg(1)\}$#T, proceed with 2 and so on. If deg$(1) > 1$, terminate. We only search for the second subroute up to two stages so that the effort will not take much processing time.

Step 5-4. If the number of second subroutes generated is more than one, then choose the least hop.

Step 6. Combine the first subroute and the second subroute forming a new route. Add to the *POP_MATRIX*.

Step 7. If any duplication of nodes exists between r_1 and r_2, discard the routes, and do not perform mutation. Otherwise, connect the routes to make up a mutated chromosome. However, nodes already included in an upper partial route should be deleted from the database so as not to include the same node twice in the new route. The upper partial route represents the surviving portion of the previous route after mutation; it is the partial chromosome stretching from the first gene to the intermediate gene at the mutation point.

6.6.6   Route selection schemes

There are a few selection schemes to be considered. They are the tournament (TOUR), elitism, stochastic universal selection (SUS), and roulette-wheel selection (RWS). Each has its own merit and can be examined to find the most useful application. The schemes are briefly described, and in the next section we examine and chose one of them. By means of an individual's fitness evaluation, we can conclude that individuals with the same Pareto dominance will have the same Pareto rank. Hence, the selection scheme is described by the following equation:

$$P\left\{T_s\left(\vec{X}\right) = X_i\right\} = \frac{f(X_i)}{\sum\limits_{k-1}^{n} f(X_k)} \tag{6.26}$$

(a)

```
Input: The Population P(τ) the tournament size t ε {1,2, ..., N}
Output: The population after selection P(τ)'

tournament (t, J₁, ..., J_N):
    for i ← 1 to N do
            J_i' ← best fit individual out of t randomly picked
            individuals from { J₁, ..., J_N } ;
    od
    return { J'₁, ..., J'_N}
```

(b)

```
Input: The Population P(τ) and the reproduction rate for each
    fitness value R_i ε [0, N]
Output: The population after selection P(τ)'

SUS (R₁, ..., R_n, J₁, ..., J_N):
    sum ← 0
    j ← 1
    ptr ← random [0,1)
    for i ← 1 to N do
            sum ← sum + R_i where R_i is the reproduction rate
                    of individual J_i
            while (sum > ptr) do
                    J_i' ← J_i
                    j ← j + 1
                    ptr ← ptr + 1
            od
    od
    return { J'₁, ..., J'_N}
```

Figure 6.8 (a) Tournament selection algorithm. (b) SUS algorithm.

1. Tournament selection (TOUR)

   In TOUR, a number of individuals are chosen randomly from the population, and the best individual from this group is selected as a parent. This process is repeated as often as there are individuals to be chosen. These selected parents produce uniform, random offspring. The algorithm for TOUR is shown in Figure 6.8a.

2. Roulette wheel selection (RWS)

   The simplest selection scheme is RWS. It is a stochastic algorithm and involves the following technique. The individuals are mapped to continuous segments of a wheel, such that each individual's segment is proportional in size to its fitness. A random number is generated, and the individual whose segment spans the random number is selected. The process is repeated until the desired number of individuals is obtained. Selection is random but biased towards individuals with the least cost or highest fitness. In fact, RWS is the first selection method where the probability P_i for each individual is defined by:

$$P\{\text{Individual i is chosen}\} = \frac{F_i}{\sum\limits_{j=1}^{F_i} F_j} \qquad (6.27)$$

where F_i equals the fitness of individual i. The use of RWS limits the maximization of the GA, because the evaluation function must map the solutions to a fully ordered set of values. Extensions, such as windowing and scaling, have been proposed to allow for minimization and negativity. In RWS, the individuals are mapped to contiguous segments of a line, such that each individual's segment is equal to its fitness. A random number is generated, and the individual whose segment spans the random number is selected. The process repeats until the desired number of individuals is obtained (called *mating population*). This technique is analogous to a roulette wheel with each slice proportionally sized to the individual's fitness.

3. Elitism
   In elitism, the individuals that are the fittest fraction μ of the population are retained for the next generation. The remaining fraction of $(1 - \mu)$ of the population is selected for crossover using TOUR.

4. Stochastic universal selection (SUS)
   A perceived drawback of RWS is its high degree of variance. An unfit individual could, by chance, reproduce more times than a fitter one. The mechanics of SUS overcomes this effect. Individuals are allocated a proportion of a circumference of a wheel according to their fitness value. Only one spin is required in order to select all reproducing individuals, and so the method is less computationally demanding compared to RWS (Blickle and Thiele 1995). Each individual is defined by Equation 6.27, and the SUS algorithm is shown in Figure 6.8b.

$$p(T_i) = \frac{f(T_i)}{\sum\limits_{j=1}^{N_p} f(T_j)} \qquad (6.28)$$

6.7 GA Parametric Evaluations and Preferences

6.7.1 Parameters of concern

Selecting GA parameters such as population size, mutation rate, and crossover rate could be a difficult task. Each combination of parameters may produce a

Table 6.2   Variables used for the simulation runs

Selection methods	RWS, TOUR, SUS, elitism
Population sizes	10–2,000 individuals
Fixed mutation rate	0.01–0.5
Crossover rate	0.80–1.00
Generations	10–10,000 generations

variety of outcomes. Haupt and Haupt (2004) outlined a general procedure for evaluating these parameters, after which suitable parameters are adopted. In this case, four selection methods are considered; RWS, TOUR, SUS, and elitism technique. Next, the parameters P_c, P_m, and population size are considered. It is necessary to examine the performance of each and select according to preferences. Matlab is used to initially design a GA-based routing algorithm without the QoS function. The route selection is based on the shortest path without considering the bandwidth, delay, or *nci*. The cost for each path is randomly generated. The main objective is to examine all the GA parameters that are useful for the protocol design. Hence, in this section a mobile network is considered, consisting of 20 nodes, randomly distributed within a perimeter of 1000 m by 1000 m with node transmission range of 250 m. In each selection, population sizes are varied to locate the best point of convergence, corresponding to the lowest measure. For each reading, 10 simulation runs are done, and the results are averaged. Table 6.2 outlines the variables for optimization of GA parameters.

6.7.2   Population size and its average minimum cost, C_{AMC}

The effect of population size is investigated by fixing the mutation rate ($P_m = 0.01$) and varying the population size. The simulation is run for 2000 generations. The minimum cost in each generation is recorded and the average minimum cost C_{AMC} evaluated over the range from 0 until the 2000th generation. Figure 6.9 shows the plots of C_{AMC} for the four different selection methods (with $\mu = 0.05$ for elitism). It shows that in RWS, a population size in excess of 700 produces a significantly low cost. With a large population, the RWS method finds it easier to choose the low-cost individual. Consequently, the probability of a low-cost individual being selected becomes high. Apart from this, with a large population size there are too many sectors within the wheel, making the probability of selecting each sector smaller. The most significant result is that of TOUR and elitism. With a population size of approximately 10, it produces very low C_{AMC}. Hence the best choice of selection method would be TOUR and elitism. Figure 6.9 is re-plotted in Figures 6.10 and 6.11, concentrating on a population size under 200 and 100, respectively. The results reinforce the view that a population size under 100 is appropriate for both the elitism and TOUR. In fact,

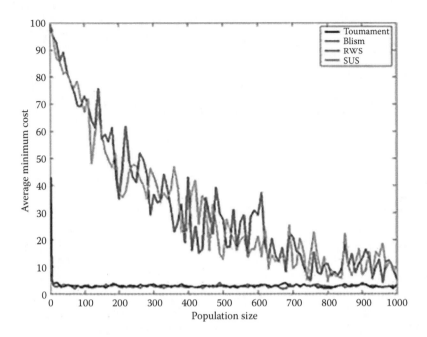

Figure 6.9 Plot of average minimum cost for various selection methods.

populations as low as 20 could be used and still produce a good fit. Thus TOUR is chosen.

6.7.3 The fitness function compared to the generations

Another necessary parametric study for the GA process is to find the most reasonable number of generations that could produce a satisfactory result. Again, this is done offline. The mutation rate is fixed at 0.1 and the crossover rate at 0.7. A simulation is run for different sized populations. The fitness measures are taken for different populations. The result is shown in Figure 6.10. It shows the lowest cost as a function of the number of generations for TOUR. This figure also shows that a larger population size generally produces lower-cost solutions. However, the rate of convergence is about the same when the population exceeds 20. The use of a larger population requires greater processing power and therefore is not desirable. From the graph it can also be inferred that convergence could occur in less than 20 generations.

6.7.4 Crossover and mutation probability

Another set of very important parameters for the GA implementation is the crossover probability P_c and the mutation probability P_m. The parameters determine how many times crossovers occurred and how many times mutations

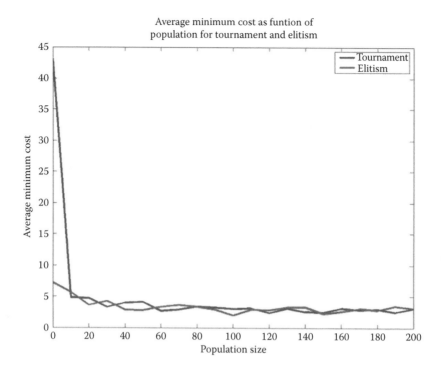

Average minimum cost as funtion of
population for tournament and elitism

Figure 6.10 Plot of C_{AMC} as a function of population size—tournament, and elitism (0.05).

occurred within a transmission period. The occurrence of crossover and mutation increases the convergence rate. De Jong (Haupt and Haupt 2004) tested various combinations of GA parameters and concluded that mutation is necessary to restore lost genes, but this should be kept at a low rate; otherwise the GA degenerates into a random search. Further study by Schaffer et al (1989) suggests that the parameters should have these recommended ranges: population size of 20 ~ 30, mutation rate of 0.005 ~ 0.1, and crossover rate of 0.75 ~ 0.95. Another study by Haupt (2000) concluded that the best mutation rate for GAs lies between 5% and 20%, while the population size should be less than 16. For this thesis, where GA operation is done online, the value of P_c and P_m is taken to be between 0.4 and 0.9 and between 0.05 and 0.2, respectively. The choice of these parameters should produce a reasonably high efficiency packet transmission. The population size is limited up to the number of routes discovered. The limit is also imposed on the number of generations. Haupt and Haupt (2004) provide useful guidelines on when to stop the algorithm. It would be better to stop the algorithm when the value of the selected route does not change or when it reaches the maximum number of generations. We designed the algorithm so that

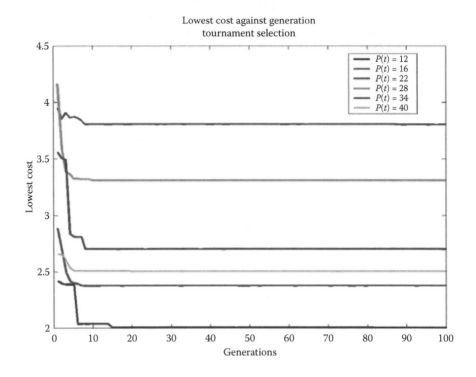

Lowest cost against generation
tournament selection

Figure 6.11 Lowest cost against generation for tournament selection.

it stops when the value of the route does not change, and we also restricted the maximum number of generations to 20.

Simulation experiments are done online by considering a MANET scenario running QOSRGA protocol, with 20 nodes placed within an area of 1000 m × 1000 m. Each node has a radio propagation range of 250 m and channel capacities of 2 Mbps. Up to 10 sources are initiated transmitting Constant Bit Rate (CBR) data with a payload of 512 bytes. The nodes move randomly within a random waypoint mobility model. In each point, 10 simulation runs are carried out and each run takes 200 s. Two sets of simulation experiments are conducted, one for calculation of crossover probability and the other for mutation probability. The experiment is done using the source traffic rate of 40, 100, and 900 kbps. The aim of the first set is to identify exactly the possible values of P_c that would give the best results. The metric is the transmission efficiency. It is defined as the ratio of average throughput of all nodes to the average load of all the nodes in the network. We varied P_c but set P_m constant as 0.1. The results are shown in Figure 6.12. For P_c with values from 0.4 to 0.8 the transmission efficiency is more than 80%. For 100 kbps source, the maximum efficiency occurred when P_c is approximately 0.65 and for 40 kbps source it is 0.4. The efficiency does not

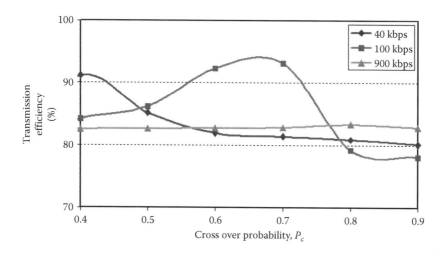

Figure 6.12 Transmission efficiency as a function of crossover probability.

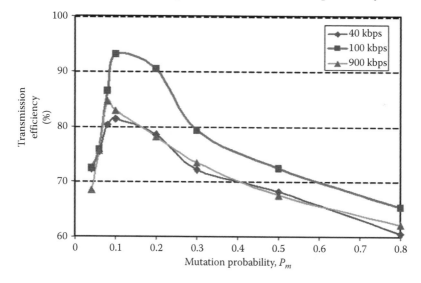

Figure 6.13 Transmission efficiency as a function of mutation probability.

deviate very much at the 900 kbps CBR source. Hence, as a general guideline the value of P_c as 0.7 was chosen for all future simulation experiments.

For mutation probability, a similar simulation is done. This time the crossover probability is fixed at 0.7, and the mutation probability varies from 0.04 to 0.8. Figure 6.13 shows that the mutation probability produces the highest transmission efficiency when it is 0.1, which is more than 80% for all three different traffic rates. Hence, it is concluded that the crossover probability and mutation probability can be taken as 0.7 and 0.1, respectively.

6.8 Performance Evaluation of GA Based QoS Route Selection

6.8.1 Qualitative properties of QOSRGA

In MANET, link quality changes rapidly and with unpredictable outcomes as each node moves randomly over a given area. Consequently, route lifetime is reduced considerably, and a QoS routing mechanism should be able to find new routes to maintain connectivity and route reliability while preserving QoS requirements. The theme of this section is about evaluating the performances of QOSRGA such that it can (i) provide for effective operations over a range of networking contexts and (ii) react effectively to topological changes and traffic demands while maintaining effective routing in mobile networking contexts. In evaluating the merit of a routing protocol, one needs both qualitative and quantitative metrics with which to measure its suitability and performances. Initially, the qualitative properties are described, and then quantitative characteristics based on the simulation results are obtained using Opnet Modeler. The aim of the simulation experiments is to use various performance metrics such as packet delivery ratio, end-to-end packet transmission delay, throughput, and normalized routing load (NRL) as a comparative evaluation of the proposed QoS routing algorithms. QOSRGA was compared with the most common MANET protocols, which are Best-Effort – Dynamic Source Routing(BE-DSR) and Best-Effort – Ad hoc On-demand Distance Vector (BE-AODV) (Haupt 2000; Johnson et al. 2007).

This section outlines the qualitative properties of QOSRGA, which are compatible within the context of a mobile networking environment. The following list describes the qualitative properties that can be attributed to QOSRGA.

1. *Distributed operation*: The overall protocol is run on a per node basis. Its operation is triggered by the arrival of packets from the next-hop neighbors. No central unit exists to organize the network. End nodes function as source or sink. The intermediate nodes function as routers, forwarding packets to the next-hop neighbors.

2. *Loop-freedom*: The protocol inherently avoids looping in order to increase the overall performances. One of the processes in the QOSRGA QoS routing operation is to check for loop occurrences. If a loop occurs, then a restoration process to remove the redundant nodes is initiated.

3. *Demand based operation*: The protocol is designed to operate on demand. Demand is initiated at the application layer by the packet generation

process, which embeds bandwidth and delay requirements. When demand is not available, the periodic connectivity packet transmission still continues to operate, in which case it maintains the information regarding neighbor nodes connectivity.

4. *Interaction with the standard IP routing protocol*: QOSRGA protocol is designed to be either a source or a sink within MANETs. The routing protocol successfully interacts with the standard IP layers by adhering to the proper IP addressing and convention. The IPv4 convention is used throughout.

5. *QOSRGA is built on three cooperative protocols*: QOSRGA consists of three cooperative protocols, which include the NDMRD protocol (Abdullah and Parish 2007), the Node State Monitoring protocol (Abdullah 2015), and QoS Route selection using GA (Abdullah et al. 2008).

6.8.2 Simulation environment model, performance metrics, and scenarios

OPNET 10.5 Modeler (Riverbed Modeler 2017) is used to create a simulation environment for the proposed QOSRGA algorithm and compares its performance with the already existing on-demand BE-AODV and BE-DSR routing protocols. The field is assumed to be a flat square shape, with the dimensions 1000 m by 1000 m. The maximum number of nodes in the network is set to 40 nodes (Lee and Gerla 2001; Zhou and Lee 2005). The physical layer parameters are as follows: maximum data rate of 2 Mbps, operating frequency of 2.4 GHz, and transmit power of 5 mW using DSSS modulation. Free space propagation with log-normal shadowing model is assumed with wireless transmission range of 250 m. Each node uses the basic configuration of Institution of Electrical and Electronic Engineering (IEEE) 802.11 (IEEE LAN MAN Standards Committee 1999), CSMA/CA operating in the DCF mode. All transmitters and receivers are configured such that they are set when performing the computation. A node is considered to be within another node's transmission range if it satisfied the distance and path loss calculation. The initial setup of the simulation is shown in Figure 6.14.

Mobility and traffic models are similar to those previously reported (Lee and Gerla 2001; Perkins et al. 2001), and the random waypoint model (RWP) (Bettstetter et al. 2003) is used to model the random motion of nodes. Each node started its journey from a random location to a random destination point with a specific speed. Once the destination is reached, it stopped for a period of time and then calculated another random destination point. The initial angle of motion for every node is set at 0 degrees. In this version of Opnet, the random waypoint

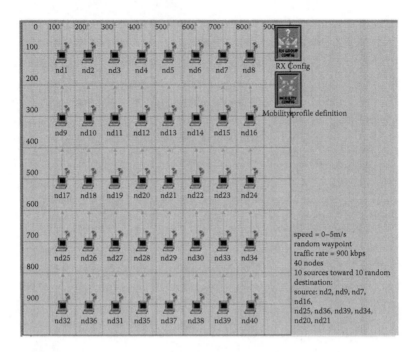

Figure 6.14 The initial setup of simulation environment with 40 nodes.

mobility model is included with the wireless module of the Opnet Modeler, configured into the model through the Mobility Profile Definition. All nodes move within the boundary of the field configuration. The pause time is constant at 1 s (Lee and Gerla 2001; Zhou and Lee 2005) for all the simulation experiments. This gives consistency in the nodes' movement for all of the scenarios. The start time is set to zero until the end of simulation. The only variation within the RWP model is the speed. The speed is configured into a uniform distribution between zero and V_{max}, where V_{max} can be set accordingly. Traffic sources with 512-byte data packets (Marina and Das 2001; Zhou and Lee 2005) are CBR in nature. The source-destination pairs are spread randomly over the network, and the number of sources is varied to change the offered load into the network. During the lifetime of a flow, the source nodes continuously generated data packets at the rate determined by the inter-arrival rate. The sending rate is varied accordingly, from between 24 and 292 pps. Nodes in all the three protocols maintain an infinite send buffer that contains queued packets. Each node buffered all data packets while waiting for a route. All packets (both data and routing) from the routing layer are queued at the buffer until the layer is able to transmit them. Routing packets are given higher priority than data packets in the buffer. Simulations are run for 200 simulated seconds (Lee and Gerla 2001; Zhou and Lee 2005). Each data point represented an average of 10 runs with identical traffic models, but different randomly-generated mobility scenarios by using different seeds to the random number generator. Another interesting aspect of the protocol design is to understand the protocol performance with various congestion levels.

The following metrics (Chlamtac et al. 2003) are used in varying scenarios to evaluate the performance of QOSRGA protocol.

Average packet delivery ratio (APDR): Since the study is essentially based on bandwidth measurement, a metric is proposed that expressed the efficiency of bandwidth, as an APDR. The APDR is defined as the ratio between the total packets received by the upper layer of every node to the total packets generated at the upper layer of every node in the MANET system. It is expressed in terms of a percentage, as in the following equation:

$$\text{APDR} = \frac{\frac{1}{N_n} \sum_{i=1}^{N_n} P_{\text{arr}(i)}}{\frac{1}{N_n} \sum_{i=1}^{N_n} P_{\text{gen}(i)}} \tag{6.29}$$

Average total end-to-end delay of data packets (ATETED): This includes all possible delays from the moment the packet is generated to the moment it is received by the destination node. The statistics of average delay of all the packets received during the simulation time are taken and then divided by the average total number of packets arrived at every receiving node. This gives the average delay of a packet.

$$\text{ATETED} = \frac{\sum_{j=1}^{N_n} \left\{ \sum_{i=1}^{m_a} T_a - T_s \right\}}{\frac{1}{N_n} \sum_{j=1}^{N_n} \left\{ \sum_{k=1}^{m_a} P_{\text{arr}(k)} \right\}} \tag{6.30}$$

Total average throughput (TAT): In this context the throughput is defined as the total number of bits (in bits/s) forwarded from the Wireless Local Area Network (WLAN) layers to higher layers in all WLAN nodes of the network. To find the average throughput of a single node, one has to divide by the number of nodes in the system.

$$\text{TAT} = \sum_{i=1}^{N_n} \left\{ \frac{\sum_{j=1}^{\text{max}} P_{\text{arr}(j)}}{\sum_{k=1}^{\text{max}} (T_a - T_s)} \right\} \tag{6.31}$$

Normalized routing load (NRL): In order to measure the cost of the QOSRGA protocol as compared to other protocols, a metric is included to evaluate the overhead necessary for successful data packet transmission. The overhead is usually measured as the number of control packets transmitted to establish and maintain the paths in the network. This is rather misleading, since the amount of resources wasted due to imprecise routing information is not considered. Hence, in this work an NRL is used, which considered all the routing packets that are dropped in other nodes in the network. The NRL is defined as follows:

$$\text{NRL} = \sum_{i=1}^{N_n} \left\{ \frac{\sum\limits_{j=1}^{\max} p_{\text{arr}(j)} - \sum\limits_{k=1}^{\max} d_{\text{arr}(k)}}{\sum\limits_{k=1}^{\max} d_{\text{arr}(k)}} \right\} \qquad (6.32)$$

The NRL is an important metric to compare the performance of different protocols, since it can give a measure of the efficiency of protocols, especially in a low bandwidth and congested wireless environment. Protocols that transmit a large number of routing packets can also increase the probability of packet collisions and waiting time of data packets in transmission buffer queues.

6.8.3 Impact of source traffic rate variation on performance

The simulation experiments are carried out by keeping the maximum node velocity constant at 2 m/s, with 40 nodes. This was to study the effect of varying the source traffic rate from 20 to 1200 kbps. In the simulation environment, 10 nodes were set as the sources to random destination nodes. The traffic rate of the sources was varied by configuring the source node with an exponentially-distributed inter-arrival rate.

1. Average Packet Delivery Ratio (APDR)
   The control packet transmission must be considered, since these packets also load the network. In order to compare the APDR, the average of total traffic sent and average of the total traffic received are recorded for each traffic rate of 20–1200 kbps. For each traffic rate of the load, the simulation is repeated for 10 runs, and the average readings are recorded. Each run used a seed with different values, so as to diversify the simulation output. Hence each point in the graph is a result of 10 runs. The ratio of average total packets received to average total packets sent is taken for each traffic rate. A plot of APDR against source traffic rates is shown in Figure 6.15. At low source traffic rate, all protocols showed similar results. The plots dropped rapidly until about 40% where each produced a different rate. BE-DSR goes further until 20%, where it stays constant. QOSRGA performed a little better than BE-AODV. When searching for the routes, QOSRGA readily acquired network information that included the bandwidth availability and the connectivity index. It had chosen the route that had less probability of being lost in the near future. It chose a route that is more reliable than the other two BE protocols. When the source traffic is more than the 400 kbps, congestion caused the ratio to stabilize at approximately 40%.

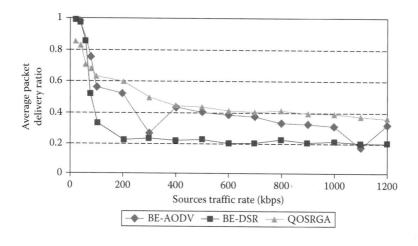

Figure 6.15 APDR against the source traffic rates.

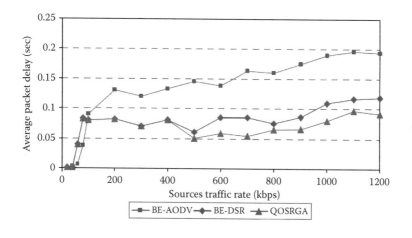

Figure 6.16 Average end-to-end delay against source traffic rate.

2. Average end-to-end delay of data packets (ATETED)

   Figure 6.16 depicts the variation of the average end-to-end delay as a function of the traffic rate. It can be seen that QOSRGA protocol has lower average delay than BE-DSR and BE-AODV under all source traffic rates. The primary reason is that the number of route discoveries is reduced in QOSRGA. Although QOSRGA has a low number of route discoveries, its delay also decreases gradually with increase in the traffic rate. This is attributed to the fact that an increase in the traffic rate leads to higher network traffic load. Since QOSRGA is configured to select bandwidth first and then, due to the GA process, the route selection always tries to locate the route with enough bandwidth first without dropping packets, as happens for BE-DSR and BE-AODV. Beyond 800 kbps, the congestion is so high that the average delay for the QOSRGA protocol is increased. From 20 until 1200 kbps, the average packet delay for QOSRGA was less than 100 ms, which is within the stated QoS requirements.

3. Total average throughput (TAT)

   Figure 6.17 shows the total average throughput against the source traffic rate. Below 100 kbps, the performance of QOSRGA is similar to BE-DSR. The rate of increase in throughput for QOSRGA and BE-AODV is steeper between 100 until 600 kbps compared to BE-DSR. Beyond 600 kbps, the throughput increase rate starts to decline. One observation is that QOS-RGA and BE-AODV delivered a similar performance, while QOSRGA was approximately 5% better than BE-AODV and almost 100% better than BE-DSR. When the source node generates packets at higher bandwidths, the GA process works better by choosing a path that can accommodate the given bandwidth. BE-DSR, on the other hand, dropped all packets that cannot be accommodated, due to limited bandwidth.

6.8.4   Impact of node mobility on the performance

The second set of simulation experiments varied the velocity of a 40 nodes network and 10 CBR sources. The mobility is varied to see how it affected the

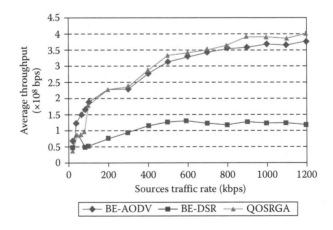

Figure 6.17 Average throughput as a function of sources traffic rate.

different metrics that are measured. The packet sending rate is fixed at 98 pkts/s (400 kbps) and 4 pkts/s (20kbps). The simulations are run with uniform velocity, where the maximum velocities are 0.5, 1, 1.5, 2, 5, 10, 15, 20, and 25 m/s. In all the simulations, the velocity is limited to 25 m/s (equivalent to 90 km/h), in order to maintain node connectivity successfully.

1. Average packet delivery ratio (APDR)

   The graph of APDR against node maximum velocity is shown in Figure 6.18. Two sets of results are obtained, one for sources at 4 packets/s and the other for 98 packets/s. At 4 packets/s, sources BE-AODV performed better than BE-DSR and QOSRGA for the whole maximum velocity range. By comparing QOSRGA and BE-DSR, QOSRGA produced a slightly better APDR. When the mobility is less than 12 m/s, QOS-RGA gives a similar reading. When node mobility is more than 12 m/s, QOSRGA performed better; in fact, 5% better than BE-DSR. For high-bandwidth sources of 98 packets/s, it is clear that QOSRGA consistently performed better than BE-DSR and BE-AODV for all the mobility ranges. Generally, it is 5%–15% better than BE-AODV and 5%–30% better than BE-DSR. In QOSRGA, multiple routes were found with the corresponding QoS metrics information B_{AVA}, D_{ETE}, D_{MAC}, and *nci*. The selection of the routes is based on the probable length of time each node pair stays connected, indicated by *nci*. The degradation of BE-DSR occurred as the mobility rate increased. In high mobility scenarios, many route reconstruction processes are invoked. When a source floods a new Route Request packet(RREQ) to recover a broken route, many intermediate nodes send

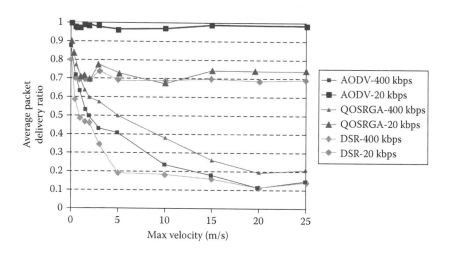

Figure 6.18 Mean packet delivery ratio as a function of maximum velocity.

RREP packets back to the source, because of the route caching mechanism of BE-DSR. However, routes overlap the existing routes, resulting in severe congestion, and it cannot deliver packets along the route. Moreover, the stale routes produce a reply to source with invalid routes. Ultimately, many packets are dropped, resulting in poor BE-DSR performance. In QOSRGA, an aging mechanism is used; hence, the stale routes will be replaced.

2. Average end-to-end delay of data packets (ATETED)
   The average end-to-end delay includes all possible delays from the moment the packet is generated to the moment it is received by the destination nodes. Generally, there are three factors affecting end-to-end delay of a packet: (i) Route discovery time, which causes packets to wait in the queue before a route is found; (ii) Buffering waiting time, which causes packets to wait in the queue before they can be transmitted; and (iii) The length of the routing path. More hops means a longer time to reach the destination node. Figure 6.19 depicts the variation of the average end-to-end delay as a function of the velocity of nodes.

   It can be seen that the general trend of all the curves is an increase in delay with the increase of the velocity of nodes. The reason is mainly that high mobility of nodes results in an increased probability of link failure that causes an increase in the number of routing rediscovery processes. This means data packets have to wait longer in the queue until a new routing path is found. The delay of BE-DSR is lower than QOSRGA and BE-AODV for 98 packets/s source data. When the source sends 4 packets/s,

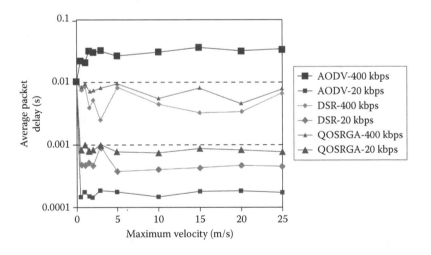

Figure 6.19 Average packet delay as a function of maximum velocity.

BE-AODV and BE-DSR are better than QOSRGA. When the velocity is more than 5 m/s, the delay for all protocols is maintained at almost the same level. QOSRGA performed the worst. This is clear, since QOSRGA was designed to collect as much information about the network as possible, so that the process of route selection using the GA is based on all possible information. However, all the delays incurred by QOSRGA are still less than 0.1 s. This is because availability of cached routes in QOSRGA eliminates route rediscovery latency that contributes to the delay when an active route fails. In addition, when a congestion state occurs in a routing path, the source node can distribute incoming data packets to the other non-disjoint routing paths to avoid congestion. This reduces the waiting time of data packets in the queue.

3. Total average throughput (TAT)

   Figure 6.20 shows the total average throughput of the QOSRGA compared to the other protocols. Throughput is the total number of bits delivered to the destination hosts. For QOSRGA, the ability of transferring the data dropped from 2.5 to 1.5 Mbps as the mobility increased from 2 to 25 m/s. The throughput is less than that of BE-AODV, but when compared to BE-DSR, QOSRGA offered an improvement of 25%–80%. Nodes with high velocity will produce small numbers of low value *nci* among the node pairs. The number of routes of longer lifetime will be less; hence, the rate of data transfer to the destination nodes will be less.

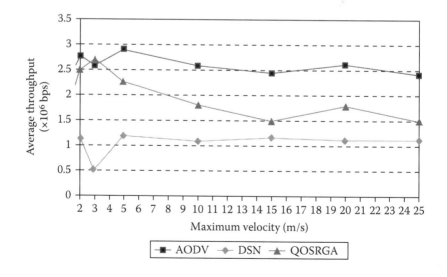

Figure 6.20 Average total throughput against mobility level.

6.8.5 Influence of node density on performance

The ability of different MANET protocol schemes to handle node density is analyzed in this set of simulations. It inherently assessed the scalability of QOS-RGA and compared its performance to BE-DSR and BE-AODV with different node densities. In this case, the area, source traffic rate, and maximum velocity are kept constant at 1000 m × 1000 m, 100 kbps, and 2 m/s, respectively. The number of source nodes is set to 10, generating towards random destinations. Each point was obtained after 10 runs with different seed numbers. The node density was defined as the ratio of the total number of nodes in the network to the area of the field configuration. The numbers of nodes are varied from 10 to 50, with the same size field configuration. The simulation experiments are based on the densities of 1×10^{-6}, 2×10^{-6}, 3×10^{-6}, 4×10^{-6}, and 5×10^{-6} nodes per square meter. The metrics measured are: (i) APDR ratio, (ii) average end-to-end packet delay, (iii) average normalized routing load (ANRL), and (iv) average throughput.

1. Average packet delivery ratio (APDR)

   Figure 6.21 illustrates the APDR for QOSRGA, BE-AODV, and BE-DSR as a function of node density. Overall the patterns of the QOSRGA graph and BE-DSR graph are normally quite similar. BE-DSR fell more rapidly from its maximum value down to a density of 2.0×10^{-6}, and then stabilized at 4.0×10^{-6} and 5.0×10^{-6}. After 5.0×10^{-6} onwards, QOSRGA produced better results. It is 10% better than the BE-DSR at high node density. The operation of QOSRGA requires fast accumulation of multiple routes. As the node density increases, a reasonable number of routes for an initial population can be obtained. A good number of multiple routes

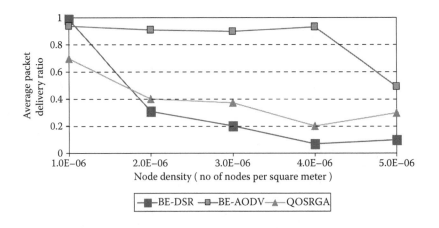

Figure 6.21 Average packet delivery ratio against node density.

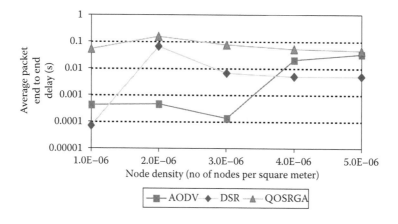

Figure 6.22 **Average end-to-end delay against node density.**

ensure better selection process by the GA. A route with very low *nci* could then be produced and selected.

2. Average end-to-end delay of packets (ATETED)

Figure 6.22 illustrates the Average end-to-end packet delay as a function of node density. Generally QOSRGA performed the worst among the other protocols. It varied from approximately 0.05 to 0.1. This can be attributed to the fact that for QOSRGA, in every intermediate node, the processing time is significant. The node had to do the monitoring of packet arrival, set, and reset all the node cache and perform the bandwidth calculation, *nci* calculation, and, most importantly, run the route selection routine using the GA. Hence, the packet delay observed from Figure 6.22 can be attributed to these node state manifestations. However, for node densities of more than 3.0×10^{-6}, the delay is less than 0.1 s, which is the maximum delay allowed.

3. Normalized routing load (NRL)

Figure 6.23 shows the NRL as a function of node density. Since QOS-RGA first performs the NDMRD protocol route discovery algorithm, the number of routing packets generated contributed significantly to the number of packets in transition within the wireless framework. The second phase of QOSRGA is to perform the route selection algorithm among the valid routes discovered by NDMRD protocol. These two phases require a significant overhead routing load. For densities less than 3×10^{-6}, the value for ANRL is less than that for BE-DSR. Beyond that BE-DSR is better. A high density configuration produces more nodal interactions for QOSRGA, resulting in very high routing packets per data packet. For low

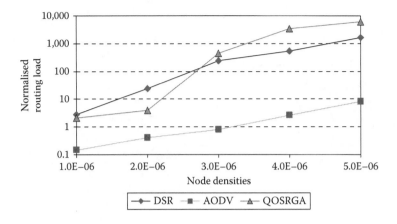

Figure 6.23 Average normalized routing load against node density.

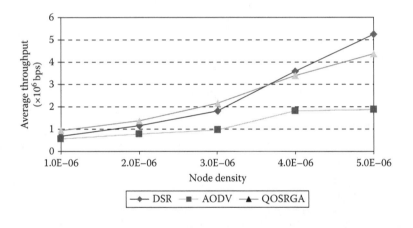

Figure 6.24 Average total throughput against node density.

densities (10 and 20 nodes) the value of routing packet per data packet is approximately equal to BE-DSR.

4. Total average throughput (TAT)

Figure 6.24 shows the total average throughput of the QOSRGA compared to other protocols. The average throughput increased as the node density was increased. In this context, QOSRGA performed on par with the BE-DSR protocol. When a node density is high, the probability of longer route lifetime can be realized.

6.8.6   Effect of congestion level on the performance

To study the impact of congestion level on QOSRGA performance, the number of active flows in the simulation is varied from 2, 5, and up until 30 in steps of 5. Each traffic flow carried traffic of 100 kbps or 24 packets/s. The mobility level is set to a normal distribution with maximum velocity 2 m/s. The number of nodes is fixed at 40. The metrics measured are: (i) average packet delivery ratio, (ii) average end-to-end packet delay, (iii) average normalized routing load, and (iv) average throughput.

1. Average packet delivery ratio (APDR)

   Figure 6.25 shows the graph of APDR as a function of network traffic load. Generally all protocols are sensitive to an increase in traffic load. Node mobility is the same for all protocols, and hence, the packet drops are caused by buffer overflow, collision, and congestion. BE-DSR is most sensitive to traffic congestion. QOSRGA performs better than BE-DSR, and at some points shows an improvement of more than 40%. The reason is that when the number of sources increases, there are more collisions in the air and congestion in node buffers. In QOSRGA, more load means more RREQ packets being sent. More route replies are generated within the RAL. The population size of the routes will increase, giving a better chance of selection of the fittest solution that is the longest connectivity pair within the routes. Furthermore, less route breakage will occur in the whole session compared to BE-DSR.

2. Average end-to-end delay of packets (ATETED)

   Figure 6.26 shows the effect of congestion level on average end-to-end packet delay for QOSRGA, compared to BE-AODV and BE-DSR.

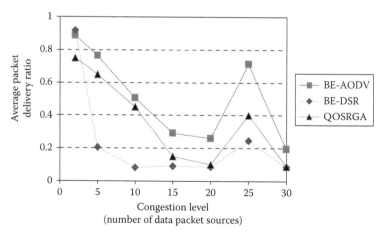

Figure 6.25 Average packet delivery ratio as a function of congestion level.

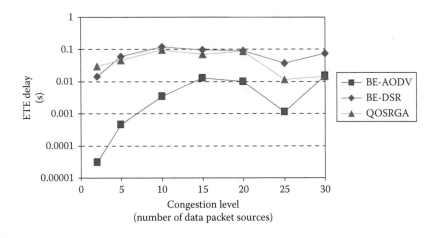

Figure 6.26 Average end-to-end delay against congestion level.

QOSRGA had a larger delay compared to BE-DSR, due to the high routing overhead as the network became more congested. QOSRGA, and also BE-DSR, had to perform route rediscovery if the existing route was broken. This may add to the delay. At the highest congestion of 30 flows, the delay for QOSRGA was equal to the delay for BE-AODV. When the number of sources was increased, the overall delay for QOSRGA packet also increased. BE-AODV used a scheme where the node upstream of the disconnected link initiated an immediate route reconstruction. Since the route rediscovery is done locally, less time is needed to search for and obtain a new route. Due to these local recovery scheme delays, BE-AODV yielded shorter delays.

3. Total average throughput (TAT)
   Figure 6.27 shows the total average throughput of QOSRGA compared to other protocols. QOSRGA performed better than BE-DSR by about 5%. As the number of sources was increased, the throughput of QOSRGA decreased. The reason is that there are more collisions in the air and congestion in node buffers when the number of sources increases.

4. Normalized routing load (NRL)
   Figure 6.28 depicts the NRL as a function of congestion level. With an increase of the number of sources, the probability of packet collision and packet congestion increases. This leads to the decrease of NRL. It is seen that QOSRGA has similar ANRL to both BE-AODV and BE-DSR for all possible numbers of sources. The ANRL in BE-AODV and BE-DSR

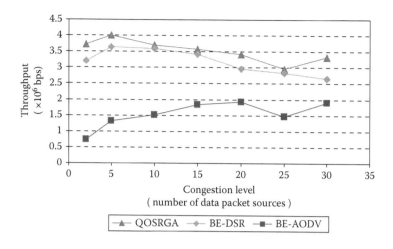

Figure 6.27 Total average throughput against the congestion level.

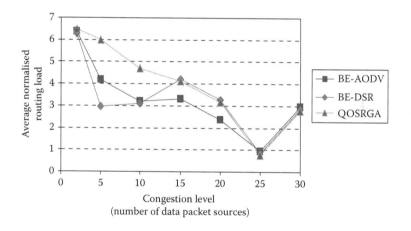

Figure 6.28 Average normalized routing load as a function of congestion level.

decreases more quickly than that in QOSRGA when there is an increase of the number of sources. The reason is that QOSRGA has multiple non-disjoint routes. When an active routing path encounters packet congestion due to high network load traffic, the source node will terminate the current route and calculate the next route. To avoid the loss of data packets, the source node can instantly select another valid non-disjoint routing path from its route table to send data packets towards the destination.

6.9 Conclusions

A scheme has been presented for multiple objective QoS routing protocol for MANETs based on GA. In the proposed scheme of QoS routing, selection of a route was based on node bandwidth availability, short end-to-end delay, and the longest node pair connectivity time indicated by *nci*. The route selection algorithm was outlined and implemented. The variable length chromosomes represented the routes and genes represented the nodes. The algorithmic process was initialized by introducing a limited population. Limited population was accumulated during the route discovery by the NDMRD protocol. The fitness calculation is done using the weighted sum approach, combining the entire objective functions into a single objective. The weight of each objective was deduced by inspection. Next, a one-point crossover was implemented, with additional restoration function and route mutation. To execute crossover and mutation process properly, it was shown that the values of P_c and P_m chosen satisfy the transmission efficiency test. Performance of the protocol was done to determine its behavior due to congestion, node density, mobility, and traffic sources. From these simulation experiments it was deduced that QOSRGA protocol performs better for most of the cases.

Bibliography

J. Abdullah. 2011. QoS routing solutions for mobile ad hoc network. *Mobile Ad-Hoc Networks: Protocol Design*, InTech. Available from: https://www.intechopen.com/books/mobile-ad-hoc-networks-protocol-design/qos-routing-solutions-for-mobile-ad-hoc-network, last accessed on 04 October 2017.

J. Abdullah. 2015. QOSRGA protocol using non-disjoint multiple routes in mobile ad hoc networks. *WSEAS Transactions on Communications* 14: 449–460.

J. Abdullah, M. Y. Ismail, N. A. Cholan and S. A. Hamzah. 2008. GA-based QoS route selection algorithm for mobile ad-hoc networks. In *6th National Conference on Telecommunication Technologies 2008 and 2008 2nd Malaysia IEEE Conference on Photonics. NCTT-MCP 2008*, pp. 339–343.

J. Abdullah and D. J. Parish. 2007. Node connectivity index as mobility metric for GA based QoS routing in MANET. In *ACM/IEEE International Conference on Mobile Technology, Application and Systems*, ISBN 978-1-59593-819-0, 10–12 September, 2007 Singapore.

G. S. Ahn, A. T. Campbell, A. Veres and L. H. Sun. 2002. Supporting service differentiation for real-time and best-effort traffic in stateless wireless ad hoc networks (SWAN). In *IEEE TMC*, vol. 1, July–September 2002, pp. 192–207.

A. Barolli, E. Spaho, L. Barolli, F. Xhafa and M. Takizawa. 2011. QoS routing in ad-hoc networks using GA and multi-objective optimization. *Mobile Information Systems* 7: 169–188, IOS Press.

C. Bettstetter, G. Resta and P. Santi. 2003. The node distribution of the random waypoint mobility model for wireless ad hoc networks. *IEEE Transactions on Mobile Computing* 2(3): 257–269.

S. Bhandari et al. 2012. High speed dynamic partial reconfiguration for real time multimedia signal processing. In *Digital System Design (DSD), 2012 15th Euromicro Conference on IEEE*, 2012.

T. Blickle and L. Thiele. 1995. A comparison of selection schemes used in genetic algorithms. *TIK-Report*.

S. Chen and K. Nahrstedt. 1999. Distributed quality of service routing in ad hoc networks. *IEEE JSAC* 17: 1488–1505.

I. Chlamtac, M. Conti and J. J.-N. Liu. 2003. Mobile ad hoc networking: imperatives and challenges. *Ad Hoc Networks* 1(1): 13–64.

D. A. Coley. 1999. *An Introduction to Genetic Algorithms for Scientist and Engineers.* World Scientific Publishing: Singapore.

M. Conti and S. Giordano. 2014. Mobile ad hoc networking: Milestones, challenges, and new research directions. *IEEE Communications Magazine* 52(1): 85–96.

R. Elbaum and M. Sidi. 1996. Topological design of local area networks using GA. *IEEE/ACM Transactions on Networking* 4: 766–778.

M. Gen and R. Cheng. 2000. *Genetic Algorithms and Engineering Optimization.* Wiley-Intersciences Publication: Canada.

R. L. Haupt. 2000. Optimum population size and mutation rate for a simple real genetic algorithm that optimizes array factors. In *Antennas and Propagation Society International Symposium, IEEE*, Vol. 2, pp. 1034–1037.

R. L. Haupt and S. E. Haupt. 2004. *Practical Genetic Algorithms.* Second Editions. Wiley-Interscience: Canada.

R. H. Hwang, W. Y. Do and S. C. Yang. 2000. Multicast routing based on genetic algorithms. *Journal of Information Science and Engineering* 16(6): 885–901.

IEEE LAN MAN Standards Committee. 1999. Wireless LAN medium access control (MAC) and physical layer (PHY) specifications. In *IEEE Std 802.11a Technology Representive IEEE Std. 802.11a*, 1999.

J. Inagaki, M. Haseyama and H. Kigajima. 1999. A genetic algorithm for determining multiple routes and its applications. In *Proceedings of the IEEE International Symposium on Circuits and Systems*, pp. 137–140.

D. B. Johnson, D. A. Maltz and Y. C. Hu. 2007. The dynamic source routing protocol for mobile ad hoc networks (DSR). In *IETF MANET Working Group, INTERNET-DRAFT,* 2007. Available from: https://datatracker.ietf.org/doc/rfc4728/, last accessed on 07 April 2017.

M. Karimi. 2011. Quality of service (QoS) provisioning in mobile ad-hoc networks (MANETs), *Mobile Ad-Hoc Networks: Protocol Design*, InTech. Available from: https://www.intechopen.com/books/mobile-ad-hoc-networks-protocol- design/quality-of-service-qos-provisioning-in-mobile-ad-hoc-networks, last accessed on 04 October 2017.

C. Y. Lee and E. K. Antonsson. 2000. Variable length genomes for evolution strategies. In *Proceedings of the Genetic and Evolutionary Computation Conference (GECCO 2000)*.

S. B. Lee, A. Gahng-Seop, X. Zhang and A. T. Campbell. 2000. INSIGNIA: An IP-based quality of service framework for mobile ad hoc networks. *Journal PADC* 60: 374–406.

S. J. Lee and M. Gerla. 2001. Split multipath routing with maximally disjoint paths in ad hoc networks. In *Proceedings of the IEEE ICC*, 11-14 June, Helsinki, pp. 3201–3205.

B. Li and K. Nahrstedt. 1998. A control theoretical model for quality of service adaptations. In *Proceedings of Sixth IEEE International Workshop on Quality of Service*; 18–20 May, 1998, Napa, CA, USA; pp. 145–153.

Y. Liu and J. Huang. 2009. A novel fast multi-objective evolutionary algorithm for QoS multicast routing in MANET. *International Journal of Computational Intelligence Systems* 2(3): 288–297.

T. Lu and J. Zhu. 2013. Genetic algorithm for energy-efficient QoS multicast routing. *IEEE Communications Letters* 17(1): 31–34.

Q. Ma and P. Steenkiste. 1997. Quality-of-service routing for traffic with performance guarantees. In *Proceedings of IFIP Fifth International Workshop on Quality of Service*: Springer. US, 115–126.

MANET IETF. 2017. Manet Working Group Charter, *Internet Engineering Task Force*, Available from: https://datatracker.ietf.org/wg/manet/documents/, last accessed on 07 April 2017.

S. Mao, Y. T. Hou, X. Cheng, H. D. Sherali and S. F. Midkiff. 2005. Multipath routing for multiple description video in wireless ad hoc network. *Proceedings IEEE* (Vol. 1, pp. 740–750) *24th Annual Joint Conference of the IEEE Computer and Communications Societies. IEEE INFOCOM2005*. IEEE.

M. K. Marina and S. R. Das. 2001. On-demand multipath distance vector routing in ad hoc networks. In *Proceedings of the 9th IEEE International Conference on Network Protocols (ICNP)*, November 2001, pp. 14–23. Los Alamitos, Calif: IEEE Computer Society.

U. Mehboob, J. Qadir, S. Ali and A. Vasilakos. 2016. Genetic algorithms in wireless networking: Techniques, applications, and issues. *Soft Computing* 20(6): 2467–2501.

P. Mohapatra, J. Li and C. Gui. 2003. QoS in mobile ad hoc networks. *IEEE Wireless Communications* 20: 44–52.

M. Mosko and J. Garcia-Luna-Aceves. 2005. Multipath routing in wireless mesh networks. In *Proceedings of the First IEEE Workshop on Wireless Mesh Networks (WIMESH'05)*, September 2005, Santa Clara, CA.

M. Munetomo, Y. Takai and Y. Sato. 1998. A migration scheme for the genetic adaptive routing algorithm. In *Proceeding of the IEEE International Conference on Systems, Man, and Cybernetics*,1998, San Diego, US, Vol. 3 pp. 2774–2779.

A. Nasipuri, R. Castaneda and S.R. Das. 2001. Performance of multipath routing for on-demand protocols in mobile ad hoc networks. *ACM/Baltzer Mobile Networks and Applications Journal* 6: 339–349.

C. E. Perkins and P. Bhagwat. 1994. Highly dynamic destination-sequenced distance-vector routing (DSDV) for mobile computers. *Computer Communications Review* , Vol. 24 No.4, 234–244. ACM.

C. E. Perkins, E. M. Royer and S. R. Das. 2001. Performance comparison of two on-demand routing protocols for ad hoc networks. In *IEEE Personal Communications*, February 2001, pp. 16–28.

G. Pravin, G. Katkar and P. Ghorpade. 2010. Mobile ad hoc networking: Imperatives and challenges. *IJCA Special Issue on MANETs* 3: 153–158.

Riverbed Modeler. 2017. Previously known as Opnet Modeler. Available from: https://www.riverbed.com/. Access on 1 Feb 2016.

J. D. Schaffer, R. A. Caruana, L. J. Eshelman and R. Das. 1989. A study of control parameters affecting online performance of genetic algorithms for function optimization. In *Proceedings of the Third International Conference on*

Genetic Algorithms, George Mason University, Morgan Kaufmann Publishers Inc., San Fransisco, 1989, pp. 51–60.

N. Shimamoto, A. Hiramatus and K. Yamasaki. 1993. A dynamic routing control based on a genetic algorithm. In *Proceedings of the IEEE International Conference on Neural Networks*, San Francisco, 26 Mac 1993, pp. 1123–1128.

Z. Wang and J. Crowcroft. 2006. Quality-of-service routing for supporting multimedia applications. *IEEE Journal on Selected Areas in Communications* 14(7): 1228–1234.

S. H. Wong and J. Wassell. 2002. Dynamic channel allocation using a genetic algorithm for a TDD broadband fixed wireless access network. In *Proceedings of the IASTED International Conference in Wireless and Optical Communications*, July 17–19, 2002, Banff, Alberta, Canada, pp. 521–526.

A. Younes. 2011. Multicast routing with bandwidth and delay constraints based on genetic algorithms. *Egyptian Informatics Journal* 12(2): 107–114.

B. Zhou, A. Marshall, T. H. Lee. 2005. A cross-layer architecture for DiffServ in mobile ad-hoc networks. In *Proceedings of International Conference on Wireless Networks, Communications and Mobile Computing*, June 13–16, 2005, Maui, Hawaii, USA.

Chapter 7

Network Flows under Thermal Restrictions

Samiksha Sarwari

Indian Institute of Technology Roorkee

Shrisha Rao

International Institute of Information Technology, Bangalore

CONTENTS

7.1 Introduction

Many systems have components that are subject to thermal degradation, and which therefore must be managed carefully to obey temperature constraints. This is particularly true of electronics [14,21], but large systems such as data centers [5,10] require extensive thermal management as well. It is therefore essential to monitor and control the flow of work through the nodes of such a system, in addition to the use of special equipment and measures for cooling.

The physics of thermal management can be quite complex [9,17], but in practical systems, heuristics are generally used. This is true of computer systems and networks [15,19] as well as industrial process systems [13].

Modeling of congestion in networks is also a well-known problem; besides computer networks, it is also studied in the context of vehicular traffic [16] and air traffic [12]. In traffic network models, congestion control is attempted using variable pricing and other changes to node characteristics [2,18].

Besides congestion caused by a surfeit of packets or other arrivals at a network node, there can also be constraints due to a node's time-varying capacity. This is preeminently seen in wireless sensor networks where nodes are subject to varying power levels [20].

Existing works on capacitated networks and flow routing [1,4,11] do not address these issues; there does not seem to be any sufficiently general way to consider thermal constraints, time-varying network characteristics such as power levels, or the like. In this chapter, we use models and results to address this.

In this chapter, we give optimal solutions to the problem of maximum flow through the following two models of a thermal network with capacity constraints on nodes. (Though we speak of temperature, the concept of a thermal network and the respective parameter of temperature can be suitably modified to model any network of nodes that exhibit the same characteristics; the thermodynamics of temperature or heat are not essential to our analyses.)

In the first model, a node that reaches a critical temperature stops functioning and can no longer be used to transmit packets; however, it can cool with time. The network in this *dissipating model* thus has the property of reviving itself over time (i.e., once the network is exhausted (because the nodes are too hot), it is possible to give it some rest (and let the nodes cool) so that we can again send more packets through it). Keeping in mind our aim to send as many packets through the network as possible, we realize that the problem now changes to sending the maximum possible packets while minimizing the time during which the network becomes dysfunctional, so that the number of packets in a given duration is maximized, which is equivalent to saying that we maximize the rate of flow through the network. This is a dynamic problem, as the nodes repair themselves with time, thereby making the state of the network depend upon one more factor (i.e., time). We study the transient state of the network, in which the minimum node-cut-sets vary with time, and move on to analyze the network to figure out if there exists a steady state. This means that we try to find out whether we need the state of the network at all time instants to obtain a maximum flow using the Ford-Fulkerson algorithm [8], or if there exists a closed-form solution that depends only upon the initial state of the network and the information about the nodes' heat dissipation. We therefore set out to prove that there indeed exists a steady state of the network, for which we prove that there exists a node-cut-set that is the minimum node-cut-set throughout after a certain amount of time. Using the results, we are able to find the value of the maximum rate of flow achievable in this network.

In the second model, the network does not have the same self-healing properties, but we have some kind of special packets called *cooling packets* at our disposal, which can decrease the temperature of nodes. So, given a dysfunctional network, we can send these cooling packets to specific critical nodes, so as to make the network functional again. One advantage of this model is that we have the liberty to send these special packets only to the nodes that need thermal repair (i.e., cooling). But this is also what makes this problem more challenging than the previous one, as now we need to figure out the optimal strategy for sending these cooling packets so as to minimize the requirement of these packets while maximizing the flow through the network (i.e., we need to minimize the repair cost, while maximizing the efficiency of the system). For this, we first find out the value of maximum flow of packets using as many cooling packets as we may require (i.e., find the maximum flow if we thermally repair all the nodes of the network completely). The question is then if we can obtain the same amount of flow using a smaller number of cooling packets, and further, what is the least number of cooling packets needed to ensure maximum flow. We analyze this scenario by finding the exact nodes that need repair, and the exact minimum possible amount of repair that will make the network work at its best. The trick used to solve this problem is based on the fact that the minimum node-cut-set determines the maximum flow. So, we do not really require any other node-cut-set to work at capacity greater than the maximum possible capacity of the minimum node-cut set. This means that we do not need to repair all nodes to their best capacities, nor do we need to repair all the nodes. The next step is to identify the nodes and the minimum capacities at which they should function, and find out the optimal routing pattern for the same. This is done by creating a set of walks such that if we send cooling packets via these walks, not only is the minimum node-cut-set revived to its maximum capacity, but the nodes of other node-cut-sets are also revived to the extent that none of them becomes the limiting node-cut-set. We prove that there exists such a set of walks, and calculate the number of cooling packets to be sent via these walks, and the corresponding maximum flow achieved.

Table 7.1 A summary of the results

Network type	Results
Static network	Theorem 7.1 (max-flow min-cut)
Uniform with dissipation	Corollary 7.1 (consequence of Theorem 7.2 and Theorem 7.1); Theorem 7.3; Theorem 7.4
Nonuniform with dissipation	Theorem 7.5 (generalization of Theorem 7.2); Theorem 7.6; (generalization of Theorem 7.4)
Nonuniform with cooling	Theorem 7.7; Theorem 7.8; Theorem 7.9; Theorem 7.10; Theorem 7.12

Table 7.1 provides an insight into the chapter in brief.

Overall, this chapter is organized as follows. In Section 7.2, first (in Section 7.2.1) we introduce some of the preliminaries and provide a background on which the subsequent sections are based. The system model (Section 7.2.2) involves a detailed description of the constraints on the Thermal Network and the results and techniques to maximize the flow of packets subject to these constraints. In Section 7.3, this system model with an additional characteristic that the nodes can cool themselves down with time is considered. This is a dynamic system, the analysis of which requires an in-depth analysis of the transient and steady states of the system. We discover some properties of the system that are used to determine the rate of maximum flow that can be achieved. In Section 7.4, the system model with another special characteristic is discussed. In this case, the nodes are not attributed with the self-cooling properties, but we have dedicated cooling packets for the purpose of repair of the network. The problem is to optimize the flow of the heating packets along with optimization of the number of cooling packets used so as to be able to reduce the maintenance cost of the network. Such networks have some special properties with respect to their minimum node-cut-sets, which determine the maximum flow due to the max-flow min-cut theorem.

7.2 Thermal Network

7.2.1 Terminology

We have a network G (also referred to as network) with nodes v_i and edges v_iv_j (directed from v_i to v_j). The temperature of a node, say v_i, cannot rise above a temperature (called critical temperature, θ_{c_i}), and cannot fall below the specified base temperature (θ_{0_i}). A node at its critical temperature cannot be traversed any more, and is called a dysfunctional node. The packets (heating packets, which shall be referred to simply as packets throughout the text) have the property that they heat up the nodes they traverse by a certain amount $\triangle T_u$. (Hence, the temperature restriction on the nodes limits the number of packets that can traverse any node. Thus, we define the capacity c_i of node v_i to be the maximum number of packets that can traverse v_i before it becomes dysfunctional.)

If some nodes of the network become dysfunctional such that there exists no path for the packets to travel from s to t, the network is said to be disconnected (or the network is said to be dysfunctional). Technically, this means that all nodes of some or the other node-cut-set are dysfunctional.

Definition 7.1   A *node-cut-set* is a set of nodes, the removal of which disconnects the network such that s and t lie in two separate blocks of the disconnected network (or equivalently, separates s from t).

The problem is essentially to maximize flow (i.e., to obtain the maximum flow, which is the maximum possible amount of flow from s to t that can be achieved through the network before the network becomes dysfunctional).

Section 2 is a special case of this basic model wherein the nodes have the capacity of cooling themselves down. We shall denote this rate by ω. This phenomenon will be referred to as *dissipation,* drawing analogy from the natural dissipation phenomenon. However, because of the base temperature constraints, a node cannot be cooled down below θ_{0_i}.

Since this network is time dependent, we are interested in maximizing the rate of flow of packets (the number of packets traveling from s to t per unit time), which shall be denoted by $\bar{f}$. This analysis will be conducted separately on a uniform and a nonuniform network.

Definition 7.2   A *uniform network* is a network in which all the nodes have identical capacities. A network that is not uniform is called a *nonuniform network*.

Section-3 is another special variant of the basic model wherein we have cooling packets (entities that decrease the temperature of a node upon traversal by an amount equal to $\triangle T_d$). This model, however, does not have the dissipating properties.

The following is a brief mention of the famous result we shall be using throughout and related definitions:

Definition 7.3   The capacity C_M of a set M is defined as the sum of the capacities of all the nodes of that set. That is,

$$C_M = \sum_{v_i \in M} c_{v_i}. \tag{7.1}$$

Definition 7.4   A min node-cut-set or minimum node-cut-set is defined as the node-cut-set whose capacity is less than or equal to the capacity of any other node-cut-set, where capacity of any node-cut-set is given by Definition 7.3.

Theorem 7.1
***Max-Flow Min-Cut Theorem** [3] : In any network, the value of a maximum flow is equal to the capacity of a minimum cut.*

Table 7.2 summarizes the notation used.

7.2.2   System model

Given a network G, with nodes denoted by v_i, having base and critical temperatures θ_{0i} and θ_{ci}, our problem is to maximize the number of packets traveling from source s to the sink t until the network becomes dysfunctional. The packets

Table 7.2   Notation

Symbol	Description
v_i	ith node of the network G.
θ_{0i}	Initial/base temperature of the node v_i, which is also equal to its minimum possible temperature.
θ_{ci}	Critical temperature of node v_i. The node v_i ceases to function above this temperature.
c_i	Capacity of node v_i.
s	Source of the flow; the packets originate from this node.
t	Sink of the flow.
$\triangle T_u$	The amount by which a packet increases the temperature of a node upon traversal.
ω	The amount by which the temperature of a node decreases per unit time.
$\triangle T_d$	The temperature by which a cooling packet decreases the temperature of a node upon traversal subject to conditions mentioned in Section 7.4.1.
$\bar{f}$	Rate of flow of packets.
M	A node-cut-set of network G. Since a network can have many node-cut-sets, we shall refer to them as M_i throughout the text.
W	The set of walks from s to t via the nodes of the node-cut-set for which the corresponding walk set is defined.
C_{M_i}	Capacity of the ith node-cut-set, which is equal to the sum of capacities of all nodes that belong to the set M_i.
τ	The amount of time for which the network is given rest to dissipate heat and become functional again.
β	Cooling capacity of a cooling packet.

have the property that they increase the temperature of a node by an amount equal to $\triangle T_u$ units upon traversal.

Constraints:

The lower and upper limits on the temperature of the nodes impose a constraint on the number of packets that can traverse that node. Let us denote the maximum number of packets that can traverse a node v_i before v_i becomes dysfunctional by c_i, the capacity of the ith node. Let n packets be able to cross node i before it gets dysfunctional. A packet increases the temperature of a node by $\triangle T_u$ upon traversal, which gives:

$$n\triangle T_u \le \theta_{ci} - \theta_{0i} \tag{7.2}$$

$$n \le \frac{\theta_{ci} - \theta_{0i}}{\triangle T_u}. \tag{7.3}$$

Since n denotes the number of packets, it has to be an integer. So, the maximum value n can attain is:

$$n_{\max} = \left\lfloor \frac{\theta_{ci} - \theta_{0i}}{\triangle T_u} \right\rfloor. \tag{7.4}$$

This $n_{\max}$ is in fact the capacity of node i by definition. So,

$$c_i = \left\lfloor \frac{\theta_{ci} - \theta_{0i}}{\triangle T_u} \right\rfloor. \tag{7.5}$$

There are no such temperature restrictions on the edges. Also, the number of packets that can be dispatched from the source or that can get into the sink at any instant does not constrain the number of packets traveling through the network. This means that as many packets as the network can allow through it at any instant can be dispatched by the source and can be absorbed into the sink. The problem—to maximize the number of packets that can travel from source s to sink t through the network with capacity constraints on nodes—has already been solved by modifying the network (which will be explained in a later section) and applying the Ford-Fulkerson algorithm [8]. Nevertheless, we mention it here in full detail as it will be referred to in further analysis of more complicated networks.

Node splitting technique [6,7]:

This technique of node splitting is often used for spot programming when solving flow questions having flow limitations on the nodes. Every node v_i is split into two nodes $v \cdot r_i$ (v_i right) and $v \cdot l_i$ (v_i left). These two nodes are joined using a directed edge from $v \cdot l_i$ to $v \cdot r_i$. All edges incident into the erstwhile node v_i are now made incident into the node $v \cdot l_i$, and all edges incident out of v_i are made incident out of the node $v \cdot r_i$. The directed edge joining $v \cdot l_i$ and $v \cdot r_i$ is given a capacity equal to the capacity of the node v_i. All edges of the original network are given an infinite capacity (assuming no limit exists on the flow through the edges). This transforms the node-limited flow problem to the familiar edge-limited flow problem, which can be easily solved using the Ford-Fulkerson or other max-flow algorithms.

7.3 Dissipating Model

The base model provides an elementary yet important starting point for the analysis of much more complicated yet interesting networks, one such model being the dissipating model. A dissipating network is fundamentally the base model with nodes exhibiting certain special characteristics. These nodes have a special property of self-repair. This is done by dissipating heat with time; that is, the nodes, if given some time, lose some of the heat, thereby cooling themselves sufficiently below the critical temperature, which makes them functional again. However, a node cannot cool itself down beyond its base temperature.

The rate of dissipation will be denoted by ω (i.e., a node cools down by ω units of temperature per unit of time).

Note 7.1 We are not discretizing time, for the sake of practicality. This means that the network can be given rest for any amount of time, not necessarily integral values. Or, equivalently, we can say that it is not necessary that a node's temperature be reduced only by an amount that is an integral multiple of ω.

The problem, to maximize the rate of flow of packets from source s to sink t through the dissipating model, is tackled in parts, wherein the first part deals with a uniform network (Definition 7.2) and the second part deals with a nonuniform network.

7.3.1 Uniform network

For a complete understanding of the dynamic behavior of the uniform network with dissipation, a complete analysis including both transient and steady state analysis will be performed for the problem of maximization of the rate of flow of packets in the following sections. Where on one hand the transient state analysis provides insight into the dynamically changing packet flow through the network, the steady state analysis illustrates the ultimate state the network achieves, which does not change with time.

7.3.1.1 Transient state analysis

As a consequence of the assumption that the flow of packets from s to t requires no time, at time $\tau = 0$, no dissipation occurs while the packets travel through the network. Therefore before any dissipation occurs, the maximum possible number of packets would have already traversed the network, thereby making it dysfunctional. This stage at $\tau = 0$ is then no different from the base model. Hence, the maximum flow is given by the max-flow min-cut theorem on the base model.

The following proposition speaks about what the minimum node-cut-set is going to be:

Lemma 7.1

The minimum node-cut-set in a uniform network is the one with minimum cardinality, where cardinality of a set refers to the number of nodes in the set.

Proof 7.1 By Definition 7.4, the min node-cut-set of network G is the node-cut-set of G with minimum capacity (the capacity of a set being the sum of capacities of the nodes in the set). Therefore, here the min node-cut-set is the set M_i, where i is such that C_{M_i} is minimized.

$$C_{M_i} = \sum_{j:v_j \in M_i} c_j = c|M_i|,$$

where $|M_i|$ denotes cardinality of the set M_i (i.e., the set of nodes in the set M_i).

Without loss of generality, let $i = k$ for which C_{M_i} is minimum (i.e., let M_k be the min node-cut-set). That is:

$$C_{M_k} = \min_i C_{M_i} = \min_i c|M_i| = c\min_i |M_i|. \tag{7.6}$$

But

$$C_{M_k} = c|M_k|. \tag{7.7}$$

Therefore, from (7.6) and (7.7):

$c\min_i |M_i| = c|M_k|$, which means $|M_k| = \min_i |M_i|$.

This tells us that the min node-cut-set is the one with minimum cardinality, which completes our proof.

After flow $f = c|M_i|$, the nodes of the node-cut-set M_k become dysfunctional, thereby disconnecting the network (M_k being a node-cut-set). To revive the network again, the network needs sufficient time to dissipate the heat and become functional again. Let the network be given rest for τ units of time.

Note 7.2 It is to be observed that the network has again reached the same state it was previously in, where we had a disconnected network, which was then given τ units of rest. Again a maximum possible number of packets pass through the network, and it becomes disconnected. We shall call this one cycle as one *stage*. So, technically, a stage of a network is the state it goes into after it has been revived after some rest, and allows a certain number of packets to pass through it before becoming dysfunctional again.

The only parameter that might vary is τ. However, it does not impact the state the network is in, as is established by the following result:

Theorem 7.2

In a uniform network with dissipation, the minimum node-cut-set is going the same throughout.

Proof 7.2   We shall prove the result using induction. We shall first prove that the min node-cut-set is the same for Stage 1 and 2 (let it be M_k). We assume that the min node-cut-set is M_k in Stage n. Then, if we are able to prove that in Stage $n+1$ as well, M_k will be the min node-cut-set, we would be done.

Let us consider these stages one by one as follows:

Stage 1: The network initially has all its nodes working at capacity c. Then, the capacity of a node-cut-set M_i being the sum of capacities of its nodes, we get:

$$C_{M_i}^{(1)} = c|M_i|.$$

Min node-cut-set being the node-cut-set with minimum capacity, can be obtained by minimizing C_{M_i} over all node-cut-sets M_i

$$\min_i C_{M_i}^{(1)} = \min_i c|M_i| = c\min_i |M_i| = c|M_k|,$$

(where M_k is assumed to be the node-cut-set with minimum cardinality without loss of generality).

So, the min node-cut-set in Stage 1 is the set M_k, which is the node-cut-set with minimum cardinality. The min node-cut-set in this stage could also have been obtained directly by applying Lemma 7.1.

After a maximum flow $f = c|M_k|$ by max-flow min-cut theorem, the network becomes disconnected. The residual capacity R_{M_i} of a node-cut-set M_i being the capacity left after a transfer of a certain number of packets through the network is given by:

$$R_{M_i}^{(1)} = C_{M_i}^{(1)} - f = C_{M_i}^{(1)} - C_{M_k}^{(1)} = c(|M_i| - |M_k|).$$

Let the network be given sufficient amount of rest (say for time τ units) so that the network becomes functional again. Then, this revived state is the Stage 2 of the network.

Stage 2: The improved capacities ($C_{M_i}^{(2)}$) of the node-cut-sets after τ units of rest are:

$$C_{M_i}^{(2)} = R_{M_i}^{(2)} + \frac{\tau\omega}{\triangle T_u}|M_i|$$

$$= c(|M_i| - |M_k|) + \frac{\tau\omega}{\triangle T_u}|M_i|, \forall i.$$

Proceeding as in Stage 1, the min node-cut-set for this state of the network is obtained for that value of i for which the capacity of the node-cut-set is minimum.

$$\min_i C_{M_i}^{(2)} = \min_i \left(c \left(|M_i| - |M_k| \right) + \frac{\tau\omega}{\Delta T_u} |M_i| \right)$$

$$= \min_i \left(c|M_i| + \frac{\tau\omega}{\Delta T_u} |M_i| \right) - c|M_k|,$$

which is minimum for $\min_i |M_i|$ (i.e., the node-cut-set with minimum cardinality, which is M_k as per the assumption made in Stage 1).

Now it is established that the min node-cut-set is the same for Stages 1 and 2. Next, assume that M_k is the min node-cut-set for Stage n.

Stage n: Assume that the min node-cut-set in this stage is M_k. Let $C_{M_i}^{(n)}$ denote the improved capacity of the node-cut-sets in this stage.

Since M_k is the min node-cut-set, the maximum flow is equal to $C_{M_k}^{(n)}$ (by the max-flow min-cut theorem) and

$$C_{M_k}^{(n)} \leq C_{M_i}^{(n)}, \forall i. \tag{7.8}$$

After this amount of flow takes place through the network, the network is disconnected with the residual capacities of the node-cut-sets being:
$R_{M_i}^{(n)} = C_{M_i}^{(n)} - C_{M_k}^{(n)}$.

Again, we revive this network by giving it sufficient time (let it be τ units) to be functional again. This increases the capacities of the nodes by $\frac{\tau\omega}{\Delta T_u}$ and makes the network functional again. In this next stage of the network

Stage $n+1$:

$$C_{M_i}^{(n+1)} = R_{M_i}^{(n)} + \frac{\tau\omega}{\Delta T_u} |M_i|$$

$$= C_{M_i}^{(n)} - C_{M_k}^{(n)} + \frac{\tau\omega}{\Delta T_u} |M_i|.$$

The minimum node-cut-set is the one with minimum capacity (minimum of all the node-cut-sets). Minimizing $C_{M_i}^{n+1}$ over i, we get:

$$\min_i C_{M_i}^{(n+1)} = \min_i \left(C_{M_i}^{(n)} - C_{M_k}^{(n)} + \frac{\tau\omega}{\Delta T_u} |M_i| \right)$$

$$= \min_i \left(C_{M_i}^{(n)} + \frac{\tau\omega}{\Delta T_u} |M_i| \right) - C_{M_k}^{(n)}.$$

We know that $C_{M_i}^{(n)}$ is minimum for $i = k$ (by the assumption in Stage n) and $|M_i|$ is also minimum for $i = k$ (by assumption).

So, $C_{M_i}^{(n)} + \frac{\tau\omega}{\triangle T_u}|M_i|$ is minimum for $i = k$. This means that $C_{M_i}^{(n+1)}$ is minimum for $i = k$, or M_k is the min node-cut-set for Stage $n + 1$. Hence, we have proved by induction that the min node-cut-set is the same in all the Stages in the case of a uniform network with dissipation.

Corollary 7.1
The rate of maximum flow in case of a uniform network with dissipation is the same in every stage.

Proof 7.3   Lemma 7.2 states that the minimum node-cut-set in case of Uniform Network is going to be the same in every stage. Let this node-cut-set be denoted by M_k. Then, by max-flow min-cut Theorem, the maximum flow f should be equal to C_{M_k}.

Next, we need to show that this capacity is the same in each stage. Consider the ith stage of the network. Since in the $(i-1)$th stage as well, the min node-cut-set would have been M_k (by Lemma 7.2), the residual capacity of this node-cut-set after maximum flow would have become zero. Therefore, after τ units of rest, the improved capacity $C_{M_k}^{(i)} = R_{M_k}^{(i-1)} + \frac{\tau\omega}{\triangle T_u}|M_k| = \frac{\tau\omega}{\triangle T_u}|M_k|$, which is also going to be the maximum flow, f, through the network in this stage (M_k being the node-cut-set):

$$f^{(i)} = \frac{\tau\omega}{\triangle T_u}|M_k|. \tag{7.9}$$

It can be easily seen that the right hand side of the equation is independent of the stage (i), which means that for the ith stage (where i is arbitrary), the capacity of the min node-cut-set, and hence the maximum flow is $\frac{\tau\omega}{\triangle T_u}|M_k|$.

So, the rate of flow in the ith stage, $\bar{f}^{(i)}$ becomes:

$$\bar{f}^{(i)} = \frac{f}{\tau} = \frac{\omega}{\triangle T_u}|M_k| \text{ (using (7.9))}.$$

It is to be noted that the right hand depends neither on the stage, nor the value of τ, and is thus a constant quantity. Hence, it has been proved that the rate of maximum flow is the same at every stage.

Note 7.3 The corollary suggests that the rate of maximum flow in case of a uniform network with dissipation is the same throughout, which means that the network has reached the steady state. However, it must be noted that the rate of this maximum flow might not necessarily be the maximum rate of flow and jumping to this conclusion might be wrong even though it seems intuitively correct. The value of flow and corresponding value of τ such that the rate of flow is maximized is discussed next.

7.3.1.2 Steady state analysis

The corollary suggests that the rate of maximum flow through the network remains the same throughout, which means that this state is indeed the steady state of the network.

Let the rate of flow in the steady state be denoted by $\bar{f}$. Then,

$$\bar{f} = c \frac{|M_k|}{\tau}.$$

The next problem is to find the value of τ so that we obtain the maximum rate of flow of packets from s to t through the network.

Note 7.4 Intuitively, it seems that τ should not be so small that the nodes of the node-cut-sets are not able to cool down even $\triangle T_u$ units, as then the capacity (the number of packets that can traverse a node) will remain zero, and the network is disconnected. Also, τ should not be so large that some of the nodes reach their base temperature from which they cannot cool down further, and, hence, the number of packets that can traverse through the network does not rise as much as the time taken, which would eventually decrease the average rate of flow. To be on a safer side, τ should be such that the capacity of any node is increased by one, which means

$$\tau\omega = \triangle T_u. \tag{7.10}$$

The following result proves this claim thus establishing the optimality of the solution:

Theorem 7.3
The rate of flow of packets in steady state through a network with dissipation is maximum for

$$\tau = \frac{\triangle T_u}{\omega}. \tag{7.11}$$

Proof 7.4 Since each packet increases the temperature of a node by $\triangle T_u$, upon traversal, when $\tau = \frac{\triangle T_u}{\omega}$, the increase in capacity, $\triangle c$, of each node is given by: $\triangle c = \lfloor \frac{\tau\omega}{\triangle T_u} \rfloor$.

Substituting $\tau = \frac{\triangle T_u}{\omega}$:

$$\triangle c = \left\lfloor \frac{\frac{\triangle T_u}{\omega}\omega}{\triangle T_u} \right\rfloor = \lfloor 1 \rfloor = 1.$$

So, the capacity of the min node-cut-set M_k is increased by $|M_k|$ units. Thus, the rate of flow in each stage is given by:

$$\bar{f} = \frac{|M_k|}{\tau} = \frac{|M_k|}{\frac{\triangle T_u}{\omega}}$$

$$\therefore \bar{f} = \frac{|M_k| \omega}{\triangle T_u}. \tag{7.12}$$

It will suffice to prove that for any other value of τ, the rate of flow cannot be greater than the rate of flow for $\tau = \triangle T_u / \omega$. Any other value of τ can either be greater than $\triangle T_u / \omega$ or less than $\triangle T_u / \omega$. Let us consider both these cases one by one:

Case 1: $\tau < \frac{\triangle T_u}{\omega}$

For this value of τ, the temperature of all nodes is decreased by $\tau \omega < \triangle T_u$ (by using: $\tau < \frac{\triangle T_u}{\omega}$).

Since a packet increases the temperature of a node by $\triangle T_u$ upon traversal, so, the increase in capacity of a node, $\triangle c$, by giving τ units of rest is:

$$\triangle c = \left\lfloor \frac{\tau \omega}{\triangle T_u} \right\rfloor. \tag{7.13}$$

But,

$$\frac{\tau \omega}{\triangle T_u} < \frac{\frac{\triangle T_u}{\omega} \omega}{\triangle T_u} = 1. \tag{7.14}$$

Using (7.13) and (7.14),

$$\triangle c = 0. \tag{7.15}$$

So, for τ units of rest, where τ is such that $\tau < \frac{\triangle T_u}{\omega}$, the capacity of nodes is not increased. This means that the capacity of no node of the node-cut-set M_k can be increased. Hence, the network remains disconnected and the rate of flow is going to be zero in every stage.

Case 2: $\tau > \frac{\triangle T_u}{\omega}$

For this value of τ, the temperature of all the nodes will be decreased by $\tau \omega$ units. But, $\tau \omega > \triangle T_u$ (given for this case). Since a packet increases temperature of a node by $\triangle T_u$ units, the increase in capacity of each node, $\triangle c$ is given by:

$$\triangle c = \left\lfloor \frac{\tau \omega}{\triangle T_u} \right\rfloor \tag{7.16}$$

But,

$$\frac{\tau \omega}{\triangle T_u} > \frac{\frac{\triangle T_u}{\omega} \omega}{\triangle T_u} = 1. \tag{7.17}$$

Using (7.16) and (7.17),

$$\triangle c \geq 1. \tag{7.18}$$

The capacity of the node-cut-set M_k is increased by $\triangle C_{M_k}$ such that:

$$\triangle C_{M_k} = \triangle c |M_k| \geq |M_k|. \tag{7.19}$$

Since M_k remains the min node-cut-set throughout (by Lemma 7.2), the maximum flow f is given by the capacity of M_k, which is given by 7.19. Hence, the rate of flow, $\bar{f}$ becomes:

$$\bar{f} = \frac{f}{\tau}$$

$$= \frac{\triangle C_{M_k}}{\tau}$$

$$= \frac{\lfloor \frac{\tau \omega}{\triangle T_u} \rfloor |M_k|}{\tau}.$$

Suppose, if possible, that this rate is greater than the rate of flow given by 7.12 when $\tau = \frac{\triangle T_u}{\omega}$:

$$\frac{\lfloor \frac{\tau \omega}{\triangle T_u} \rfloor |M_k|}{\tau} > \frac{|M_k| \omega}{\triangle T_u}$$

$$\lfloor \frac{\tau \omega}{\triangle T_u} \rfloor > \frac{\tau \omega}{\triangle T_u}$$

which is not possible. Hence, our supposition is wrong. Therefore, the rate of flow in this case will always be less than or equal to the case when $\tau = \frac{\triangle T_u}{\omega}$

The results from Cases 1 and 2 prove that the rate of flow is maximum when $\tau = \frac{\triangle T_u}{\omega}$.

Theorem 7.4
The maximum rate of flow in any stage is given by

$$\bar{f} = \frac{|M_k| \omega}{\triangle T_u}. \tag{7.20}$$

Proof 7.5 The maximum rate of flow in any stage = (maximum flow f)/(minimum value of τ for obtaining the maximum flow f). Therefore,

$$\bar{f} = \frac{|M_k|}{\frac{\triangle T_u}{\omega}} = \frac{|M_k| \omega}{\triangle T_u} \text{ (using (7.9) and Theorem 7.3).}$$

7.3.2   Nonuniform network

A nonuniform network differs from a uniform network in that in this case the capacities of all the nodes may be different. Let c_i denote the capacity of node v_i.

Since the network is the same, the node-cut-sets are going to be the same, denoted by M_i, $i = 1, 2, \ldots, m$. The capacity of the set M_i is the sum of the capacities of all its nodes, and is denoted by $C_{M_i}^j$ for stage j. However for the sake of convenience, the initial capacities of the nodes and node-cut-sets (which are also the same in Stage 1), will be used without the superscript. Also, after max-flow has taken place through the network, the remaining capacities of the node-cut-sets are denoted by $R_{M_i}^j$ for stage j and those of the node by r_i^j and are called *residual capacities*.

On similar lines as the uniform network, the analysis for the transient and the steady states is done separately as follows.

7.3.2.1 Transient state analysis

Using a similar argument as in Section 7.3.1.1, at $\tau = 0$, no dissipation has taken place so far and hence the network in Stage 1 is the same as the base model:

Stage 1: We have a network having nodes v_i with capacities c_i, respectively. Applying the max-flow-min-cut theorem, the maximum flow is given by: $f^{(1)} = \min_i C_{M_i} = C_{M_{k_1}}$ (without loss of generality, the min node-cut-set in stage 1 is assumed to be M_{k_1}).

After the flow has taken place, the residual capacities of the node-cut-sets would be: $R_{M_i}^{(1)} = C_{M_i} - f = C_{M_i} - C_{M_{k_1}}$. After f packets travel from s to t, the network becomes disconnected (as all the nodes of M_{k_1} become dysfunctional). For the network to become connected, at least one of the nodes $\in M_{k_1}$ should become functional (i.e., the capacity of at least one node in M_{k_1} should be at least $\triangle T_u$). So, we give the network τ units of rest such that: $\tau \omega = \triangle T_u$. (This value of τ gives the maximum flow as well as the maximum rate of flow, shown in Theorem 7.3.)

Stage 2: After giving the network τ units of rest, the new capacity of the ith node-cut-set (denoted by $C_{M_i}^{(2)}$) becomes:

$$C_{M_i}^{(2)} = R_{M_i}^{(1)} + |M_i| = C_{M_i} - C_{M_{k_1}} + |M_i|$$

(i.e., the capacity increases by 1 unit for each vertex, thereby increasing the capacity by $|M_i|$ units).

For this stage, applying the max-flow-min-cut theorem gives the maximum flow $f^{(2)}$ as follows:

$$\begin{aligned}
f^{(2)} &= \min_i (C_{M_i}^{(2)}) \\
&= \min_i (C_{M_i} - C_{M_{k_1}} + |M_i|) \\
&= \min_i (C_{M_i} + |M_i|) - C_{M_{k_1}} \\
&= C_{M_{k_2}} + |M_{k_2}|) - C_{M_{k_1}},
\end{aligned}$$

where we have assumed the set k_2 to be such that $C_{M_i} + |M_i|$ is minimized.

After flow of $f^{(2)}$ units from the network, the residual capacity of the ith node-cut-set becomes:

$$
\begin{aligned}
R_{M_i}^{(2)} &= C_{M_i}^{(2)} - f^{(2)} \\
&= C_{M_i} - C_{M_{k_1}} + |M_i| - \left(\min_i(C_{M_i} + |M_i|) - C_{M_{k_1}}\right) \\
&= C_{M_i} - C_{M_{k_1}} + |M_i| - \left(C_{M_{k_2}} + |M_{k_2}| - C_{M_{k_1}}^2\right) \\
&= C_{M_i} + |M_i| - C_{M_{k_2}} - |M_{k_2}|.
\end{aligned}
$$

Stage 3: Proceeding in a similar way, after giving the network τ units of rest, the new capacity of the ith node-cut-set(denoted by $C_{M_i}^{(3)}$) becomes:

$$
\begin{aligned}
C_{M_i}^{(3)} &= R_{M_i}^{(2)} + |M_i| \\
&= C_{M_i} + |M_i| - C_{M_{k_2}} - |M_{k_2}| + |M_i| \\
&= C_{M_i} + 2|M_i| - C_{M_{k_2}} - |M_{k_2}|.
\end{aligned}
$$

For this stage, applying the max-flow-min-cut theorem gives the maximum flow, say f^3 as follows:

$$
\begin{aligned}
f^3 &= min_i(C_{M_i}^3) \\
&= min_i(C_{M_i} + 2|M_i| - C_{M_{k_2}} - |M_{k2}|) \\
&= min_i(C_{M_i} + 2|M_i|) - (C_{M_{k_2}} + |M_{k_2}|) \\
&= (C_{M_{k_3}} + 2|M_{k_3}|) - C_{M_{k_2}} - |M_{k_2}|,
\end{aligned}
$$

where we have assumed the set k_3 to be such that $C_{M_i} + 2|M_i|$ is minimized. After flow of f^3 units from the network, the residual capacity of the ith node-cut-set becomes:

$$
\begin{aligned}
R_{M_i}^{(3)} &= C_{M_i}^{(3)} - f^{(3)} \\
&= C_{M_i} + 2|M_i| - C_{M_{k_2}} - |M_{k_2}| - (C_{M_{k_3}} + 2|M_{k_3}| - C_{M_{k_2}} - |M_{k_2}|) \\
&= C_{M_i} + 2|M_i| - C_{M_{k_3}} - 2|M_{k_3}|.
\end{aligned}
$$

This analysis of transient state suggests that the min node-cut-set might vary from stage, unlike the case of a uniform network. The next question is whether or not there exists any steady state for this kind of network, for which we analyze the state of the network as the number of stages increases under the steady state analysis.

7.3.2.2 Steady state analysis

Continuing in the same way as in the previous section, suppose we reach the nth stage, where n is some large number. Then, we have the following result, which proves that there indeed exists a steady state for a nonuniform network with dissipation.

Theorem 7.5
The min node-cut-set is the same throughout after n stages, where

$$n = \begin{cases} 0 & \text{if } C_{M_k} \le C_{M_i}, \forall i \text{ and} \\ \max_i \left(\frac{C_{M_k} - C_{M_i}}{|M_i| - |M_k|} \right) & \text{otherwise,} \end{cases}$$

where M_k denotes the node-cut-set with minimum cardinality.

Proof 7.6   Continuing as above till the nth stage(where n is some number), we get:

Stage n:
In the nth stage, the flow will be given by:

$$f^{(n)} = \min_i((C_{M_i} - C_{M_{k_1}}) + n|M_i| - |M_{k_{n-1}}|)$$
$$= \min_i(C_{M_i} + n|M_i|) - C_{M_{k_1}} - (n-1)|M_{k_{n-1}}|.$$

Now we claim that as n becomes large, the minimum given by $\min_i(M_i)$, and C_{M_i} is negligible in comparison with $n|M_i|$.

To prove this claim, consider a set M_k such that $|M_k|$ is less than $|M_i|$ for all i except k. Also, C_{M_k} may or may not be the least. We wish to prove that there exists an n such that M_k is going to be the min node-cut-set for all stages after stage n. For that, we need to show that for all stages after n, $\min_i(C_{M_i} + n|M_i|)$ occurs at $i = k$.

Equivalently, we need to show the existence of n such that

$$C_{M_k} + n|M_k| \le C_{M_i} + n|M_i|, \forall i \ne k. \tag{7.21}$$

Case 1: $C_{M_k} \le C_{M_i}$, $\forall i$. Also, $|M_k| \le |M_i|$.
Combining the equations, we get:

$$C_{M_k} + n|M_k| \le C_{M_i} + n|M_i|, \ \forall n \ge 0,$$

which proves the result for this case.

Case 2:
$C_{M_k} > C_{M_i}$, for some or all i.
Then, $C_{M_k} + n|M_k| \le C_{M_i} + n|M_i|$

$$\Rightarrow C_{M_k} - C_{M_i} \le n(|M_i| - |M_k|), \forall i$$

$$n \ge \frac{C_{M_k} - C_{M_i}}{|M_i| - |M_k|}, \forall i. \tag{7.22}$$

So, for $n \geq \max_i \left(\frac{C_{M_k} - C_{M_i}}{|M_i| - |M_k|} \right)$, the minimum node-cut-set is always M_k as it satisfies (7.21). And there exists an i, which maximizes the RHS in equation 7.22. With this, we establish the existence of such a value n, thus proving our claim.

So, after a considerable time has elapsed, the min node-cut-set is the same throughout (i.e., M_k such that $M_k = \min_i(|M_i|)$).

Note 7.5 This model (nonuniform network with dissipation) is, in fact, a generalization of the uniform network with dissipation. The Case 1 of Theorem 7.5 is a general case of the uniform network. We obtain the result that the minimum cut set is the same throughout for this case, which is concurrent with the result 7.2 of the uniform network.

Theorem 7.6
The maximum rate of flow in every stage in the steady state is given by:

$$\bar{f} = \frac{|M_k| \omega}{\triangle T_u}. \tag{7.23}$$

Proof 7.7   Since in all subsequent stages, the min node-cut-set is the same (i.e., M_k) using Lemma 7.2 and Lemma 7.3, the rate of flow, $\bar{f}$ in any subsequent stage is the same as in the previous case and is equal to

$$\bar{f} = \frac{|M_k| \omega}{\triangle T_u}.$$

7.4   Nonuniform Network with Cooling

The uniform networks, being a subset of nonuniform networks, do not require separate analyzation. So, here in this section, we consider a general network (nonuniform network) with a cooling mechanism. The model description is as follows.

We have a network, with the nodes at their maximum capacities. The problem is the same—to send as many heating packets from s to t via the network as possible. The only way this model differs from the basic model is that we have some "cooling packets" for the repair and maintenance of the network. *Cooling packets* are the packets that can travel via the network such that they cool a node they traverse by an amount $\triangle T_d$. Also, a node that is already at its maximum capacity (meaning that the node is already operating at the lowest temperature possible) does not require any cooling (in fact, it cannot be cooled any further because of restrictions on the lower bound of the temperature for each node), so the cooling packet does not cool such a vertex, which essentially means that it does not lose its cooling capacity (see Definition 7.5) while traversing that node.

Definition 7.5 *Cooling capacity* (β) is defined as the amount by which the cooling packet can cool the nodes before getting exhausted. This means a cooling packet can cool at most n nodes before getting exhausted, where n is such that:

$$n \triangle T_d \leq \beta, \text{i.e.}, n \leq \left\lfloor \frac{\beta}{\triangle T_d} \right\rfloor.$$

We consider identical cooling packets (i.e., all of them must be of the same capacity β). The cooling packets are meant only for cooling purposes and are distinct from regular packets whose flow from s to t is sought to be maximized.

When the cooling capacity of a cooling packet is exhausted, it is assumed to simply disappear from the network (the assumption is concurrent with the assumptions on the cooling packet, viz a cooling packet shall only be used for cooling purpose s, which it fails to, once its cooling capacity is exhausted).

7.4.1 Maximizing flow through the network using cooling packets

Our problem is to find the dispatch pattern (of heating and cooling packets) such that the flow (of heating packets) from source s to sink t is maximized. Initially, we have a network, with all the nodes at their maximum possible capacity (since every node is at its minimum possible temperature initially and hence maximum possible number of heating packets can traverse that node before it reaches its upper bound and becomes dysfunctional). Since this model is exactly similar to the base model for calculating the maximum flow via this network, we can follow the same approach as in the base model (applying Ford-Fulkerson Algorithm on the equivalent network (modified using the node-splitting technique)).

Once maximum flow has been achieved via this network, it becomes disconnected. Let M_k denote the corresponding min node-cut-set, which is the node-cut-set that has actually disconnected the network. It should be noted, however, that there might exist other nodes that have become dysfunctional, but do not belong to the node-cut-set M_k. However, those nodes need not be identified, as they will not play a decisive role in further analysis.

To connect this disconnected network, it is obvious that we need to repair the nodes in the node-cut-set M_k. The only option available to us is using the cooling packets for this purpose.

Also, we know that the network will become functional if at least one of the nodes in M_k is repaired. However, the maximum flow possible in this case would be less than or equal to the initial capacity of set M_k. In cases where nodes of set M_k are the limiting nodes, flow can be maximized if all nodes of M_k work at their maximum capacities (the other case will be handled later in the analysis).

The cooling packets will obviously have to be sent via directed paths/walks to the target nodes in M_k.

Definition 7.6 A *walk* is a directed path from a vertex v_1 to another vertex v_2 such that a node may be traversed more than once, but any edge is traversed just once. Specifically for this chapter, walk is used to refer to a directed walk from s to t.

Let W denote the set of walks via which we have sent the cooling packets to the nodes in M_k.

Definition 7.7 Walk W_S to a set of nodes S is defined as a set of walks from s to t such that the walks traverse all the nodes of the set S once. The set S is then said to be *entirely spanned* by W_S. If the set of walks W_S spans only some of the nodes of S and not all, S is said to be *partially spanned* by W_S.

So, the set of walks to the node-cut-set M_k refers to a set of walks that pass through all nodes of the set M_k. Then, we have the following result.

Lemma 7.2
Let the cooling packets be sent to M_k, M_k being the min node-cut-set of the network, via the set W and let the resulting network (with increased capacity of nodes) be denoted by G^*. Then, the min node-cut-set of G^* will either be M_k again or M_i, where M_i is the node-cut-set partially spanned by W.

Proof 7.8 For this, we prove that the min node-cut-set can never be the set M_i such that M_i is spanned entirely by W and $i \neq k$. Let the increase in capacity of a node-cut-set M_i in G^* be denoted by $\triangle C^*_{M_i}$. Then, if the set M_i is spanned by W entirely, the capacity of the set M_i is increased by at least as much as that of M_k, i.e., $\triangle C^*_{M_i} \geq \triangle C^*_{M_k}$.

Also, since M_k was the min node-cut-set, the residual capacity (R_{M_k}) of M_k after flow of f units would have become zero, whereas that of M_i will be greater than or equal to zero. So, after the cooling packets have been sent through the network,

$$C^*_{M_i} = R_{M_i} + \triangle C^*_{M_i}$$
$$\geq \triangle R_{M_k} + C^*_{M_k},$$
$$\therefore C^*_{M_i} \geq C^*_{M_k}. \tag{7.24}$$

Therefore, in the next stage, the node-cut-set M_i cannot be the min node-cut-set, where i is such that M_i is entirely spanned by W and $i \neq k$. (Even if the equality holds in 7.24, we can assume M_k to be the min node-cut-set for the sake of preserving the generality of the result.) Thus we have proved that the min node-cut-set can never be the set M_i such that M_i is spanned entirely by W and $i \neq k$.

Lemma 7.3
The maximum possible flow via the network G^* is f, where f' is the flow obtained by applying the max-flow-min-cut theorem on the initial network G.

Proof 7.9 In the initial network G, all the nodes are at the temperature θ_{0i}, which is the minimum possible temperature that the node can attain. So, the capacity of each node is the maximum, and let the maximum flow be f. Let G^* denote the network, the capacity of nodes of which has been improved by employing the cooling packets. We wish to prove that in no case can the maximum flow through the network G^* exceed f. Let M_i, $i = 1, 2, \ldots, m$ be all the possible node-cut-sets in the network G^*, and let C_{M_i} denote their respective capacities (the capacity of a node-cut-set is equal to the sum of the capacities of its nodes).

The maximum flow f^* is given by the max-flow-min-cut theorem as

$$f^* = \min_i C_{M_i}.$$

Let us assume, without loss of generality, that the node-cut-set M_k is the one with minimum capacity. Then, since all its nodes are at their maximum possible capacities (because they have been cooled to their respective base temperatures by using cooling packets), it follows that C_{Mk} is working at its maximum capacity. This directly implies that

$$\max f^* = \max \min_i C_{M_i} = \max C_{M_k} = C_{M_k}.$$

This is equal to the value of maximum flow through the initial graph G. Hence, the result.

Lemma 7.4

For the network to yield the maximum flow, every node-cut-set must work at least at the capacity κ, where κ is given by $\kappa = \min_i C_{M_i} = f$.

Proof 7.10 We reason as follows.

(a) Suppose it is not necessary. That is, there exists a node-cut-set, say, M_a, which works at the capacity C'_{M_a} less than f. Then, by applying the max-flow-min-cut theorem, $C'_{M_a} < f$.

So, the maximum flow in this case will be $f' = C'_{M_a} < f$.

But we know that the maximum capacity of the set M_k is C_{M_k}, which can be attained by increasing the capacity of its nodes to their maximum capacities, which is not impossible. If we increase the capacities of all the node-cut-sets in this way, it is easy to see that the maximum flow will then be C_{M_k} only, because M_k has the minimum capacity of all the node-cut-sets when all node-cut-sets are working at their maximum capacities.

(b) Even if we do not increase the capacity of the node-cut-sets to their respective maximum, and only up to κ, even then the max-flow-min-cut theorem says we can attain the flow equal to f. And by Lemma 7.3, f is in fact the maximum possible flow via this network G.

So, we deduce that we can attain the maximum possible flow via G if the capacity of every node-cut-set is at least $\kappa = \min_i C_{M_i}$.

Note 7.6 Our objective now becomes: to send cooling packets through the network in such a way that all node-cut-sets work at at least the capacity given by $\min_i C_{M_i}$. It has already been proved why the maximum flow cannot exceed the value f. So, it is established that we can attain maximum flow of f via the network. So, now we have a disconnected network, say G', which we have to repair by sending cooling packets so as to make it functional again so that it yields the maximum flow.

Theorem 7.7
To attain the maximum flow f via network G', the cooling packets must be sent via directed walks such that the walks span the entire network.

Proof 7.11   The result follows immediately from Lemmas 7.2, 7.3, and 7.4.

Note 7.7 The theorem does not state how many cooling packets are to be sent. We can safely send as many cooling packets as required to make the set M_k work at its maximum capacity. The number of packets required for the same is given by Theorem 7.9.

To send cooling packets so as to span the entire network, it is necessary that such a set of walks spanning the entire network exists. This is what we shall prove next.

Note 7.8 By *entire network*, we mean the entire functional network. This means that the nodes that cannot be traversed by heating packets should not be considered as part of the network. Therefore, we define only those nodes that allow the flow of heating packets to be part of the network.

Theorem 7.8
There exists a set of walks that spans the entire network.

Proof 7.12   Suppose there exists a node v that the heating packets traverse, but $\nexists$ any walk from s to t via that node. This is self contradictory as the node being traversible by heating packet itself implies that there exists a path from s to t via v. A path is also a walk. So, there exists a walk from s to t via v, which directly implies that there exists a walk from s to v, thereby contradicting our supposition. Hence, there exists a set of walks that spans the entire network.

Theorem 7.9

For achieving the maximum flow in a network with cooling packets by sending the cooling packets via the walks spanning the entire network, we need

$$n \geq \sum_{i:v_i \in M_k} \left\lceil \frac{c_i * \triangle T_u}{\triangle T_d} \right\rceil$$

cooling packets per "max flow" number of heating packets.

Proof 7.13 We have to bring the nodes in the set M_k to their maximum capacities by sending cooling packets. Now, the number of cooling packets to be sent to node $v_i \in M_k$ with capacity c_i is given by n_{v_i} such that: $n_{v_i} \triangle T_d \geq c_i \triangle T_u$ so that
$n_{v_i} = \left\lceil \frac{c_i \triangle T_u}{\triangle T_d} \right\rceil$.

Since we need to repair all the nodes in M_k, the total number of cooling packets to be sent per f number of heating packets would be:

$$n = \sum_{i:v_i \in M_k} \left\lceil \frac{c_i \triangle T_u}{\triangle T_d} \right\rceil .$$

But the set of walks spanning the set M_k might not span the entire network. So, there may exist other nodes that the walks did not cover. To span the entire network, we need to employ more cooling packets for such unspanned nodes. This results in am increase in the number of cooling packets to be sent per f heating packets, and hence,

$$n \geq \sum_{i:v_i \in M_k} \left\lceil \frac{c_i \triangle T_u}{\triangle T_d} \right\rceil .$$

7.4.2 Reducing the number of cooling packets and the cooling capacity required

Do we really need to send cooling packets so as to span the entire network? Perhaps not. It seems a little counter-intuitive, but we have the following results to substantiate the realization.

Lemma 7.5

Every walk in W that spans the set M_k traverses at least one node of each node-cut-set M_i, $i = 1, 2, \ldots, m$.

Proof 7.14 We prove the result by contradiction. Let us assume that there exists a set, say M_a and a walk w_i such that w_i does not traverse any vertex of M_a. Then, if all the nodes of M_a become dysfunctional, there still exists a path w_i from s to t, which contradicts the fact that M_a is a node-cut-set. Hence, every walk $w_i \in W$ traverses at least one vertex of every node-cut-set.

For the next result, we need to define what we mean by a walk through a node:

Definition 7.8   *A walk via a node v_i is a walk from s to t such that the node v_i lies on that walk (or equivalently, the walk traverses the node v_i).*

Theorem 7.10
Maximum flow f in G^* can be achieved by sending cooling packets to the nodes of M_k via walks from s to t via nodes in M_k such that the set of walks, say W, spans the entire set M_k.

Proof 7.15   We are given a set W of walks that span M_k, and via which we are sending the cooling packets. Now, let the residual capacity of the node-cut-sets in the network G be denoted by R_{M_i} and let the increase in capacity of a node-cut-set M_i be denoted by $\triangle C_{M_i}$. Let the resultant capacity of the node-cut-set M_i in the network G^* be represented by $C_{M_i}^*$. Now, when we send cooling packets via nodes in M_k such that the capacity of the node-cut-set is increased by f, using Lemma 7.5, the capacity of all other node-cut-sets is increased at least by f (i.e., $\triangle C_{M_i} \geq \triangle C_{M_k}$).
Also, since M_k was the min node-cut-set, $C_{M_k} \leq C_{M_i}$.
And after flow f has taken place, in the resultant network G',

$$C_{M_k} - f \leq C_{M_i} - f$$

(i.e., $R_{M_k} \leq R_{M_i}$).

Therefore,

$$R_{M_k} + \triangle C_{M_k} \leq R_{M_i} + \triangle C_{M_i}$$

(i.e., $C_{M_k}^* \leq C_{M_i}^*$).

So, using max-flow-min-cut on the network G^*, we obtain the maximum flow $C_{M_k}^*$, which is equal to f.
Since f is the maximum possible flow that can ever be achieved via G, we have thus obtained an improved approach to obtain the maximum flow through G^*.

Theorem 7.11
For achieving max flow in a cooling network, we need

$$n = \sum_{i:v_i \in M_k} \left\lceil \frac{c_i \triangle T_u}{\triangle T_d} \right\rceil$$

number of cooling packets per maximum flow number of heating packets.

Proof 7.16   The previous result implies that now we do not need to send cooling packets to span the entire network G. Rather, our objective is to span the entire set M_k, and send cooling packets so that all nodes of M_k function at their respective maximum capacities so that the set M_k, which is going to be the min node-cut-set in the subsequent stage, works at its maximum possible capacity that yields the maximum flow.

The number of cooling packets that span M_k entirely so that all nodes of M_k work at their respective maximum capacities is given by Theorem 7.9 to be:

$$n = \sum_{i:v_i \in M_k} \left\lceil \frac{c_i \triangle T_u}{\triangle T_d} \right\rceil$$

Note 7.9   As per Theorem 7.10, our objective is just to send cooling packets via M_k such that all nodes of M_k work at their respective maximum capacities. Since a cooling packet loses $\triangle T_d$ of its cooling capacity upon traversing a node, we can save this cooling capacity by sending these cooling packets via shortest possible walks such that they span M_k and make its nodes work at their respective maximum capacities. So now, we not only reduce the number of cooling packets required but also the cooling capacity required.

We now give two results on the value of β, first to make the network functional and then to make the network functional such that it yields maximum flow.

Theorem 7.12
The capacity of a cooling packet required for making a dysfunctional network functional should at least be equal to the minimum of the shortest distances between s and t via $v_i \in M_k$ (the min node-cut-set of the network), (i.e., $\beta \geq \min_{i:v_i \in M_k} d(s, v_i, t)$).

Proof 7.17   For the network to become functional again, we need to repair at least one vertex of M_k. To reduce the cooling capacity requirement, we would send the cooling packet to the vertex $v_i \in M_k$ such that v_i is nearest to s. Hence, if we denote by $d(s, v_i, t)$ the distance from s to t via node v_i, the minimum possible value of β required would be:

$$\min_{i:v_i \in M_k} d(s, v_i, t).$$

Theorem 7.13
The capacity of a cooling packet required to obtain maximum possible flow through the network should at least be equal to the maximum of the shortest distances between s and t via $v_i \in M_k$ (the min node-cut-set of the network), (i.e., $\beta \geq \max_{i:v_i \in M_k} d(s, v_i, t)$).

Proof 7.18   According to Theorem 7.10, to obtain maximum possible flow through the network, we need to send cooling packets via the set of walks W such that W entirely spans M_k. Also, we need to repair all the nodes of M_k to their respective maximum capacities, so as to be able to obtain a maximum flow f. For this, we need the capacity of the cooling packets to be such that even the node ($\in M_k$) farthest from s is also traversed. For that, we need capacity to be such that $\beta = \max_{i:v_i \in M_k} d(s, v_i, t)$, which proves the result.

Note 7.10   We are not maximizing or minimizing over the distances from s to v_i; rather, we are maximizing or minimizing over distance from s to t via v_i because if we do not traverse from s to t, in the subsequent stage M_k might not necessarily be the min node-cut-set.

7.5   Conclusion

This chapter defines a thermal network, and gives results for the maximum flow that is achievable through a thermal network. In many networks, there are restrictions on the nodes, which may be repair constraints or pollution level constraints (in road networks) or the amount of data to be transferred through a node that is already stressed (in computer networks). This work is a generalization to all such problems, whose systems may thus be regarded as real-life thermal networks. An aspect of the model we have discussed is that it is dynamic in nature, thereby capturing the temporal properties of the nodes. There are infrastructures that are to be maintained and used for very long durations. In such networks, we have to maximize the flow while maintaining nodes in a manner that does not contribute to their breakdowns.

Also, our results give the maximum flow values through the network under such constraints that can be used to measure the amount of error in heuristic algorithms developed for similar problems.

This chapter also opens up the scope of developing exact algorithms for such networks using the approach we have implemented. The models can also be extended (e.g., to get a dissipating model with the rates of dissipation being different for different nodes). Algorithmic work is also possible, especially with real-life system data; for instance, algorithms for coolant problems, optimizing the capacity of the coolants used, and similar applications and implementation to practical problems.

Acknowledgment

The work of the first author was partially supported by an INSPIRE research fellowship from the Department of Science and Technology, Government of India.

Bibliography

[1] J. Berger and M. Barkaoui. A new hybrid genetic algorithm for the capacitated vehicle routing problem. *Journal of the Operational Research Society*, 54(12):1254–1262, 2003.

[2] W. R. Black. *Sustainable Transportation: Problems and Solutions*. The Guilford Press, New York, 2010.

[3] J. A. Bondy and U. S. R. Murty. *Graph Theory with Applications*. North-Holland, New York, 1976.

[4] V. Campos and E. Mota. Heuristic procedures for the capacitated vehicle routing problem. *Computational Optimization and Applications*, 16(9):265–277, 2000.

[5] A. Capozzoli, G. Serale, L. Liuzzo, and M. Chinnici. Thermal metrics for data centers: A critical review. *Energy Procedia*, 62:391–400, 2014.

[6] T. H. Cormen, C. Leiserson, R. Rivest, and C. Stein. *Introduction to Algorithms, 3rd edition*. The MIT Press, Cambridge, MA 2009.

[7] N. Deo. *Graph Theory with Applications to Engineering and Computer Science*. Prentice-Hall, Englewood Cliffs, NJ, 1974.

[8] L. R. Ford and D. R. Fulkerson. Maximum flow through a network. *Canadian Journal of Mathematics*, 8:399–404, 1956.

[9] P. Fraundorf. Thermal capacity in bits. *American Journal of Physics*, 71(11):1142–1151, 2003.

[10] Y. Fulpagare and A. Bhargav. Advances in data center thermal management. *Renewable and Sustainable Energy Reviews*, 43:981–996, 2015.

[11] G. Gallo, M. D. Grigoriadis, and R. E. Tarjan. A fast parametric maximum flow algorithm and applications. *SIAM Journal on Computing*, 18(1), 30–55, 1989.

[12] G. Glockner. Effects of air traffic congestion delays under several flow-management policies. *Transportation Research Record: Journal of the Transportation Research Board*, 1517:29–36, 1996.

[13] G. P. Henze, M. Laguna, and M. Krarti. Heuristics for the optimal control of thermal energy storage. In I. H. Osman and J. P. Kelly, editors, *Meta-Heuristics: Theory and Applications*, pp. 183–202. Springer, Boston, MA, 2012.

[14] A. Hosseini and V. Shabro. Thermally-aware modeling and performance evaluation for single-walled carbon nanotube-based interconnects for future high performance integrated circuits. *Microelectronic Engineering*, 87(10):1955–1692, 2010.

[15] Y. Joshi and P. Kumar, editors. *Energy Efficient Thermal Management of Data Centers*. Springer, New York, 2012.

[16] B. S. Kerner. Modeling approaches to traffic congestion. In R. A. Meyers, editor, *Encyclopedia of Complexity and Systems Science*, Springer, New York, pp. 9302–9355. 2009.

[17] C. Kittel and H. Kroemer. *Thermal Physics*. Freeman, San Francisco, CA, 2000.

[18] K. M. Kockelman and S. Kalmanje. Credit-based congestion pricing: a policy proposal and the public's response. *Transportation Research Part A: Policy and Practice*, 39(7–9):671–690, 2005.

[19] M. R. V. Kumar and S. Raghunathan. Heterogeneity and thermal aware adaptive heuristics for energy efficient consolidation of virtual machines in infrastructure clouds. *Journal of Computer and System Sciences*, 82(2):191–212, 2016.

[20] N. A. Pantazis and D. D. Vergados. A survey on power control issues in wireless sensor networks. *IEEE Communication Surveys & Tutorials*, 9(4):86–107, 4th Quarter 2007.

[21] Y. Zhan, S. V. Kumar, and S. S. Sapatnekar. Thermally aware design. *Foundations and Trends in Electronic Design Automation*, 2(3):255–370, 2008.

Chapter 8

Bio-Inspired Solutions and Network on Chip (NoC) Fault Tolerant Algorithms

Muhammad Athar Javed Sethi

University of Engineering and Technology, Peshawar

Fawnizu Azmadi Hussin and Nor Hisham Hamid

Universiti Teknologi PETRONAS, Malaysia

CONTENTS

System on Chip (SoC) has seen a continuous increase of computational nodes on it. This has increased the communication requirement between them. Complexity of the communication has also increased due to these nodes. This leads scientists and researchers to think over a new communication paradigm. Time to market access is another major factor that forced the scientists to rethink the design of SoC architecture. Because of these requirements, scientists presented the idea of Multiprocessor System on Chip (MPSoC). This improved the performance and processing power of embedded systems. Philip Nexperia, TI OMAP, and ST Nomadik are a few MPSoC platforms available in the market today. As the numbers of devices increased, the interconnections of internal components of MPSoC became a major issue. The bus-based systems between these internal components were also not able to handle the Globally Asynchronous Locally Synchronous (GALS) concept. In the GALS approach every component works in a different clock domain. This further triggered the research for on-chip communication. Scientists started thinking over the reliable and efficient communication between MPSoC components. Later, they came up with the idea of Network on Chip (NoC) architectures to address the complex communication requirements of MPSoC. NoC have brought the packet switching concept into the on-chip communication. Today, a few of the commercially available NoC's are Arteris, Silistrix, and INoC [1,35].

NoC solved the communication issues between multiple cores present on the chip, but it also has encountered various faults. Fault tolerance is the key concept, which differentiates the NoC from the traditional data network. As the devices are shrinking, more and more devices are coming on per unit area of SoC. Due to this, devices and interconnects between them are suffering from cross talk and permanent and temporary faults. In order to overcome the faults, routing algorithms have been used in the NoC. These routing algorithms help to achieve a reliable communication between multiple devices on NoC by bypassing the faulty nodes and links [32].

8.1 Introduction

Network on Chip (NoC) is a communication framework for on-chip communication. It has replaced the traditional bus and crossbar interconnections with a network of wires. These buses have drawbacks of congestion, delayed communication, and latency problems. NoC has the benefit of higher bandwidth, concurrency, modularity, scalability, and effective reuse of resources. NoC is constructed by routers, processing elements (PEs), and interconnects. Routers are connected together through point to point links with each other. The PEs are connected to the routers through network interface (NI). Routers in turn are connected with neighbor routers. NI separates the data communication from network communication. NI transforms the message generated by PE to packets and vice versa. NI sends these packets to routers, which route them in the particular direction of NoC according to the routing algorithm [5]. Figure 8.1 shows 4 × 4 mesh NoC having 16 routers and PEs. PEs can be a homogeneous resource or they can be a heterogeneous resource. PEs can be a processor, memory, cache, reconfigurable block, digital signal processing (DSP) core, or any other device, as shown in Figure 8.2.

In NoC, PEs can be connected using any topology as regular and irregular are two broad categories of NoC topologies. In regular topology, routers and PEs are connected in some structured and organized manner while in irregular topology

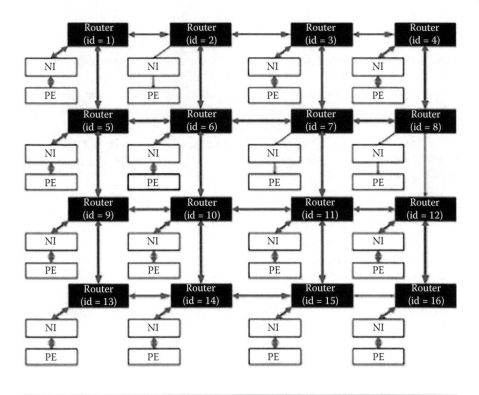

Figure 8.1 Network on Chip (NoC).

routers and PEs are not connected in any fixed pattern. Figures 8.3 and 8.4 show the NoC torus, tree, ring, and irregular topologies.

8.2 NoC Topologies

In mesh topology, the routers are connected with each other through point to point connection in a mesh structure. The routers at the first and last column and row of NoC are not connected with any other neighbor routers. The torus topology is similar to mesh except that the first and last column and row elements are also connected with each other. Torus increases and eases the routing decisions. In tree structure topologies the routers are connected in a hierarchical design such that the parent routers have child routers connected with them. The PEs are connected with leave routers. In star topology, the routers are connected with the centralized arbiter. The arbiter manages the communication between multiple routers. The arbiter may be a specialized device or it can be a router as well. In octagon topology, the eight routers are connected in a crossbar fashion

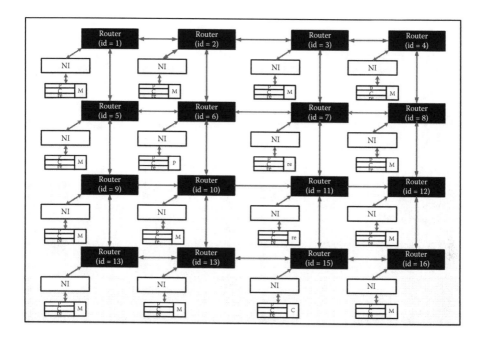

Figure 8.2 NoC with heterogeneous resources.

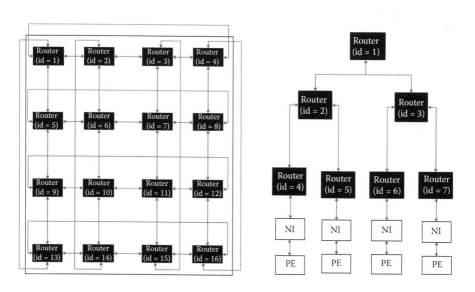

Figure 8.3 NoC torus and tree topology.

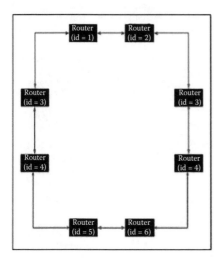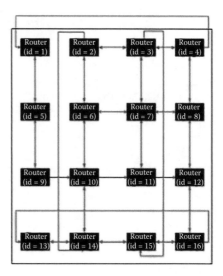

Figure 8.4 NoC ring and irregular topology.

with bidirectional links [13]. In crossbar topology the routers are connected with a number of wires connected in a crossbar fashion. Ring topology refers to the (*n*) routers connected in a ring. That is, all routers are connected with each other through interconnects in a ring shape. In Spidergon topology, a number of routers (>2) are connected together in a bidirectional ring in both clockwise and anti-clockwise direction. Spidergon topology is similar to a spider web. In subnet topology, there are small groups of PEs connected with the router. This group is called the subnet. These subnets are in turn connected with each other [13].

NoC, with the help of the network of wires between PEs, have solved the communication problem between them. Faults in NoC arise as the sizes of the devices were continuously reduced to cope with the communication requirement of PEs. The fault tolerance is a very important concept nowadays to cope with these faults.

8.3 Fault Tolerance

Fault tolerance is a key issue nowadays as devices on the chip are reducing in size using nanotechnology. There are two broad categories of faults: permanent and temporary faults. Permanent faults are due to manufacturing defects, electromigration, and dielectric breakdown or due to any other physical damage. Permanent faults can only be removed by extra hardware (i.e., spare routers, wires, or resources) available on NoC [10,23,24]. The temporary faults may occur due to changes in voltage, temperature fluctuations, and congestion at routers. Temporary faults can be dealt with using fault tolerant routing techniques [11,12,16,22,27–30,33,34,46–48].

Fault tolerant routing algorithms are used to recover from the faulty router(s) and interconnect(s) in the NoC. These techniques ensure reliable and efficient communication between source PE and destination PE. Deterministic, stochastic, fully adaptive, and partially adaptive, routing algorithms are broad categories of fault tolerant routing algorithms. In the deterministic routing algorithm, the packet follows a particular fixed path from source to destination while in the stochastic routing algorithm, the packet is sent in all or a particular direction based on the type of routing algorithm. In the fully adaptive routing algorithm, the routing depends on the routing table at the router. These algorithms are adaptive as compared to the stochastic routing algorithm. The partial adaptive routing algorithm puts restrictions on certain turns in NoC. These algorithms are less complex but lack the adaptiveness as compared to fully adaptive routing algorithms [39,41,42].

8.4 Fault Tolerant Routing Algorithms

Fault tolerant algorithms are used to recover from the faulty router(s) and interconnect(s) in the NoC. These techniques ensure reliable and efficient communication between source PE and destination PE.

8.4.1 *Deterministic routing algorithms*

In deterministic routing the packet routes from the certain point to another using a fixed path. These algorithms lack the adaptiveness. Xy, yx, xyz, and zyx [12,16,22,28,33,34,48] are a few examples of dimension order routing (DOR) algorithms. DOR algorithms are the simplest algorithm of deterministic routing algorithms. These DOR algorithms are deadlock free. In minimal path routing, the packet can traverse using multiple (shortest) paths to reach the destination. Minimal path routing algorithms are prone to deadlock as compared to DOR [2].

In xy routing algorithm, the packet is first routed to the row of the destination node and then to the column of the destination node. In yx, the packet is first routed to the column of the destination node and then to the row of it. These algorithms are static, deadlock free, and deterministic algorithms. These algorithms lack the adaptiveness and are not suitable for a complex network of resources. Zyx is the 3D NoC routing algorithm. In this algorithm the packet is first sent to the NoC layer, then it is routed to the column of the destination, and later it is sent to the particular row of the destination [28]. The zyx algorithm is a static, deterministic, and deadlock free algorithm, but it lacks adaptiveness.

8.4.2 Stochastic routing algorithms

Packets are broadcasted in all or a particular direction of NoC depending on the type of algorithm. These algorithms are simple to implement, but they consume a large amount of energy. They have congestion, deadlock, livelock, and high bandwidth utilization problems. These algorithms do not perform well even at low traffic rate because they broadcast the packet in all directions. These techniques lack the adaptiveness, and they are not dynamic in nature. A few types of these stochastic algorithms are probabilistic gossip flooding scheme, directed flooding, N-random walk [28–30,48], and connection oriented stochastic routing (COSR) [27].

8.4.2.1 Probabilistic gossip flooding scheme

In this algorithm packets are sent multiple times to multiple paths of NoC. These algorithms do not have the information of the destination. When one of the copies of a packet arrives at the destination then all other packet copies are deleted. This algorithm consumes a lot of bandwidth and energy. This algorithm lacks the adaptiveness and it is not dynamic in nature. Packets in this scheme may get lost or collapse during the traversal of NoC [30,48].

8.4.2.2 Directed flooding scheme

Packets are replicated in particular or directed paths in directed flooding scheme. The performance of this algorithm is better than probabilistic algorithm as it consumes less bandwidth and other network resources [28]. This algorithm is less affected by high error rates. The performance is better than probabilistic algorithm when the gossip rate is low [22].

8.4.2.3 N-Random walk

(N) number of packets are sent to the network in a particular direction in the N-random walk algorithm. This technique is efficient as compared to the directed flooding algorithm as it consumes less bandwidth and energy [48]. Random walk has less communication overhead as compared to probabilistic gossip flooding scheme, directed flooding scheme and N-Random walk and it also provides a useful level of fault tolerance [6].

8.4.2.4 Connection oriented stochastic routing (COSR)

This scheme is a hybrid technique that combines circuit and packet switching. A connection is established using circuit switching between source and destination. After connection establishment, packets are sent over the established connection using packet switching [27]. This algorithm is easy to implement and fault tolerant. The drawback of this technique is the connection that is established for a certain period of time may lead to underutilization of the bandwidth, congestion, and denial of services to some PEs.

8.4.3 Fully adaptive routing algorithms

The routing at fully adaptive algorithms depends upon the routing table or on the routing information collected from the neighbor nodes at the router. Based on this information the direction of the packet is decided at run time. Routers constantly communicate with each other to update the routing table or neighbor nodes information. Updating of routing information takes a lot of power, energy, and time, which affects the throughput of the NoC [28,46]. Source routing for NoC (SRN) and force directed wormhole routing (FDWR) are two examples of a fully adaptive routing algorithm [22,34]. Fully adaptive routing algorithms are very dynamic, but updating of routing information consumes a lot of area, energy, and power, which sometimes degrades the performance of NoC. The flow of control messages between routers at times creates a congested, deadlocked situation in NoC. The techniques [36–38] are fully adaptive routing algorithms, but they do not have the routing table. They collect the neighbor information through certain control messages. Based on this information, the router makes a decision in which direction the packet should be routed.

8.4.3.1 Source routing for NoC (SRN)

Router discovery and route maintenance are two main components of this algorithm. First, the algorithm dynamically discovers its path from source to destination. Later, it maintains the path by adapting another path to bypass the faulty router. Due to its limited number of broadcasts, SRN has less communication overhead. SRN is not able to handle complex networks. SRN has a restriction on source routing table, which limits adaptiveness [22].

8.4.3.2 Force directed wormhole routing (FDWR)

FDWR divides the traffic over the network by force. It follows the wormhole switching mechanism. In wormhole switching the packet is divided into header, body, and tail flits. Header flit finds and contains the path until the destination. Body and tail flits follow the header flit specified path. Body flits contain the data that is to be transferred from source to destination while tail flits inform the routers and destination about termination of communication. In FDWR, the header flit finds the path by avoiding the faulty routers. The neighbor routers that are working properly exchange messages about their availability and working status. If the routers receive the messages, and with the help of the routing table, FDWR finds the shortest path from source to destination [34]. The advantage of FDWR is that it evenly distributes the traffic across the network with the help of the routing table. It is fault tolerant as it avoids the faulty router by taking another path. In this algorithm, flits of the packets may take different path on the NoC. With the help of the sequence number, flits and packets are reordered at the destination. The drawback of this technique is that the destination has to reorder a packet, which consumes energy, power, and time.

The advantages of fully adaptive routing techniques are that they are dynamic in nature. The router decides at run time on which link it should route the packet. However, constantly updating the routing table requires a lot of time, area, energy, and power. The communication between routers in fully adaptive routing algorithms creates a deadlock and congestion situation in NoC [28,29].

8.4.4 Partial adaptive routing algorithms

As the name suggests, partial adaptive algorithms are partially adaptive. They put some restrictions on the routes that can be taken by a router in NoC [25]. These algorithms solve the problem of deadlock and also consume less energy and power as there are no routing tables. These algorithms limit the adaptiveness of the NoC, and latency of the packets increases due to restrictions [28]. West first, negative first, north last, south last, odd-even, and planar adaptive [11,12, 16,28,33,46–48] are a few examples of these algorithms.

8.4.4.1 West first, negative first, north last, south last routing algorithms

In west first, packets transmitted to the west should be the first because later in this algorithm a packet cannot be transmitted in a westward direction. Similarly, the negative first algorithm allows all other turns except turns from the positive direction to negative direction. Packet routing at negative direction must be done first and all other turns in NoC are done later. In north last, as the name suggests, packets that need to be routed to north must be transferred there last. In the south last algorithm, a packet can be routed to the south direction as a last turn [46].

8.4.4.2 Odd-even routing algorithm

In the odd-even routing algorithm no deadlock occurs. This algorithm restricts turns at every column so there are no turns from east to north or from east to south. Similarly, in odd columns there should be no turn from north to west or from south to west. The advantage of this algorithm is that it avoids livelocks and consumes less energy due to restriction on turns [12,46].

One of the advantages of turn based models is that no routing table is required. The router has to determine which turn algorithm it is routing the packets. The router can incorrectly send the packet in the wrong direction. In turn based algorithms, routers become more complex. Among all partial adaptive routing algorithms, the odd-even routing algorithm provides more adaptiveness as there are less restricted turns in it.

8.4.4.3   *Planar adaptive routing algorithm*

A constant number of virtual channels is the basic requirement of the planar adaptive routing algorithm, and it does not depend on the size or dimension of the network [47]. This algorithm routes the packet in two dimensions until it reaches the destination. This algorithm avoids deadlock and also consumes less routing time because there is no routing table in it. The planar adaptive algorithm can be applied to (*n*) dimensional planes by allocating three virtual channels [11]. The disadvantage of this technique is that it requires extra virtual channels for communication.

All these traditional algorithms have drawbacks of congestion, livelocks, deadlocks, latency, efficient bandwidth, and throughput utilization. To overcome these drawbacks bio-inspired fault tolerant techniques are proposed. These techniques are inspired by the biological brain fault tolerant and robust techniques. In order to get inspiration from nature, certain rules should be taken into consideration.

8.5   Bio-Inspired Solutions

Scientists and researchers are always naturally attracted to solve the complex and difficult engineering world problems. In order to implement the biological concept to solve the real world problems, it should be clear that the engineering problem is similar in nature. The bio-inspired solution is feasible to implement and the tools and mechanism required to implement it are available. In the past, people have adapted different biological concepts to solve real world issues. Swarm intelligence and behavior of insects have been used in solving the routing problems in computer networks. Network security software has been inspired by the immune system of the human body to catch abnormal behaviors and usage of resources on the network [14].

The robust and dynamic nature of biological solutions is being used by the scientists in the NoC. Self-configuration, self-healing, and self-optimization are a few of the dynamic nature characteristics of the immune system that are being used in NoC to make it autonomic. The bio-inspired NoC (BNoC) behaves and reacts like an immune system. Just like immune systems, the BNoC detects the pathogens (application behavior and system state change) that enter the body (system) and delivers a response to heal (adapt to changes) it. In BNoC, the self-configuration, self-healing, and self-optimization are performed at the application, communication, and architecture layer. BNoC can react like an immune system against pathogens that have entered the body. It detects the infection (applications behavior or system state changes) and delivers a response to eliminate it (i.e., adapt to changes) [3].

We are getting inspiration from biological brain fault tolerant techniques to make NoC communication reliable as the biological brain is highly robust and fault tolerant. The brain tries to work properly, even if some neurons, synapse, or some other part of the brain is damaged. Synaptogenesis and sprouting [7] are two biological brain techniques used by us to implement the self-adapt and self-heal concepts in NoC.

8.6 Biological Brain Characteristics

The biological brain is a complex organ having a network of neurons interconnected with the help of synapses. It works faster than any supercomputer. A neuron is the basic building unit of the brain. There are around 80—120 billion neurons in a human brain. Figure 8.5 shows the biological neuron [19].

Old brain and new brain (the neocortex) are two broad categories of the biological brain. The neocortex is 77% of the whole brain. This part is related to learning and processing capabilities of humans. Neurons, columns, and hierarchies of neurons are the three parts of the neocortex. The basic block of neocortex is a neuron. Neurons are connected through synaptic junction. This junction is formed by the connection of dendrites of one neuron and axon terminal of another neuron as shown in Figure 8.6. The input to neurons comes from dendrites and the output follows through axons to the axon terminal. The axon is covered by a myelin sheath that protects the signals travelling inside the axon. This sheath also increases the speed of the electrical signal traveling through the axon. The neuron's cell body is called the soma and contains the nucleus. The neuron on one side of the junction is the presynaptic junction neuron, while

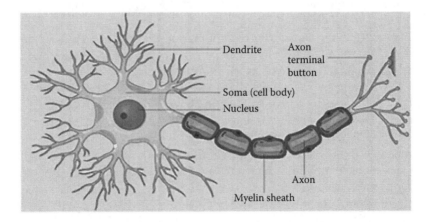

Figure 8.5 Biological neuron.

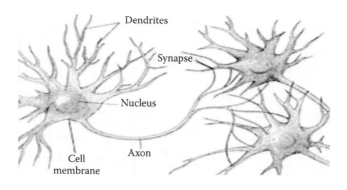

Figure 8.6 Biological synapse.

the other side is called the postsynaptic junction neuron. The synaptic cleft separates axon terminal and dendrites. Neurons only receive signals from those neurons to which they are connected. A neuron triggers an output if the signal strength is above a certain threshold after summing all of the inputs coming from various neurons. This output is in chemical form while an electrical signal flows though the axon [7].

The brain is highly adaptable and new connections (synapse) are dynamic in nature. There are between 1000 and 10,000 new connections per neuron in the brain. These connections can change very rapidly. Plasticity (changes in brain) reduces with age. Damaged neurons self-destruct (a process called apoptosis) and are reabsorbed by the body. There are around five support cells for every neuron, called neuralgia (or just glia) and these cells take care of maintenance issues. Some damaged neurons cannot be repaired in humans; for example, the severing of the spinal cord is one such damage. If the firing rate of neuron is higher than the resting rate then some neurons get fatigued and stop contributing to information flow. Then, due to the death of neuron, other neurons take over and communication between them remains the same with a new path [21].

8.7   Biological Brain Fault Tolerant Techniques

The basic building unit of the brain is a neuron. The biological brain contains neurons and synapses. Neurons are connected with each other through synapse. Synapse connection depends on the firing rate of the neuron. Neurons that fire together are wired together. The biological brain is very complex to model as neurons and synapses are very dynamic in nature. Two biological brain fault tolerant techniques have been adopted in the NoC (i.e., synaptogensis and sprouting).

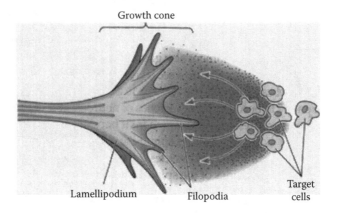

Figure 8.7 Synaptogenesis concept.

8.7.1 *Synaptogenesis*

Synaptogenesis is the concept of the biological brain. It is a self-adapting mechanism of the biological brain when two neurons want to connect and communicate with each other. In this phenomenon, the growth cone (having lamellipodium and filopodia) present at the top of the axon and dendrites terminal finds the path to the target neuron. The filopodia actually finds the path for connection with the target neuron. Chemical attractant is released by the target neuron to attract the growth cone. A synapse is formed between the source and target neuron due to this biological method. The synaptogensis process is shown in Figure 8.7.

8.7.2 *Sprouting*

Sprouting is the self-healing concept of biological brain. Due to various biological reasons that include blood clotting, hypertension, and high blood pressure, neurons and synapse get damaged in the biological brain. Then, a new synapse or sprout emerges from the axon that is already connected with the source neuron. With the help of chemical adhesion material (CAM) this sprout is connected with the target neuron as shown in Figure 8.8. After various experiments on fishes and frogs (amphibians), scientists concluded that this feature is not present in humans.

8.8 Bio-Inspired NoC Techniques

These two biological brain techniques are adopted in NoC to make it fault tolerant and efficient. In the first phase, the synaptogenesis algorithm was

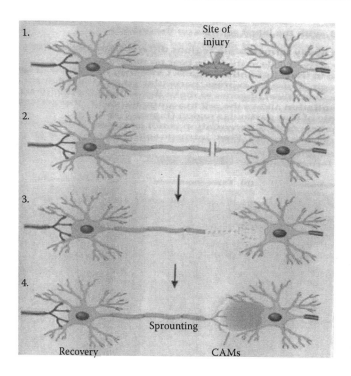

Figure 8.8 Sprout concept.

implemented in NoC; later, the algorithm was improved by adopting the sprouting concept. The performance of the sprouting based algorithm was better as compared to the synaptogenesis algorithm. In the upcoming sections, we will name and call the improved algorithm the "sprouting" based algorithm.

8.8.1 Synaptogenesis based NoC

This algorithm helps establish an optimal synapse (connection) between source PE and destination PE. It is able to detect the static faults during the connection establishment. The router exchanges two hop information during the connection establishment using multiple signals. If an interconnect or router becomes faulty during the communication between source and destination, a newer synapse is formed from the neighbor router. This is possible as the destination PE and source PE are constantly communicating with each other. The destination address in the flits helps to recover from the faults. Synaptogenesis algorithm takes 3 ns to detect one fault. In order to improve the fault detection and performance of NoC, a sprouting algorithm was proposed.

8.8.2 Sprouting based NoC

With the help of the sprouting algorithm, a synapse (sprout) emerged from the synapse already connected with the source PE when a router or interconnect becomes faulty during the communication. This algorithm detects the runtime faults during the communication and creates a shorter synapse to bypass the faulty interconnects or routers. This shorter synapse avoids the unnecessary traversal of routers and tries to connect with an older synapse, or directly with the destination if an older synapse does not exist.

These techniques recover from the static and runtime faulty routers and interconnects by creating a shorter synapse. Multiple synapses are formed between source PE and destination PE, which efficiently utilizes the bandwidth, maximizes the throughput, and recovers from the faults quickly. A sprout takes 1 ns to detect a fault. This helps the algorithm to quickly create a shorter synapse as compared to synaptogenesis algorithm. The sprout emerged as the destination address and is saved in every flit. Because of this, the overall performance of the sprouting algorithm is better as compared to the synaptogenesis algorithm.

These two biologically inspired techniques are implemented using two connection setups or communication services.

8.9 Connection Setups

Guaranteed throughput service (GT) and best-effort services (BE) are two broad categories of connection services in NoC.

8.9.1 Guaranteed throughput services (GT)

In GT connections, resources are reserved for a particular time period between a specific source and destination pair. The resources include channel bandwidth, routers and PEs time slot. In GT connections, the bandwidth is reserved for guaranteeing the throughput to particular connections between source and destination pair. GT connections may underutilize the network resources at certain times. This makes the GT connections an expensive option. At certain times, routers and PEs send the burst of data on these GT connections and then it remains silent for certain periods of time. This leads to the underutilization of the network resources. That is why BE services complement the GT services by utilizing the unused bandwidth. Video processing is an example of GT connection. This implies that GT connections are usually preferred for real time critical traffic applications [17,31].

8.9.2 Best-effort services (BE)

Bandwidth is not reserved in BE connections. BE uses the unused bandwidth from the GT connections. BE connections do not provide any guarantee of the

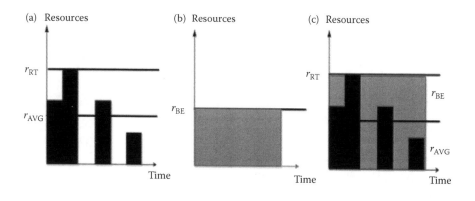

Figure 8.9 GT and BE services mechanism [31].

bandwidth. BE connections efficiently utilizes the bandwidth and resources as they are designed for average case scenarios as compared to the GT worst case mechanism. An example of BE connections are cache updates. This shows that BE connections are preferred for non-critical traffic [17,31].

Figure 8.9 shows the mechanism of GT and BE connections. In Figure 8.9a the bandwidth of the NoC is not properly utilized by the GT connections. The white space shows the unused bandwidth of NoC. BE connections are utilizing this unused bandwidth as shown in Figure 8.9c. Figure 8.9b shows the average bandwidth requirement of BE connections.

The performance of the sprouting algorithm was better as compared to the synaptogenesis algorithm due to quick fault detection and recovery. This affects the overall performance of the NoC. The synaptogenesis algorithm was implemented on router and later on interconnects faulty cases. The performance of interconnect base faulty mechanism was better. Later, the bio-inspired algorithm was improved by adopting the mechanism of sprouting. The sprouting was applied on interconnect faulty cases. The sprouting based bio-inspired algorithm was performing better as compared to the synaptogenesis algorithm.

The improved bio-inspired algorithm (combination of synaptogenesis and sprouting technique) is also implemented using GT communication setups by time division multiplexing (TDM). The bandwidth utilization of BE based connections was better as compared to GT based connections. This is normal as a packet switching mechanism utilizes the completely available bandwidth. The throughput utilization of GT based connections was better as compared to BE based algorithms due to the dedicated path between source and destination PE. Similarly, due to the dedicated path between source and destination the inter-flit arrival time of GT based connections is lesser as compared to BE based connections.

In this chapter, synaptogenesis and sprouting algorithms having BE based communication services are explained.

8.10   Bio-Inspired NoC Framework

Heterogeneous Network on Chip simulator (HNOCS), an OMNET++ based simulator [4], is used to implement the bio-inspired algorithms. HNOCS is a state of the art simulator as it supports both synchronous and asynchronous communication. HNOCS is open source and it supports all topologies. It supports parallelism and heterogeneous devices.

The bio-inspired NoC technique works on a per link basis. Whenever a faulty router or interconnect is encountered at the scheduler of the port, the scheduler initiates a new synapse connection (sprout) from the current router to the destination and tries to connect with the older synapse. Whenever a fault occurs, the destination address, routing path, and routers traversed are saved in the new synapse. These parameters help the newer synapse to connect with the older synapse. This bio-inspired algorithm is robust as it tries to connect with the older synapse and also avoid the traversal of unnecessary routers. This also decreases the latency of the packets even when faults occur. The bandwidth of the NoC is also efficiently utilized and throughput is slightly decreased for shorter periods of time as the NoC recovers from the faulty router. During the network recovery time the destination is still receiving the flits from other synapses. These flits have already traversed the faulty router. The general framework of the bio-inspired algorithm is shown in Figure 8.10.

Initially, the synapse is initiated from the source PE towards the destination PE. The synapse signal is received at the router. In return the router sends signals to neighbor routers to have two hop information. Upon receiving all information, the router priorities the path based on the direction of the destination. The synapse is routed from the NoC and reached at the destination. The destination in return sends the route reply signal. After receiving the route reply signal at source PE, the flits are sent from source PE to destination PE. If no fault is detected during the communication then flits are reached at the destination or the neighbor router detects the fault and initiates the temporary connection between current router to the older synapse or to the destination router. If there is no older synapse to connect with, the sprout is directly connected with the destination PE. Detailed information about each step is provided in Section 8.12.

8.11   Bio-Inspired NoC Network

In bio-inspired NoC, routers are connected with each other through bi-directional links. Each PE is connected with the router through the network interface. There

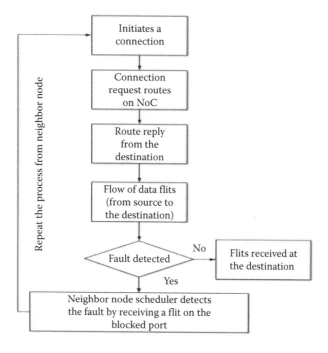

Figure 8.10 Bio-inspired NoC architecture.

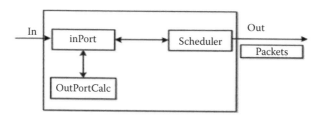

Figure 8.11 Port architecture.

are five ports in the router. Four ports of a router are connected with the neighbor routers (i.e., north, east, west and south), while the fifth port is connected with the PE. Every port has an "inPort" that receives packets from the neighbor router or from the PE. The "OutPortCalc" is used to calculate the output port for "synap" signal. The "Scheduler" is used to control the output port as shown in Figure 8.11. The "Scheduler" is connected with the "inPort" for disseminating the fault detection information and data flits to the "inPort" of that particular port. Later, these data flits are sent on the new shorter synapse initiated by "inPort".

The five ports in the router are connected with each other through cross bar interconnections. Every port is connected with every other port in the router to share control information and data flits. The bio-inspired algorithm is implemented on routers, NI, and interconnects.

8.12   Bio-Inspired NoC Algorithm

In NoC, routers discover themselves at the start of the simulation. The network discovery is initiated by the "Scheduler" of the router port. The scheduler sends the request signals to the neighbor router, which in return replies with response control signals. The response signal is sent by the "inPort" module of the particular router port. After sharing, the control signal routers know about the status of their neighbor routers (i.e., either they are working or not working). If after 1ns, the response is not received at the "Scheduler" port of the router then that port is blocked as either the interconnect or neighbor router is faulty. A bio-inspired NoC algorithm only needs one network discovery at the start of the simulation. This efficiently utilizes the bandwidth of the NoC and improves the overall performance. After network discovery, the source PE initiates the synapse "synap" signal to connect with the destination PE. Static and run time faults are detected by the "synap" signal. The static faults are detected by the synaptogenesis algorithm during the connection establishment, while the sprouting algorithm detects the run time faults. This makes the bio-inspired algorithm fault tolerant and reliable.

When the router receives the synap signal, it sends the information set (IS) request signals to its neighbors through the "Scheduler." Neighbor routers reply with the IS response signals having their information (direction neighbor, or DN) and their immediate neighbor's (IN) through the "inPort." The immediate neighbors for an ISW signal are east, north, and south, while west, north, and south are the IN's of ISE signals. Similarly, north, east, and west are the immediate neighbors of ISS signal, while ISN has IN's of south, east, and west. The neighbor router sends back the information in ISW (west), ISE (east), ISS (south), and ISN (north) response packets as shown in Figure 8.12. The neighbor router's information is gathered through the network discovery at the start of the simulation. Every port of a particular router has this information visible. The information in IS response packet is 1(working), 0 (not working), or -1 (neighbor doesn't exist). Based on this information the router decides at which direction the packet should be sent [26].

When an interconnect or router become faulty, a neighbor router detects the fault through "Scheduler." A new sprout (synapse) emerges from the neighbor router "inPort" to by-pass the faulty interconnect or router. The following code is used to implement the previously mentioned concept. The code is executed whenever a faulty router or interconnect is encountered. The port of the router

ISW

DN	East	North	South

ISE

DN	West	North	South

ISS

DN	North	East	West

ISN

DN	South	East	West

Figure 8.12 IS packet format.

connected with that particular faulty router or interconnect is blocked through "Scheduler" and "sourceInform" is sent to "inPort." The "sourceInform" is used to inform and initiate a new shorter synapse from the current router to bypass the faulty interconnect or router.

```
cModule* curRouter=getParentModule()->getParentModule();
cModule* curPort=getParentModule();
int portN=curPort->getIndex();
EV<<"Blocking particular port due to fault"<<endl;
if (portN==0)
{
    curRouter->par("southPort_W")=0;
}
else if (portN==1)
{
    curRouter->par("westPort_W")=0;
}
else if (portN==2)
{
    curRouter->par("northPort_W")=0;
}
else if (portN==3)
{
    curRouter->par("eastPort_W")=0;
}
```

```
else if (portN==4)
{
   curRouter->par("corePort_W")=0;
}
EV<<"Informing inport to initiate a new shorter synapse"
<<endl;
sourceInform=new NoCInformSourceMsg("Inform Source");
sourceInform->setKind(NOC_INFORMSOURCE_MSG);
sourceInform->setSrcId(srcId);
sourceInform->setDstId(dstId);
sourceInform->setTotalCount(totalCount);
sourceInform->setSynpseId(synpseID);
for (int i=0; i<totalCount;i++)
{
int outPort=msg->getSynpOutPort(i);
sourceInform->setSynpOutPort(i,outPort);
}
for (int i=0; i<totalCount;i++)
{
int routerIndex=msg->getRouterIndex(i);
sourceInform->setRouterIndex(i,routerIndex);
}
EV<<"Sending a signal to inPort from Scheduler"<<endl;
send(sourceInform,"sched_link$o");
```

The difference between the "synaptogenesis" and "sprouting" algorithm is the detection and informing of the "inPort" about fault occurrence. In synaptogenesis, it takes 3 ns to detect fault and to send a "sourceInform" signal to "inPort" and initiate a new synapse signal. This concept was improved in sprouting by changing the logic of fault detection code. The "sourceInform" signal is sent to "inPort" as soon as detection of the fault occurs.

In the synaptogenesis algorithm, fault is detected after the flit is sent to the neighbor router, and credit is not received from the neighbor router, in 3 ns. The timer (popRouterFaulty) is used to wait 3 ns for the credit response from the neighbor router. If the credit is not received then it means either the neighbor router or interconnect is faulty. In the sprouting algorithm, this timer is not used due to optimization and sprout concept implementation. Rather than detecting a fault, after sending the flit and waiting for credit response, the sprouting algorithm checks for the faulty interconnect and router before sending the flit to out link. This helps the sprouting algorithm avoid wasting 3 ns for the timer.

In BE services, the credit based flow control technique is used to control the flow of flits between routers and PEs. In this mechanism, the scheduler looks at the credit counter before sending the flit to the output port. If the credit is more than zero, it can send the flits at that particular port; zero credit means the port

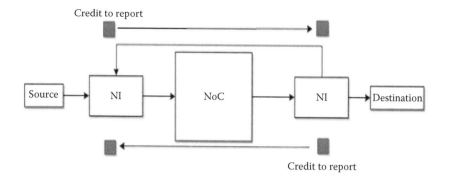

Figure 8.13 Credit based buffer management.

is busy and waiting for a feedback (credit) from its neighbor. The exchange of credit information also controls the links from congestion, collision, and avoids the deadlock situation. Figure 8.13 shows the credit based buffer management scheme. This technique is employed between adjacent routers and NI [18].

After receiving a flit at "inPort", it sends the credit back to the "Scheduler" to increment the virtual channel (vc) credit counter. The credit counter is incremented by one, as only one flit is sent by the scheduler per clock cycle. Currently, there are two vc's supported between routers to send flits on it.

```
char credName[64];
sprintf(credName, "cred-\%d-\%d", vc, numFlits);
NoCCreditMsg *crd = new NoCCreditMsg(credName);
crd->setKind(NOC_CREDIT_MSG);
crd->setVC(vc);
crd->setFlits(numFlits);
crd->setSchedulingPriority(0);
send(crd, "in$o");
```

At the receiver side, this credit message is received at the "Scheduler" and the credit counter is increased by one as specified by "numFlits" variable.

```
int vc = msg->getVC();
int num = msg->getFlits();
credits[vc] += num;
```

Algorithm 1 is the pseudo code of the overall bio-inspired algorithm. This algorithm only specifies one case when the destination is towards the east. Details about other cases are presented in the upcoming paragraphs. As previously mentioned, various changes were also made in the other components of NoC, which include "inPort," "Scheduler," and "OutPortCalc" to efficiently implement the code.

Algorithm 1: Bio-inspired algorithm(when destination is towards east)

1. Synap signal initiated from source to destination.

2. If (Destination towards East)

3.   Check East DN and INs condition

4. Else if //East DN or IN's Faulty

5.   Check North DN and INs condition

6. Else if //North DN or IN's Faulty

7.   Check South DN and INs condition

8. Else if //South DN and IN's Faulty

9.   Check West DN and INs Condition

10. Synapse connection is formed between source and destination.

11. Data flits move from source to destination over the synapse.

12. Upon router or interconnect failure neighbor router detects the fault.

13. The flits are saved in the "Scheduler" and later at "inPort" module of the router.

14. Router initiates a synap connection (sprout) and tries to connect the new synapse with the older synapse.

15. If older synapse does not exist then the new synapse directly connects with the destination.

16. Data flits are sent from the shorter newer synapse.

The "synap" signal is initiated from the source having destination address. The router sends ISW, ISE, ISS, and ISN signals to neighbor routers upon reception of the synap signal. Upon receiving all IS signals from neighbors, the router priorities the path based on the information present in the IS signals. If DN is zero (not working) then that path is blocked immediately. If the DN is 1 (working) and all three IN's are faulty, still the path is blocked as the "synap" will not be able to pass through the neighbor routers. If the DN is 1 and any one of the IN's is working the synap still sends signals in that particular direction. This algorithm can be optimized further by controlling the flow of the signal when more INs — rather than only one — are working.

Following are the five broad cases that may occur based on the information received and processed by a router using the information present in ISW, ISE, ISS, and ISN signals.

Case 1

If the destination is towards the east, then the first priority is to send the "synap" signal towards the east. If the east router or interconnect is faulty then the next priority will be to send the flit towards the north. If the north router is faulty, then the next priority is to send the "synap" signal towards the south. The least priority will be to send the "synap" signal towards the west, which is in the opposite direction from the destination.

Case 2

If the destination is towards the west, then the first priority is to send the "synap" signal towards the west. If the west router or interconnect is faulty then the next priority will be to send the flit towards the north. If the north router is faulty then the next priority is to send the "synap" signal towards the south. The least priority will be to send the "synap" signal towards the east, which is in the opposite direction of the destination.

Case 3

If the destination is towards the north, then the first priority is to send the "synap" signal towards the north. If the north router or interconnect is faulty then the next priority will be to send the flit towards the east. If the east router is faulty then the next priority is to send the "synap" signal towards the west. The least priority will be to send the "synap" signal towards the south, which is in the opposite direction from the destination.

Case 4

If the destination is towards the south, then the first priority is to send the "synap" signal towards the south. If the south router or interconnect is faulty then the next priority will be to send the flit towards the east. If the east router is faulty then the next priority is to send the "synap" signal towards the west. The least priority will be to send the "synap" signal towards the north, which is in the opposite direction of the destination.

Case 5

If current router x-axis and y-axis are equal to the target router (destination PE) x-axis and y-axis, then then the "synap" signals are sent towards the local core (PE).

After the establishment of synapse between source PE and destination PE, data flits are sent from source to destination over the path specified by the "synap" signal. The path is saved in every flit to address the fault tolerance. If

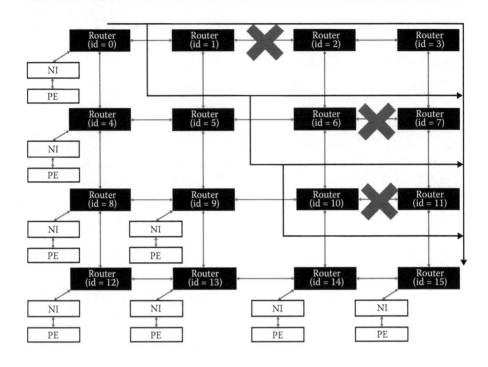

Figure 8.14 Multiple connections between source and destination.

there is no faulty router or interconnect on the way then the communication will be moving in the same manner. If during the communication a router or interconnect become faulty the scheduler of the neighbor router detects the fault and initiates a shorter synapse (sprout) from the current router to the older synapse or until the destination is reached. The old synapse information can be gathered from the flit trying to traverse through the faulty router or interconnect.

Figure 8.14 shows the implementation of the bio-inspired algorithm when interconnect failure occurs. The synapse was constructed from source (0) to destination (15) at the start of the communication between them. During the simulation the links between various routers became faulty and a shorter synapse was initiated to bypass the faulty interconnects. The temporary synapse was formed between router (1) and router (7) to bypass the faulty interconnect between router (1) and router (2). As shown in the figure, the shorter synapse is bypassing router (2) and router (3) and connecting directly with the old synapse at router (7). The same mechanism is adopted for any number of faults in the NoC. The crosses in the figure show the faulty interconnects. The bio-inspired algorithm is applicable to multiple parallel faults and the same recovery mechanism is used to recover from faults.

8.13   Results and Discussion

The bio-inspired algorithm is tested with various faulty routers and interconnects introduced during the simulation. The faults were introduced using the uniform random distribution formula. The multiple faults were introduced at the same time. The seed value of the random distributed formula was changed in each iteration to make it more randomized. In order to get the actual results without transient effects, the simulation was run 35 times for a 95% confidence interval. The simulation was repeated 35 times per fault and for every injection rate. The simulation was run multiple times based on the confidence interval measurement [8].

The injection rate of the packet was changed from 10% to 100%. The injection rate is the percentage utilization of the interconnect bandwidth. The injection rate is linked to the bandwidth of the interconnect connected with the PE. In this simulation, the interconnect bandwidth is 16Gbps. The injection rate can be varied by changing the number of idle cycles between the packets of wormhole switching [43].

In order to test the bio-inspired NoC algorithm, we took 4×4 NoC. The topology of the NoC is mesh, although this algorithm is scalable to multiple NoC sizes and topologies. Another reason to select 4×4 NoC is that most of the techniques had been implemented in this topology. The clock frequency of the NoC is 500 MHz. The flit size is 4 bytes and the packet length is 10 flits. In each iteration, different sources are injecting a various number of packets. A total of 100,000 packets or 1,000,000 flits are injected in the NoC to fully utilize the bandwidth of the NoC. The algorithm is tested using uniform and random traffic patterns. The simulation is run 50 ms per iteration. A wormhole switching mechanism is used in which a packet is divided into header, body, and tail flits. The result shows that the algorithm is fault tolerant as the NoC quickly recovers from the faults. The bandwidth, throughput and latency degrade gracefully during the network recovery from faults.

We first implemented the synaptogenesis algorithm on the faulty router and interconnect basis. The synaptogenesis algorithm was performing well on the faulty interconnect basis as compared to a faulty router basis. Later, the algorithm was improved by adopting the sprouting concept in it. The algorithms were implemented using BE communication service levels. The improved algorithm was applied on a faulty interconnect basis and its performance was better as compared to the synaptogenesis algorithm.

8.13.1   *Synaptogenesis algorithm using BE services on per faulty router basis*

The bio-inspired algorithm was implemented using BE communication service. In this NoC architecture, the algorithm does not reserve any resources on the

NoC. The packet switching concept is used along with the credit based flow control technique. The routers decide based on the situation of the NoC in which direction the packet should be sent. The bio-inspired algorithm performed well during the recovery from faulty routers. The algorithm efficiently utilized the throughput, and bandwidth during the recovery of faults. The bandwidth, throughput, and latency of the NoC were degrading gracefully during the recovery from faults.

In this case, the complete router is blocked when it becomes faulty. Due to this all five ports of the router are blocked, and they are not accessible for traversing the packet to the neighbor routers. Due to this, the number of possibilities for routing is also reduced, as all five ports are blocked. This also reduces the performance of the NoC as compared to the interconnect faulty case. In this simulation scenario, we only tested the code for one source and one destination at a time. The faults were introduced on the particular path to test the algorithm. In order to initially test algorithm, the simulation scenarios were predefined.

The average bandwidth of the NoC decreases very slightly as the number of faulty routers increases, as shown in Figure 8.15. The figure shows the gradual graceful degradation of the bandwidth as the number of faults is increased. The graceful degradation is due to quick recovery from the faulty routers. The average bandwidth utilization dropped by 1.37% during the recovery of faults.

Figure 8.16 shows a slight increase of the latency of the packets as the number of faulty routers increases. The average latency of the flits is increased by 25% during the recovery of faults.

Figure 8.17 shows the increase of interflit arrival time at the destination when the numbers of faults are increasing. The interflit arrival time is increased from

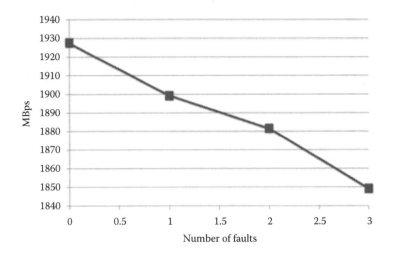

Figure 8.15 Bandwidth vs faults.

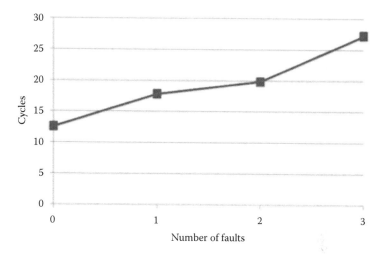

Figure 8.16 Latency vs faults.

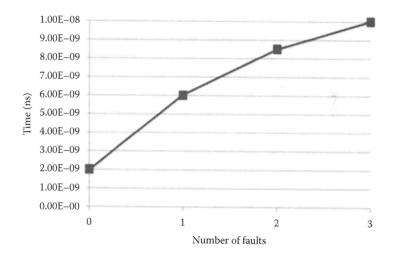

Figure 8.17 Interflit arrival time.

2 to 10 ns as faults increase from zero to seven. The percentage increase is 86.44%. The increase of the interflit arrival time is due to the blockage of complete router and its ports. This in turns limits the possible number of routes from source PE to destination PE.

Figure 8.18 shows that the packet network latency increases as the number of faults increases. Different amounts of faults were introduced during the

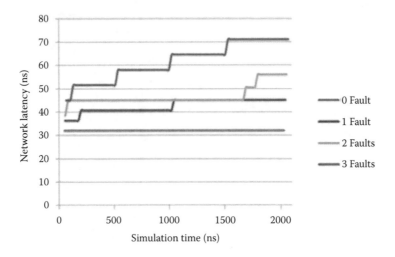

Figure 8.18 Packet network latency.

simulation. The packet network latency increases as the router has to construct a new synapse for every fault. The average packet network latency gradually increases for various amounts of faults. The figure shows the only particular case of faulty routers introduced during the simulation. The packet network latency is 32 ns for all the packets when there was no fault on the NoC. It increases to a maximum of 45 ns when one fault was introduced. When two faults were introduced during the simulation, then the packet network latency increased to a maximum of 58 ns. Similarly, when three faults were introduced the packet network latency increased to a maximum of 71 ns. The overall increase of packet network latency is 30.64%. The transition in the graph shows the increase in the packet network latency due to faults.

Figure 8.19 shows the throughput of the NoC when various routers were made faulty. The throughput decreases for a very short period of time as the network constructs the new synapse. The fault recovery time depends on the number of routers needed to traverse to establish a new synapse multiplied by the time period of a clock. The formula is shown in Equation 8.1.

$$\text{Fault recovery time} = \text{Clock Period} \times \text{Number of routers} \qquad (8.1)$$

The time at which the flit is received at the destination after fault recovery depends on the clock period and total number of routers to traverse, as shown in Equation 8.2.

$$\text{Time to receive flit at destination} = (\text{Fault recovery time})$$
$$+ \text{Clock period} \times \text{Number of routers}$$
$$(8.2)$$

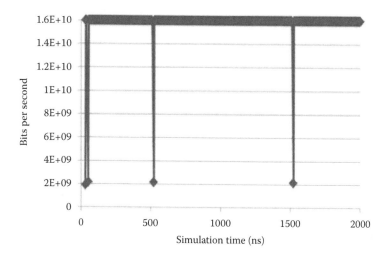

Figure 8.19 Throughput of NoC when there were three faulty routers.

For the scenario mentioned in Figure 8.19, it takes 15 ns to recover from faults and for the first flit to be received at the destination. This shows the quick recovery from the fault and the reception of flits at the destination. Moreover, during the recovery from faults, the flits are traversing on the older synapse; that is why the throughput degrades slightly and quickly recovers the maximum data rate when the new synapse is formed. The throughput is dropped to 2.133 Gbps when three faults occur during the communication. The throughput is dropped to 2.133 Gbps at 54, 523, and 1522 ns, while the NoC quickly regains the maximum throughput of 16 Gbps after recovery of the faults. This shows the robustness of the bio-inspired algorithm, and that it quickly recovers from faults. The throughput is dropped by 86.87% during the recovery from faults. The throughput is dropped so much because of complete blockage of the router and all of its ports. This in turns limits the possible number of routes from source PE to destination PE and degrades the throughput utilization.

8.13.2 Synaptogenesis algorithm using BE services on per faulty interconnect basis

The bio-inspired algorithm performed well during the recovery from faulty interconnects. The algorithm efficiently utilized the throughput and bandwidth during the recovery of faults. The bandwidth, throughput, and latency of the NoC were degrading gracefully during the recovery from faults. In this simulation scenario, we only tested the code for one source and one destination at a time. The faults

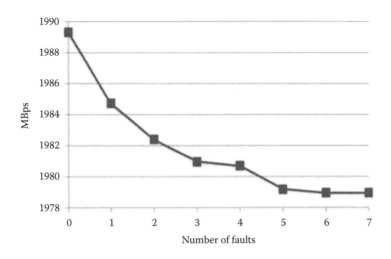

Figure 8.20 Bandwidth vs faults.

were introduced on the particular path to test the algorithm. In order to initially test the algorithm, the simulation scenarios were predefined.

The average bandwidth of the NoC decreases very slightly as the number of faulty interconnects increases, as shown in Figure 8.20. The figure shows the gradual graceful degradation of the NoC when faults occur due to the efficient algorithm. When the number of faults is higher than five, the bandwidth drop stabilizes. The reason is that the number of available paths between source and destination is being reduced as the number of faults is increased. The source and destination are connected with more direct and shorter paths. The average bandwidth utilization drops by 0.07% during the recovery of faults.

As can be seen in Figures 8.15 and 8.20, when there were no faulty routers and interconnects, respectively, the bandwidth utilization of the interconnect based algorithm was better as compared to the router based bio-inspired algorithm. The bandwidth utilization is increased to 1989 Mbps from 1927 Mbps. Similarly, when one fault occurred during the simulation, the bandwidth utilization of the interconnect based algorithm is better. The bandwidth utilization is even better when seven interconnects become faulty, as seen in the three routers faulty case. The overall bandwidth utilization increase is 5.03% from router to interconnect, based on the bio-inspired routing algorithm. The reason is that in the interconnect based algorithm there are more routing paths as compared to the faulty router algorithm. Therefore, the bio-inspired algorithm can adapt to any alternate path to bypass the fault.

Figure 8.21 shows the slight increase of the latency of the packets as the number of faulty interconnects increases. The average latency of the flits is increased by 14.07% during the recovery of faults.

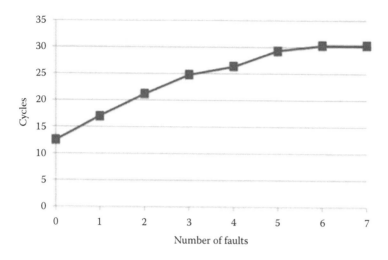

Figure 8.21 Latency vs faults.

Similarly, the latency of the interconnect based bio-inspired algorithm is less as compared to the router based bio-inspired algorithm. This is normal behavior as all the ports of the router are blocked, and no routing is possible through complete router. This increases the latency of the packet and flits. This is shown in Figures 8.16 and 8.21. The latency of the flits in the interconnect based algorithm was reduced to almost 24 cycles from 27 cycles (router based algorithm) when three faults were introduced during the simulation. The overall latency decrease is 35.16% from the router to interconnect based bio-inspired routing algorithm.

Figure 8.22 shows the gradual increase of interflit arrival time at the destination when the number of faults is increasing. The interflit arrival time is increased from 2 to 8.78 ns when going from zero to seven faults. The percentage increase is 32.35%.

Similarly, the interflit arrival time is reduced to 7.4 ns from 10 ns when three interconnects and routers become faulty in the interconnect and router based bio-inspired algorithm, respectively. The overall interflit arrival time decreases 19.89% from the router to interconnect based bio-inspired routing algorithm. This is shown in Figures 8.17 and 8.22. This is the normal behavior as the interflit arrival time should be more for complete router blockage as compared to one interconnect blockage.

Figure 8.23 shows that the packet network latency increases as the number of faults increases. Different amounts of faults were introduced during the simulation. The packet network latency increases as the router has to construct a new synapse for every fault.

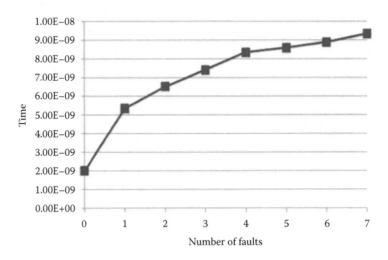

Figure 8.22 Interflit arrival time.

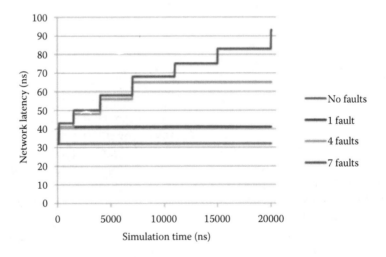

Figure 8.23 Packet network latency.

In Figure 8.23, few cases are shown for clarity. The average packet network latency gradually increases for various amounts of faults. The packet network latency is 32 ns for all the packets when there was no fault on the NoC. It increases to a maximum of 41 ns when one fault was introduced. When two faults were introduced during the simulation, then the packet network latency is increased to a maximum of 50 ns. Similarly, when three faults were introduced the packet network latency increases to a maximum of 58 ns. The packet network latency increases to 65, 75, 83, and 93 ns for four, five, six, and seven faults, respectively. The overall increase of packet network latency is 11.64%. The transition in the graph shows the increase in the packet network latency due to faults.

Figures 8.18 and 8.23 show the packet network latency graphs of the bio-inspired algorithm. The average packet network latency of the improved bio-inspired interconnect based algorithm reduced from 13 to 8.4 ns as compared to the router based algorithm. The overall packet network latency decreases 8.22% from the router to interconnect based bio-inspired routing algorithm.

Figure 8.24 shows the throughput of the NoC when various interconnects were made faulty. The throughput decreases for a very short period of time as the network finds the newer synapse. The flits are traversing on the older synapse; that is why the throughput degrades slightly and quickly recovers the maximum data rate when the new synapse is formed.

The throughput is dropped to 2.46 Gbps when the first fault occurs during the communication. The throughput is dropped to 3.56, 3.20, 2.67, 3.56, 3.20, and 2.67 Gbsp for two, three, four, five, six, and seven faulty interconnects, respectively, which occurred during the simulation. The average recovery time from faults and reception of flits at the destination is 10.72 ns. After this time the NoC quickly regains the maximum throughput of Chabe to 16 Gbps. The recovery time depends on the summation values of Equations 8.1and 8.2, as mentioned before. This shows the robustness and quick recovery from faults of the bio-inspired algorithm. The throughput is dropped by 39.32% during the recovery from faults.

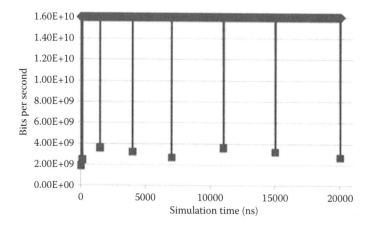

Figure 8.24 Throughput of NoC when there were seven numbers of faults.

The throughput graphs of Figures 8.19 and 8.24 show that the throughput of the improved interconnect based bio-inspired algorithm also increases from the previous algorithm of router basis. The recovery time from the fault is also reduced, as can be seen in the figures. The overall throughput increases 44.02% from the router to interconnect based bio-inspired routing algorithm.

The synaptogenesis algorithm was updated by adapting the concept of sprouting as explained in Section 8.8.2. The fault detection and recovery was improved as it takes 1ns to recover from fault as compared to 3 ns in the synaptogenesis algorithm. The performance of the synaptogenesis based bio-inspired algorithm was better when applied to the interconnect faulty case as compared to the router fault case. Therefore, the improved bio-inspired algorithm was applied to the interconnect base faulty cases.

8.13.3 Sprouting algorithm using BE services on a faulty interconnect basis

The synaptogenesis algorithm was improved by adopting the sprouting concept in NoC. The improved algorithm is the combination of synaptogenesis and sprouting. The improved bio-inspired algorithm is implemented in NoC using BE communication service. The improved sprouting based bio-inspired algorithm is performing well as compared to the synaptogenesis algorithm. The fault recovery time in the improved algorithm is better as compared to the synaptogenesis algorithm. This in turns affects the bandwidth, throughput, and latency of the NoC.

The bandwidth of the NoC degrades gracefully as the number of faults increases as shown in Figure 8.25. When the number of faults increases above

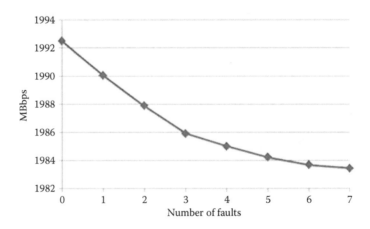

Figure 8.25 Bandwidth vs faults.

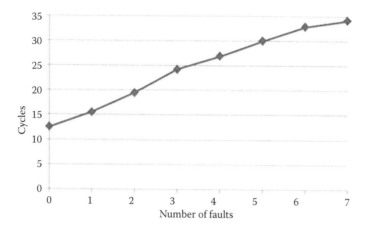

Figure 8.26 Latency vs faults.

five, the bandwidth drop stabilizes. The reason is that the number of paths available between the source PE and destination PE is reduced as the number of faults increases. The source and destination is connected with more directed and smaller paths. The average bandwidth utilization of the NoC is dropped by 0.05% during the recovery of faults.

Similarly, the latency of the bio-inspired algorithm is increased slightly as the number of faults increases. The average latency is increased by 12.91%. This is shown in Figure 8.26.

Figure 8.27 shows the gradual increase of interflit arrival time at the destination PE when the number of faults increases. The interflit arrival time is increased from 2 to 8.78 ns for zero to seven faults. The percentage increase is 28.27%.

Figure 8.28 shows that the packet network latency increases as the number of faults increases. Different numbers of faults were introduced during the simulation. The packet network latency increases as the router has to construct a new synapse for every fault. In Figure 8.28 a few cases are shown for clarity. The average packet network latency gradually increases for various numbers of faults. The packet network latency is 32 ns for all the packets when there was no fault on the NoC. It increases to a maximum of 38 ns when one fault was introduced. When two faults were introduced during the simulation, the packet network latency increased to a maximum of 46 ns. Similarly, when three faults were introduced the packet network latency increases to a maximum of 56 ns. The packet network latency increases to 62, 70, 80, and 86 ns for four, five, six, and seven faults, respectively. The overall increase of average packet network latency is 10.88%.

The throughput is dropped to 4 Gbps when the first fault occurs during the communication. The throughput is dropped to 3.2, 2.67, 4, 3.2, 2.67, and 4 Gbps

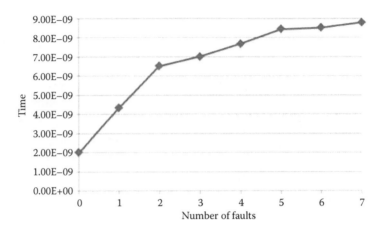

Figure 8.27 Interflit arrival time.

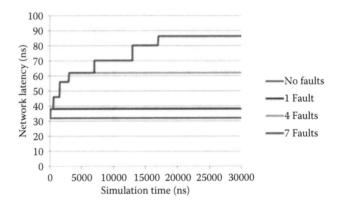

Figure 8.28 Packet network latency.

for two, three, four, five, six, and seven faulty interconnects, respectively, which occurred during the simulation. The average recovery time from faults and reception of flits at destination is 9.71 ns. After this time the NoC quickly regains the maximum throughput of 16 Gbps. The recovery time depends on the summation values of Equations 8.1 and 8.2 as mentioned before. This shows the robustness and quick recovery from faults of the bio-inspired algorithm. The transition in the graph shows increase in the packet network latency due to faults.

Figure 8.29 shows the throughput of the NoC when various interconnects were made faulty. The throughput decreases for a very short period of time as the network finds the newer synapse when fault occurs. The flits are traversing on the older synapse. That is why the throughput degrades slightly and quickly

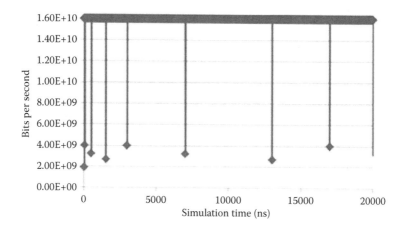

Figure 8.29 Throughput of NoC when there were seven numbers of faults.

recovers the maximum data rate when the new synapse is formed. The throughput is reduced by 37.22% during the recovery from faults.

While comparing the synaptogenesis algorithm with sprouting, we concluded that the sprouting algorithm is performing better in terms of the bandwidth and throughput utilization. In the sprouting algorithm, the average throughput and average packet latency were increased by 10.41% and 4.66%, respectively, while the average bandwidth was efficiently utilized by 0.24% as compared to synaptogenesis algorithm.

8.13.4   Synaptogenesis vs sprouting algorithm

In this section, the synaptogenesis algorithm is compared with the sprouting algorithm. Figure 8.30 shows that the average bandwidth of the synaptogenesis algorithm drops more as compared to the sprouting algorithm. The synaptogenesis algorithm takes more cycles to detect the fault while the sprouting algorithm quickly discovers the fault and the sprout emerges from the neighbor node. In synaptogenesis, as the number of faults increases from five, the bandwidth almost becomes stable because of direct and shorter paths between the source PE and destination PE, while in sprouting the decrease is continuous but slow.

Similarly, the latency of the bio-inspired techniques increases slightly as the number of faults is increasing, as shown in Figure 8.31. As the number of faults increases from four, the latency of the sprouting is increasing as compared to synaptogenesis; at the same time the bandwidth increase is higher for the sprouting algorithm.

Figure 8.32 shows the graceful degradation of average throughput as the number of faults increases. The figure shows the impact of the faults on the

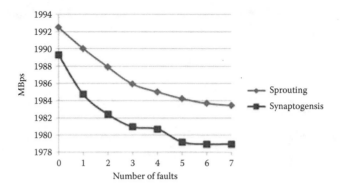

Figure 8.30 Bandwidth vs number of faults.

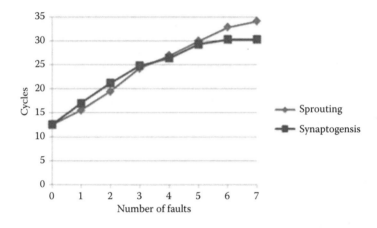

Figure 8.31 Latency vs number of faults.

throughput. The overall average throughput of the sprouting algorithm is better as compared to synaptogenesis. The quick recovery time of the sprouting algorithm from faults during the simulation is shown in Figure 8.33. The figure shows the inter flit arrival time for seven faulty interconnect cases. The interflit arrival time of the BE based synaptogenesis based algorithm is 5.88% greater as compared to the BE based sprouting algorithm.

8.13.5 Bio-Inspired algorithm vs literature techniques

The bio-inspired algorithms are also compared with the literature techniques. Table 8.1 shows the accepted traffic and average throughput values for the different algorithms. The values show that the bio-inspired algorithms are performing

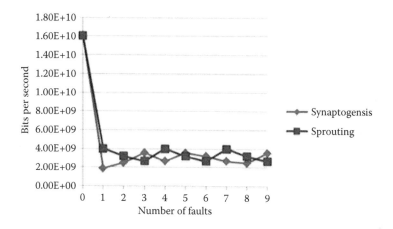

Figure 8.32 Throughput vs number of faults.

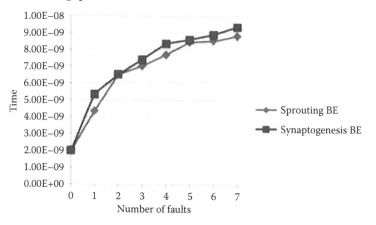

Figure 8.33 Inter flit arrival time vs simulation time.

better. Similarly, as the number of faults is increasing from four, the accepted traffic and average throughput of the synaptogenesis algorithm is better as compared to sprouting; still, they are performing better as compared to the literature technique mentioned.

The saturation point is analyzed in Table 8.1. The saturation point of the bio-inspired algorithm is better as compared to other fault tolerant techniques. The BE based synaptogenesis and sprouting algorithm have saturation points of 0.39 and 0.42 at 0 fault, which shows the efficiency and simplicity of the algorithm. This shows that the bio-inspired algorithm quickly finds the path from source PE to destination PE when there is no fault and data flits are sent over it. This saturation point is better as compared to all other techniques in Table 8.1. The

Table 8.1 Saturation points of techniques (4 × 4 NoC)

Techniques	Number of faults							
	0	*1*	*2*	*3*	*4*	*5*	*6*	*7*
Synaptogenesis algorithm [37,38,40]	0.39	0.31	0.29	0.28	0.27	0.26	0.25	0.24
Sprouting algorithm [36,38]	0.42	0.40	0.34	0.30	0.22	0.21	0.19	0.18
DBP network [23]	0.22	0.18	0.17	0.16	0.15	0.15	0.14	0.14
ODD-EVEN [9]	—	0.4	—	0.4	—	0.4	—	—
Inverted ODD-EVEN [9]	—	0.4	—	0.4	—	0.4	—	—
FTCAR [45]	0.31	—	—	0.25	—	0.24	—	—
HiPFaR [15]	0.37	0.35	—	—	—	—	—	—
Baseline [15]	0.38	0.32	—	—	—	—	—	—
RAFT2 algorithm using VOPD traffic [44]	0.26	—	0.23	—	—	—	—	—
RAFT2 algorithm using uniform traffic [44]	0.37	—	0.29	—	—	—	—	—
Fault aware dynamic routing algorithm [20]	0.2	0.2	0.2	—	—	—	—	—
Flooding algorithm [20]	0.10	0.10	0.10	—	—	—	—	—

accepted traffic rate of the bio-inspired algorithm is better as compared to the technique mentioned in Ref. [23] for different numbers of faults. Technique [23] is the most thoroughly analyzed, tested, and presented technique in the literature. The saturation point of the techniques mentioned in [9] is slightly higher in two faulty cases. The faulty cases are three and five. The saturation point of 0.4 for three different faulty cases is not common behavior in the literature. Usually the saturation point decreases as the number of faults is increased. The network saturates early as the network has less paths to traverse through, and they are occupied by all the PEs available on the NoC. Due to this NoC saturates early and it leads to overflow of buffers and queues. The bio-inspired NoC accepted traffic better than techniques [15,20,44,45] for different faulty cases.

8.14   Conclusion

These two biological brain techniques are adopted in NoC to make it fault tolerant and efficient. In the first phase the BE based synaptogenesis algorithm was implemented in NoC and later the algorithm was improved by adopting the sprouting concept. The synaptogenesis algorithm was tested by making routers and interconnects faulty. The performance of the sprouting algorithm was better as compared to synaptogenesis algorithm. In sprouting the fault detection mechanism was improved from the synaptogenesis algorithm. The synaptogenesis algorithm takes 3 ns to detect the faulty router and interconnect, while the sprouting algorithm detects the faults within 1 ns as the flits arrive at the faulty port of a particular router. The bio-inspired algorithm was also compared with the literature techniques and the accepted traffic rate (flit/cycle/node) was better as compared to other techniques.

Bibliography

[1] D. Atienza, F. Angiolini, S. Murali, A. Pullini, L. Benini, and G. De Micheli. Network-on-chip design and synthesis outlook. *INTEGRATION, the VLSI Journal*, 41(3):340–359, 2008.

[2] P. Bahrebar and D. Stroobandt. Adaptive and reconfigurable fault-tolerant routing method for 2D networks-on-chip. In *2014 International Conference on ReConFigurable Computing and FPGAs (ReConFig)*, pp. 1–8. IEEE, 2014.

[3] M. Bakhouya. Towards a bio-inspired architecture for autonomic network-on-chip. In *2010 International Conference on High Performance Computing and Simulation (HPCS)*, pp. 491–497. IEEE, 2010.

[4] Y. Ben-Itzhak, E. Zahavi, I. Cidon, and A. Kolodny. HNOCS: Modular open-source simulator for heterogeneous NoCs. In *2012 International Conference on Embedded Computer Systems (SAMOS)*, pp. 51–57. IEEE, 2012.

[5] L. Benini and G. De Micheli. Networks on chips: A new SoC paradigm. *Computer*, 35(1):70–78, 2002.

[6] T. Bjerregaard and S. Mahadevan. A survey of research and practices of network-on-chip. *ACM Computing Surveys (CSUR)*, 38(1):1, 2006.

[7] S.M. Breedlove, N. V. Watson, and M. R. Rosenzweig. *Biological Psychology*. Sinauer Associates, Incorporated Publishers, 2007.

[8] M. D. Byrne. How many times should a stochastic model be run? An approach based on confidence intervals. In *Proceedings of the 12th International Conference on Cognitive Modeling*, Ottawa, 2013.

[9] H. S. Castro and O. A. de Lima. A fault tolerant NoC architecture based upon external router backup paths. In *2013 IEEE 11th International New Circuits and Systems Conference (NEWCAS)*, pp. 1–4. IEEE, 2013.

[10] Y.-C. Chang, C.-T. Chiu, S.-Y. Lin, and C.-K. Liu. On the design and analysis of fault tolerant NoC architecture using spare routers. In *Proceedings of the 16th Asia and South Pacific Design Automation Conference*, pp. 431–436. IEEE Press, 2011.

[11] A. A. Chien and J. H. Kim. Planar-adaptive routing: Low-cost adaptive networks for multiprocessors. *Journal of the ACM (JACM)*, 42(1):91–123, 1995.

[12] G.-M. Chiu. The odd-even turn model for adaptive routing. *IEEE Transactions on Parallel and Distributed Systems*, 11(7):729–738, 2000.

[13] N. Choudhary. Network-on-chip: A new SoC communication infrastructure paradigm. *International Journal of Soft Computing and Engineering (IJSCE)*, 1(6):332–335, 2012.

[14] F. Dressler and O. B. Bio-inspired networking: From theory to practice. *IEEE Communications Magazine,* 48(11):176–183, 2010.

[15] M. Ebrahimi, M. Daneshtalab, and J. Plosila. High performance fault-tolerant routing algorithm for NoC-based many-core systems. In *2013 21st Euromicro International Conference on Parallel, Distributed and Network-Based Processing (PDP),* pp. 462–469. IEEE, 2013.

[16] C. J. Glass and L. M. Ni. The turn model for adaptive routing. In *ACM SIGARCH Computer Architecture News*, vol. 20, pp. 278–287. ACM, 1992.

[17] K. Goossens, J. Dielissen, and A. Radulescu. Æthereal network on chip: Concepts, architectures, and implementations. *IEEE Design & Test of Computers,* 22(5):414–421, 2005.

[18] A. Hansson, K. Goossens, and A. Rădulescu. Avoiding message-dependent deadlock in network-based systems on chip. *VLSI Design*, 2007, Article ID 95859, 10 p, 2007.

[19] A. Hashmi, H. Berry, O. Temam, and M. Lipasti. Automatic abstraction and fault tolerance in cortical microachitectures. In *ACM SIGARCH Computer Architecture News*, vol. 39, pp. 1–10. ACM, 2011.

[20] A. Hosseini, T. Ragheb, and Y. Massoud. A fault-aware dynamic routing algorithm for on-chip networks. In *IEEE International Symposium on Circuits and Systems, 2008. ISCAS 2008.* pp. 2653–2656. IEEE, 2008.

[21] J. W. Kalat. *BiologicalPsychology.* Wadsworth/Thomson Learning, Belmont, CA, 2001.

[22] Y. B. Kim and Y.-B. Kim. Fault tolerant source routing for network-on-chip. In *22nd IEEE International Symposium on Defect and Fault-Tolerance in VLSI Systems, 2007. DFT'07.* pp. 12–20. IEEE, 2007.

[23] M. Koibuchi, H. Matsutani, H. Amano, and T. M. Pinkston. A lightweight fault-tolerant mechanism for network-on-chip. In *Proceedings of the Second ACM/IEEE International Symposium on Networks-on-Chip*, pp. 13–22. IEEE Computer Society, 2008.

[24] T. Lehtonen, D. Wolpert, P. Liljeberg, J. Plosila, and P. Ampadu. Self-adaptive system for addressing permanent errors in on-chip interconnects. *IEEE Transactions on Very Large Scale Integration (VLSI) Systems,* 18(4):527–540, 2010.

[25] F. Moraes, N. Calazans, A. Mello, L. Möller, and L. Ost. Hermes: An infrastructure for low area overhead packet-switching networks on chip. *INTEGRATION, the VLSI Journal,* 38(1):69–93, 2004.

[26] C. Nicopoulos, V. Narayanan, and C. R. Das. *Network-on-Chip Architectures: A Holistic Design Exploration,* vol. 45. Springer, 2009.

[27] J. L Nunez-Yanez, D. Edwards, and A. M. Coppola. Adaptive routing strategies for fault-tolerant on-chip networks in dynamically reconfigurable systems. *IET Computers & Digital Techniques,* 2(3):184–198, 2008.

[28] S. Pasricha and Y. Zou. A low overhead fault tolerant routing scheme for 3D networks-on-chip. In *2011 12th International Symposium on Quality Electronic Design (ISQED),* pp. 1–8. IEEE, 2011.

[29] A. Patooghy and S. G. Miremadi. Complement routing: A methodology to design reliable routing algorithm for network on chips. *Microprocessors and Microsystems,* 34(6):163–173, 2010.

[30] M. Pirretti, G. M. Link, R. R. Brooks, N. Vijaykrishnan, M. Kandemir, and M. J. Irwin. Fault tolerant algorithms for network-on-chip interconnect. In *Proceedings of the IEEE Computer society Annual Symposium on VLSI, 2004.* pp. 46–51. IEEE, 2004.

[31] E. Rijpkema, K. Goossens, A. Rădulescu, J. Dielissen, J. van Meerbergen, P. Wielage, and E. Waterlander. Trade-offs in the design of a router with both guaranteed and best-effort services for networks on chip. *IEE Proceedings-Computers and Digital Techniques,* 150(5):294–302, 2003.

[32] F. Safaei and M. ValadBeigi. An efficient routing methodology to tolerate static and dynamic faults in 2-D mesh networks-on-chip. *Microprocessors and Microsystems,* 36(7):531–542, 2012.

[33] T. Schönwald, O. Bringmann, and W. Rosenstiel. Region-based routing algorithm for network-on-chip architectures. In *Norchip, 2007,* pp. 1–4. IEEE, 2007.

[34] T. Schonwald, J. Zimmermann, O. Bringmann, and W. Rosenstiel. Fully adaptive fault-tolerant routing algorithm for network-on-chip architectures. In *10th Euromicro Conference on Digital System Design Architectures, Methods and Tools, 2007. DSD 2007.* pp. 527–534. IEEE, 2007.

[35] M. R. Seifi and M. Eshghi. Clustered NoC, a suitable design for group communications in network on chip. *Computers & Electrical Engineering*, 38(1):82–95, 2012.

[36] M. A. J. Sethi, F. A. Hussin, and N. H. Hamid. Implementation of biological sprouting algorithm for NoC fault tolerance. In *2013 IEEE International Conference on Circuits and Systems (ICCAS)*, pp. 39–44. IEEE, 2013.

[37] M. A. J. Sethi, F. A. Hussin, and N. H. Hamid. Synaptogenesis based bio-inspired NoC fault tolerant interconnects. In *2013 IEEE International Conference on Control System, Computing and Engineering (ICCSCE)*, pp. 46–51. IEEE, 2013.

[38] M. A. J. Sethi, F. A. Hussin, and N. H. Hamid. Bio-inspired NoC fault tolerant techniques. In *2014 5th International Conference on Intelligent and Advanced Systems (ICIAS)*, pp. 1–6. IEEE, 2014.

[39] M. A. J. Sethi, F. A. Hussin, and N. H. Hamid. Survey of network on chip architectures. *Science International (Lahore)*, 5(27):4133–4144, 2015.

[40] M. A. J. Sethi, F. A. Hussin, and N. H. Hamid. Implementation and analysis of biological synaptogenesis technique on nodes and interconnects for NoC fault tolerance. *Maxwell Scientific Publication Corporation*, 2015(22):483–489, 2016.

[41] M. A. J. Sethi, F. A. Hussin, and N. H. Hamid. Noc architecture: A closer look. *Asian Journal of Information Technology*, 14(15):2531–2541, 2016.

[42] M. A. J. Sethi, F. A. Hussin, and N. H. Hamid. Review of network on chip architectures. *Recent Advances in Electrical & Electronic Engineering*, 10(1):4–29, 2017.

[43] L. Tedesco, A. Mello, D. Garibotti, N. Calazans, and F. Moraes. Traffic generation and performance evaluation for mesh-based NoCs. In *18th Symposium on Integrated Circuits and Systems Design*, pp. 184–189. IEEE, 2005.

[44] M. Valinataj. Evaluation of fault-tolerant routing methods for NoC architectures. In *2011 14th Euromicro Conference on Digital System Design (DSD)*, pp. 446–449. IEEE, 2011.

[45] M. Valinataj, S. Mohammadi, J. Plosila, and P. Liljeberg. A fault-tolerant and congestion-aware routing algorithm for networks-on-chip. In *2010 IEEE 13th International Symposium on Design and Diagnostics of Electronic Circuits and Systems (DDECS)*, pp. 139–144. IEEE, 2010.

[46] R. Ville, L. Teijo, and P. Juha. Network on chip routing algorithms. TUCS Technical Report, 2006.

[47] J. Wu. A fault-tolerant adaptive and minimal routing approach in n-D meshes. In *Proceedings of 2000 International Conference on Parallel Processing.*, pp. 431–438. IEEE, 2000.

[48] H. Zhu, P. P. Pande, and C. Grecu. Performance evaluation of adaptive routing algorithms for achieving fault tolerance in NoC fabrics. In *2007 IEEE International Conference on Application-Specific Systems, Architectures and Processors (ASAP)* pp. 42–47. IEEE, 2007.

Chapter 9

Bio-Inspired Network on Chip (BNOC)

Muhammad Athar Javed Sethi

University of Engineering and Technology, Peshawar

Fawnizu Azmadi Hussin and Nor Hisham Hamid

Universiti Teknologi PETRONAS, Malaysia

CONTENTS

Due to its nanoscale manufacturing process and the complex communication requirements of Network on Chip (NoC), various faults occur. Various fault tolerant techniques have been proposed in the literature to address the permanent and temporary faults. Permanent faults can only be removed by the redundant hardware, while temporary faults can be addressed by fault tolerant techniques.

Deterministic, stochastic, fully adaptive, and partially adaptive routing algorithms are four broad categories of fault tolerant techniques. These algorithms were not completely addressing the faults in NoC. The bio-inspired fault tolerant technique is proposed to address these faults by mimicking the fault tolerant techniques of the biological brain. In order to have a reliable, efficient, and robust communication in NoC, the interconnection mechanism of the human brain is adopted. Human brain is highly robust and fault tolerant. Two biological brain fault tolerant techniques are adopted in NoC to make it fault tolerant and robust. Biologically inspired techniques offer novel ways of making NoC fault tolerant.

Synaptogenesis and sprouting are two biological brain techniques that help the brain recover from damaged neurons and synapses. Synaptogenesis implements the self-adopt mechanism present in the brain while sprouting implements the concept of self-healing. These two techniques were implemented in NoC. The synaptogenesis algorithm was used to find the optimal and fault free path from source to destination in NoC, while the sprouting algorithm is used to create a shorter synapse to bypass the faulty router or interconnect.

The improved bio-inspired algorithm (a combination of synaptogensis and sprouting technique) is implemented using guaranteed throughput (GT) communication setup by time division multiplexing (TDM). The bandwidth utilization of best effort (BE) service based connections is better as compared to GT based connections. This is normal behavior as packet switching mechanisms utilize all of the available bandwidth. The throughput utilization of GT based connections was better as compared to BE based algorithms due to a dedicated path between source and destination. Similarly, due to a dedicated path between source and destination the interflit arrival time of GT based connections is lesser as compared to BE based connections. The performance of both GT and BE based NoC architecture and algorithms is also thoroughly analyzed.

9.1   Introduction

There are two broad categories of quality of service (QoS) parameters in NoC, guaranteed throughput (GT) service and BE. In GT the resources on the network are reserved for a particular time period between source and destination processing elements (PE's). Circuit switching or the connection oriented mechanism of packet switching is mainly used to allocate the resources and guarantee the throughput. GT connections are usually allocated to real time traffic because of high traffic and less latency requirements. GT connections address the worst case scenario because the resources can be silent for particular instances of time, which makes the bandwidth utilization at times low. BE services compliment the GT connections because of their underutilization of channel bandwidth. BE is the packet switching concept that utilizes the unallocated bandwidth to sources and other routers. BE traffic does not allocate any resources on the NoC. BE services represent the average case scenario. The channel is always utilized when the source and destination need it. Video processing and multimedia applications are an example of GT critical traffic, while cache updates are an example of BE non-critical traffic [33,75].

The bio-inspired algorithm was implemented using these connections setups. The GT connections were implemented using TDM. TDM connections help divide the bandwidth of interconnect among multiple connections using slots. The BE connections are implemented using packet switching. In packet switching, routers decide which direction the packet should be sent based on the situation of neighbor's routers, interconnects, and routing table. In this chapter, a bio-inspired NoC algorithm using a GT connection is discussed.

9.2   NoC Architectures and Fault Tolerance

Some of the NoC architectures provide GT and BE services while others provide only GT and BE communication mechanisms. Architectures that provide GT based communication services are aSOC [58], OCTAGON [46], Nexus [59], QNoC [13], spatial division multiplexing NoC [57], PNoC [40], cross road interconnection architecture [20], HIBI [76], ProtoNoC [16], CDMA NoC [101], TTNoC [71], MoCSYS [45], Aelite [37],and dAElite [92].

MicroNetwork [103], CLICH [53], SPIN [36], Dally et al. [23], PROTEO [90], CHAIN [2], RAW [94], HERMES [63], BIDI-MIN [68], OCN [39], SoCIN [108], Xpipes [8], R^2NoC [78], RaSoC [107], NoCGen [19], topology adaptive NoC [3], Arteris [88], a low latency router [49], INoC [65], XGFT [47], low latency on chip network [64], GEXPolygon and GEXSpidergon [111], Low-Power Network on Chip [56], NocMaker [17], TILEPro64 [6], UT TRIPS [34], Intel TeraFLOPS [100], SCC [41], MoCRes [44], Generalized de Bruijn Graph NoC [42], Polaris [91], EVC [52], ReNoC [93], HT-OCTAGON [26], XHiNoC [77], Network on Chip in a Three Dimensional [29], BiNoC [54], RAMPSoC [32], DRNoC [51], Dynamic Reconfigurable Network on Chip [105], Ramos et al. [72], Kilo-NoC [35], WiNoC [31], Custom Network on Chip Architecture [62], DANoC [89], WaveSync [106], AdNoC [28], Mesh based NoC [21], RecMIN [60], and SWIFT [74] are a few architectures that provide BE based communication.

There are some Network on Chip (NoC) architectures that provide both GT and BE based communication services. They are Æthereal [75], SoCBUS [102], Nostrum [61,73], Spider [27], Spidergon [22], Asynchronous on Chip Network Router with Quality of Service [30], Mango [9–11], Wolkotte et al. [104], Asynchronous NoC Architecture [5], Kavaldjiev et al. [48], DSPIN [67], art-NoC [79], and ALPIN [4]. None of these architectures are fault tolerant except artNoC, Nexus, and TTNoC. Details about these and other fault tolerant NoC architectures are mentioned in Table 9.1.

Table 9.1   **NoC Architectures providing fault tolerance**

S.No.	Architecture	Fault tolerance	Year
1	artNoC [79] (Adaptive Real-Time Network on Chip)	Fault tolerance using broadcasting of packets (concept not explained)	2007
2	Nexus [59]	-No physical architecture is provided for fault tolerance -Fault test pattern generated is for fault and delay testing	2004
3	TTNoC [71] (Time Triggered Network on Chip)	Fault Tolerant (no detail provided)	2008
4	Dally et al. [23]	Transient faults (error in bits) detection mechanism	2001
5	PROTEO [90]	Fault Tolerant (no detail provided, trying to include)	2002
6	Xpipes [8]	Distributed error detection bits	2004
7	XGFT [47] (extended generalized fat tree)	Fault diagnosis and repair (FDAR) detects and repairs static, dynamic and transient faults (bit errors and blocking of packets)	2006
8	Low latency on chip network [64]	-Error detection code -Flits ordering checks	2006
9	GEXPolygon and GEXSpidergon [111]	-Cyclic Redundancy Code (CRC) for error detection -Faulty node/interconnect detection but no detail provided	2006
10	Dynamic reconfigurable NoC for adaptive reconfigurable MPSoC [1]	Error detection and correction	2006
11	Generalized de Bruijn Graph NoC [42]	Distributed error detection bitsReliable routing algorithm to detour the faulty interconnect	2007
12	AdNoC [28] (Adaptive Network on chip)	-Detects faults during NoC adaptation at architectural level -No architectural detail provided -No details about run time node and interconnect faults detection	2012

9.3 Bio-Inspired Techniques and NoC

The basic building unit of the biological brain is a neuron. Neurons are connected together through synapse (connection). There are 80–120 billion neurons. Neurons are connected together through an axon terminal of one neuron and dendrites of another neuron. There is a network of neurons connected together through synapse. There are between 1000 and 10,000 synapses per neuron [38]. This shows the complexity and robustness of the biological brain. The interconnection of this neuron network is similar to the NoC. In NoC routers are connected to each other through interconnects. The router corresponds to the neuron while the interconnect is similar to the synapse. In order to get the inspiration from biological brain, the real world problem should be similar to the natural phenomenon [24].

Synaptogenesis and sprouting algorithms [14] are two biological brain techniques, among others, which make the brain fault tolerant and robust. The synaptogensis and sprouting algorithm are the self-adopted and self-healing concept of the biological brain. In synaptogenesis the neuron is attracted by the chemical attractant released by the target neuron. The growth cone at the top of the axon terminal having filopodia and lamellipodium is basically attracted by these chemical attractants. Later, a synapse is formed between source and destination neuron through which communication occurs. The concept is shown in Figure 8.7

Sprouting is the self-healing mechanism of the biological brain, when a synapse gets damaged due to any biological reason. The damage may be because of blood clotting or hypertension, which affects the biological brain. In the sprouting technique, a sprout emerges from the axon already connected with the source neuron and it advances towards the destination neuron. The chemical adhesion material is used by the axon to discover the new path to the destination. The concept is shown in Figure 8.8. Experiments were done on fishes and frogs (amphibians). Scientists discovered that this feature is not available in humans.

Our bio-inspired algorithm is adopting these two biological brain techniques to make the NoC fault tolerant and robust. The self-adapting mechanism of the biological brain helped the NoC algorithm find the optimal path between source and destination. This algorithm detects the faulty interconnects and nodes during connection establishment. While sprouting, a self-heal concept of the biological brain helps the NoC bypass the faulty nodes or interconnects and tries to connect with the older synapse to make it robust and fault tolerant. These algorithms help the NoC establish multiple synapses between source and destination to maximize the throughput and efficiently utilize the bandwidth.

These two bio-inspired techniques are implemented using BE and GT based NoC architectures. The architecture and algorithm is mentioned in Section 9.4.

9.4 Bio-Inspired NoC Using GT Architecture

The bio-inspired NoC algorithms are implemented using BE and GT services separately. In both service levels, we have used 4×4 NoC mesh topology. The routers are connected with each other while PEs are connected with routers through Network Interface (NI). The overall port architecture of routers in both NoCs is the same except for the architecture of "scheduler," "inPort," and some control/data pins. In GT architecture, there are slots at the "scheduler" and multiple queues at the "inPort," while in BE architecture there is a credit based flow control technique to manage the communication between routers. In GT communication architecture, no flow control technique is required. The port architecture of BE and GT routers is similar to Figure 9.1.

The "SW_in," "SW_out," "SW_Ctrl_in," "SW_Ctrl_out," and "Ctrl" pins are used to control the flow of data flits between routers. Control messages are shared through "SW_Ctrl_in" and "SW_Ctrl_out" pins between ports of a particular router. Similarly, "SW_in" and "SW_out" pins are used to send data flits between various ports of the router.

The "out" pins of the "inPort" are connected to the "SW_in" pins of the port, while "Ctrl" pins are connected with the "SW_Ctrl_in" pins. The "in" pins of the "Scheduler" are connected with "SW_out" pins, while "Ctrl" pins are connected with "SW_Ctrl_out" pins. Each "SW_in" pin is connected with every port of the router. Similarly, "SW_out" pins of the particular port are connected with every other port of the router. The "SW_Ctrl_in" pins of the router are connected with every other port of the router through "SW_Ctrl_out." Similarly,

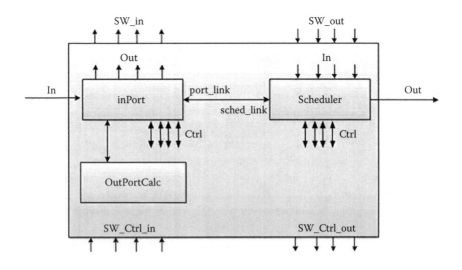

Figure 9.1 Detail port architecture.

"SW_Ctrl_out" pins are connected with "SW_Ctrl_in" pins of every other port of the particular router.

In GT architecture, each "inPort" has a separate queue for every virtual channel. This avoids the head of line blocking whenever faults occur. The scheduler has four slots that are allocated run time to various virtual channels. The optimum four slots are reserved based on the literature [37]. The slots are managed and read by the TDM clock. The TDM clock period is 2 ns. The slots help to efficiently utilize and divide the bandwidth of the NoC among four connections. The TDM concept is shown in Figure 9.2.

Different slots are allocated to four virtual channels A, B, C, D. The slot umber of the routers is increased by an increment of one, as stated in Equation 9.1. At every slot, the router is sending one packet out to link from any virtual channel. This is a contention free routing as there are no collisions between packets from different virtual channels.

$$slotAllocated = (previousAllocatedSlot + 1)\%4 \qquad (9.1)$$

Contention free routing is achieved with the help of slots. In contention free routing, there is no collision between packets of multiple connections when accessing one output port. With the help of slots, time to access out a link is divided between multiple virtual channels. Every virtual channel equally shares the bandwidth of the out link. The slots are allocated by the scheduler of the router. Every router in the NoC increases one slot from the previous slot number as shown in Figure 9.3. In the figure, slot 2 is dedicated to virtual channel B. While in the next router the slot is increased by one and slot 3 is assigned to virtual channel B. This helps to efficiently utilize the bandwidth, and one flit is routed from four routers in the consecutive four clock cycles. The scheduler skips those slots that are not allocated to any of the virtual channels. This helps the connection efficiently utilize the bandwidth of the interconnect and maintain the high throughput for

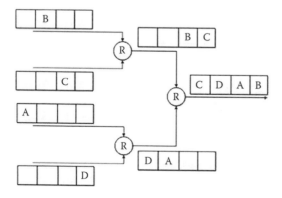

Figure 9.2 TDM mechanism for contention free routing.

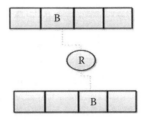

Figure 9.3 Router delay slot allocation by one slot during connection setup.

that particular connection. The efficient utilization and reading of the slots also increases the performance of a particular port as it dedicates the available bandwidth to already allocated active connections.

If the bandwidth of the out link is (b_w) and there are virtual channels (vc) transmitting the flits (data), then each vc will have a throughput of TH_{min} as in the following equation 9.2 [48].

$$TH_{min} = \frac{b_w}{vc} \tag{9.2}$$

In the bio-inspired NoC, we have currently allocated four slots per port. So, there can be four virtual channels per port. In this technique, we are statistically allocating the slots to the scheduler so the throughput (TH_{min}) is guaranteed to each and every virtual channel. Based on the TH_{min}, the time required for the message to travel from the source to the destination using the virtual channel connection is TH_{max} as shown in equation 9.3 [37].

$$TH_{max} = N * D_t + \frac{L}{TH_{min}} \tag{9.3}$$

In the above equation, N is the total number of routers traversed by the packets during the communication. D_t is the time required by each packet per router and L is the length of message in bits.

The scheduler architecture of a router is shown in Figure 9.4. The slots are reserved during the synapse establishment in the "OutPortCalc" module before deciding a direction for the "synap" signal. The "OutPortCalc" checks whether any slot is available for this port. If yes, it will allocate the slot to the virtual conection; if a slot is not available, the "synapse" is directed to another port based on the information present in the Information Set (IS) messages. Details about IS messages are explained in Section 9.5. As shown, there are four virtual channels accessed by one router at the same time. The slot table is used to keep track of the slots allocated to each virtual channel.

The following code shows the allocation and checking of a slot for "corePort." A virtual connection can only be established with a "corePort" if a slot is

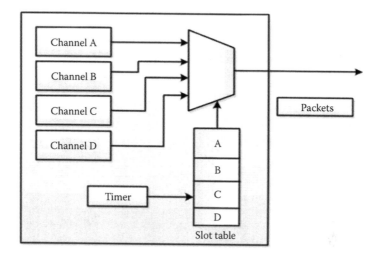

Figure 9.4 Scheduler architecture of a router having slot table.

available. The "corePort" is connected with the PEs. The slots are checked for all five ports before making the decision about the "synap" signal direction.

```
for (int i=0;i<numberOfSlot;i++)
{
    //Checking slots of current router of port 4
    accessSlot=common::routerSlots[routerIndex][4]
    [slotIndex];
    if (accessSlot==-1)
    {
    //slot is free allocated it to current port and
    //make it 1
    common::routerSlots[routerIndex][4][slotIndex]=
    portIndex;
    //Allocate this port and make it status as waiting
    //for route reply so that it cannot send "request
    //to send data" signal to which it is allocated
    common::routerPortsStatusWaiting[routerIndex][4]
    [slotIndex]=1;
    slotAllocated=slotIndex;
    swOutPortIdx = corePort;
    break;
    }
}
```

There are four virtual channels (A, B, C, D) that are allocated to slot A, B, C and D. The timer reads every slot after 2 ns on the scheduler. The scheduler sends the signal to read data from that port for a particular virtual channel. This helps with multitasking and avoids contention between multiple connections.

```
NoCReqTDMMsg* reqTDM=new
NoCReqTDMMsg("Request To Send Data");
reqTDM->setKind(NOC_REQTDM_MSG);
reqTDM->setPortIndex(portIndex);
reqTDM->setCurrentSlot(slotRead);
send(reqTDM, "ctrl$o", outPort );
```

In the above code, the scheduler is sending a "reqTDM" message to a particular port to read a particular queue allocated to particular slot and port. This signal is sent on the "ctrl" port of the "Scheduler" as this is the control signal. In reply to this signal, "inPort" reads a particular queue allocated to this slot and sends the flit to a particular "Scheduler" of a port.

For every vc, a separate queue is allocated at the "inPort" of the router as shown in Figure 9.5. This helps to avoid the head of the line blocking problem when fault(s) occurs.

When the "inPort" reads a signal from the scheduler to send the data, a particular queue is read from the "inPort," which is allocated to that slot. The "queue to read" parameter is provided by the scheduler. With the help of the TDM based router, there is no collision between packets or deadlock scenario as the scheduler only allows one flit per router during one clock cycle. The following lines of

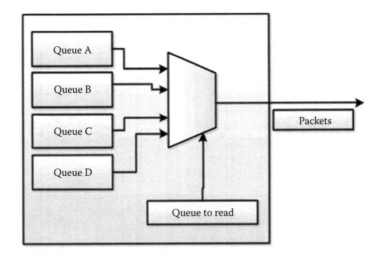

Figure 9.5 inPort architecture of a router.

code specify that "inPort" receiving a request from "Scheduler" read a particular queue. The "tdmRequestFromPort" and "tdmSlotFromPort" information is provided by "Scheduler" when a particular slot is read. Later, the queue allocated to a particular slot is read. The flit that is saved in the queue is read and sent to the output port of the "inPort." This flit is received at the particular "Scheduler" of the port and the scheduler will send it out to the link.

```
queueToRead=queueAllocatedToSlotPort[tdmRequestFromPort]
[tdmSlotFromPort];
NoCFlitMsg* tempFlit=(NoCFlitMsg*) Q[queueToRead].pop();
send(tempFlit, "out", outPort);
```

The flow of messages in the bio-inspired NoC using GT architecture is shown in Figure 9.6. The flow of signals is similar to the BE architecture except for the

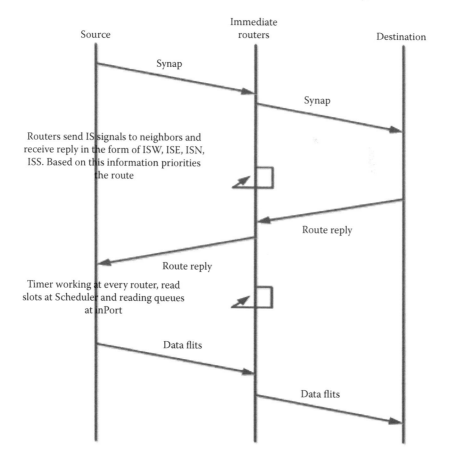

Figure 9.6 Flow of signal between source and destination.

use of the slots table at "Scheduler" and "Queues" at "inPort." At every clock cycle the slot is read at the scheduler. The scheduler sends the "Request to send" signal to the port allocated in the slot. This signal is used to read a particular queue at the "inPort," which is allocated to this slot using vc. The communication between source and destination goes like this until the end flit is sent by the source to terminate this connection.

9.5 Bio-Inspired Framework and Algorithm

Bio-inspired NoC is implemented on an OMNET++ based simulator. Heterogeneous Network on Chip simulator (HNOCS) is an open source simulator providing heterogeneous, multiple topologies, synchronous, and asynchronous NoC support [7].

The scheduler of the port constantly monitors the interconnects using the bio-inspired NoC algorithm. When a faulty node or interconnect is detected by the scheduler it initiates a new synapse connection from the current node to the destination node. During the sprout (connection) establishment the priority is always to connect with the older synapse to avoid the unnecessary routers traversal. This makes the algorithm highly robust and reduces the latency of communication between source PE and destination PE. During the recovery of fault the throughput is slightly degraded for a short period of time, and it tries to regain the original throughput after bypassing the faulty interconnect or node. The flits are still traversing on the older synapse as those flits have already bypassed the faulty node or interconnect. This allows the bio-inspired algorithm to efficiently utilize the bandwidth of the NoC. There are multiple parallel synapses between the source and destination. The shorter synapses can be formed for as many faults that may occur during the simulation.

The network discovers itself at the start of the simulation. The routers share requests and reply control signals. After sharing, the control signal routers know about the status of their neighbor routers (i.e., working or not working). The bio-inspired NoC algorithm only needs one network discovery at the start of the simulation. This efficiently utilizes the bandwidth of the NoC.

After network discovery, the source PE initiates the synapse "synap" signal to connect with the destination PE. Static and run time faults are detected by the synap signal. The static faults are detected by the synaptogenesis algorithm during the connection establishment, while the sprouting algorithm detects the run time faults. This makes the bio-inspired algorithm fault tolerant and reliable. When the router receives the "synap" signal it sends the IS request signals to its neighbors. Neighbor routers reply with the IS response signals having its information (direction neighbor) and its immediate neighbors information. The neighbor node sends back the information in ISW (west), ISE (east), ISS (south), and ISN (north) response packets as shown in Figure 9.7. The information in the IS

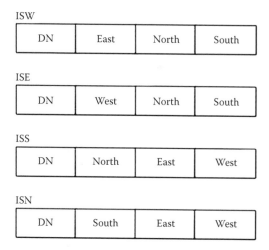

Figure 9.7 Information Set (IS) packet.

Synapse

Synapse Id	Source Id	Destination Id	Output port numbers	Router Id's	Slot numbers	Queue numbers	Connection Id

Figure 9.8 Synap packet format.

response packet is either 1 (working), 0 (not working), or −1 (neighbor doesn't exist). Based on this information the router decides in which direction the packet should be sent [66].

The bio-inspired algorithm was developed in two phases. In the first phase the algorithm was run and tested on a node basis in which the complete node was made faulty. Later, the algorithm was modified and tested on a node and interconnect basis. The algorithm performance in the latter case was better as bandwidth and throughput were efficiently utilized and the network recovery time from faults was less.

In GT based NoC there is a slight change in the packet format of "synap" and "flit" as compared to the BE services, due to the addition of queue and slot numbers as shown in Figure 9.8 and Figure 9.9, respectively.

"The synapse ID" is unique for every connection between source and destination. Whenever a fault occurs the synapse ID is changed such that it is similar to the original synapse ID. This corresponds to the sprout behavior of the original synapse. "Source ID" and "destination ID" correspond to the unique source PE and destination PE in NoC. Output port numbers are saved in Synapse and Flit packet. This is used by the flits to send data on these ports. "Router IDs" are used

Flit

Packet ID	Flit Id	Connection Id	Flit type	Message length	Source Id	Destination Id	Slot numbers	Queue numbers	Output port numbers	Synapse Id

Figure 9.9 Flit packet format.

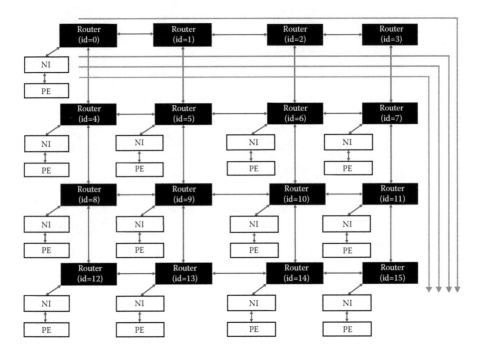

Figure 9.10 Multiple synapses between a source and a destination using TDM.

by the route reply signal. The route reply signal is sent by the destination after connection is established. "Slot numbers" are saved in this field during the connection establishment. These slot numbers are used by flits. "Queue numbers" are also used by the flit to avoid head of the line blocking. "The connection ID" corresponds to the number of connections a source can have. Currently, four parallel connections are possible between a source and a destination. "The packet ID" corresponds to the packet number in a message. "The flit ID" is the flit number in the ten flits packet. "The flit type" corresponds to the header, body, and tail flit. "The message length" is the total number of flits in the packet.

In GT based NoC, with the help of slots and queues at the "scheduler" and in "inPort" multiple parallel connections are possible between multiple sources and multiple destinations. Figure 9.10 shows the case when multiple synapses are formed between source (0) and destination (15). The source (0) initiated a connection to create a synapse with destination (15). As the ports are shared among multiple virtual channels, there can be multiple connections between one source and one destination. Connections between multiple sources and multiple destinations are also possible as shown in Figure 9.11. Port 2 (south) of router 6 is shared between two connections. Two different slots are allocated to the connections. Similarly, ports 2 (south) of routers 7 and 9 are shared by two connections. Port 4 (core port) of router 0 and router 15 is shared among 2 and 3 connections, respectively.

Figure 9.12 shows that the NoC have faulty interconnects. The cross shows the faults that occur during the communication between source and destination. Initially, source PE (0) launched a connection to connect with the destination PE (15). That is why a synapse is formed between them. But, during the communication faults were introduced among random interconnects. The temporary synapse was formed to bypass the faulty interconnect and to connect with the older synapse. As shown, a temporary synapse is formed between router 1 and router 3. This temporary synapse also avoided the traversal of unnecessary routers on the way, which made the communication faster. The same concept was

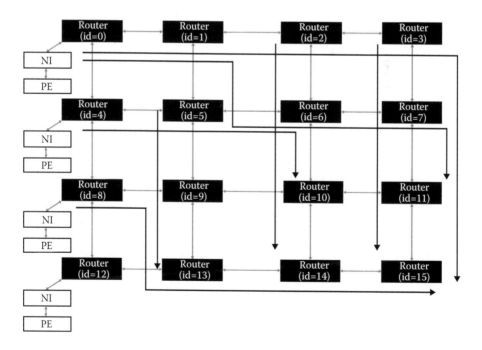

Figure 9.11 Multiple synapses between multiple sources and multiple destinations using TDM.

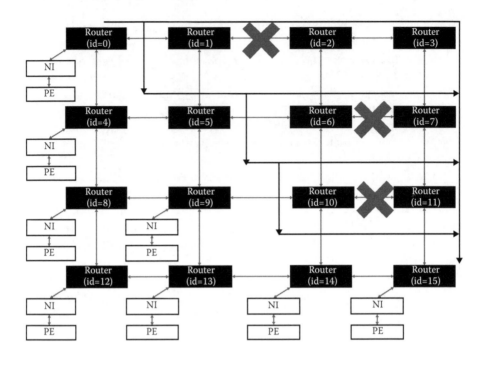

Figure 9.12 Synapse connections between source (0) and destination (15) when multiple faults occurred.

used when faults were introduced between router (6), router (7) and router (10), and router (11).

9.6   Results and Discussion

The bio-inspired algorithm is tested with a number of faults. These faults were introduced at the same time. The faults were introduced using the uniform random distribution formula. The faults were random in multiple simulations as the seed value was constantly changed. The algorithm is tested with a number of faulty routers and interconnects. The simulation was run 35 times to have a 95% confidence interval and to remove the transient effects. The simulation was repeated 35 times per fault and for every injection rate [15]. We choose the observed proportion (p) to be 90%, width of percentage error (w) to be 10%, and the $z_{\alpha/2}$ value is 1.96 for a 95% confidence interval. The values were put in the formulas to measure the number of iterations.

$$\text{C.I} = p \pm z_{\alpha/2} \sqrt{\frac{p(1-p)}{n}} \tag{9.4}$$

where C.I is the confidence interval. The width of the confidence interval can be determined by the following formula:

$$w = z_{\alpha/2} \sqrt{\frac{p(1-p)}{n}} \tag{9.5}$$

Equation 9.5 is rearranged for n. So, we got $n = 35$ iterations for a 95% confidence interval.

$$n = p(1-p) \left(\frac{z_{\alpha/2}}{w} \right)^2 \tag{9.6}$$

The injection rate depends on the percentage of utilization of the interconnect bandwidth. The interconnect bandwidth of the NoC is 16 Gbps. The injection rate of the packet was changed from 10% to 100%. In wormhole switching, by changing the number of idle cycles between the packets, the injection rate can be changed. The formula used to calculate the number of idle cycles is [96]:

$$\text{cycles} = \left(\frac{CTR}{PETR} \right) * \text{PSize} * \text{CyclesPerFlit} \tag{9.7}$$

where cycles are the number of idle cycles between the packet from PE. CTR is channel transmission rate and PETR is PE transmission rate. PSize is packet size in terms of number of flits, and cycles per flit is the number of cycles required to transmit per flit in a wormhole switching.

In order to test the bio-inspired NoC algorithm, we took 4×4 NoC. The topology of the NoC is mesh, although this algorithm is scalable to multiple NoC sizes and topologies. Another reason to select 4×4 NoC is that most of the techniques have been implemented in this topology. Table 9.2 summarizes a few of the papers using other size networks [55,69,98,109,110]. Table 9.3 summarizes the techniques implemented in 4×4 NoC. There are more papers [18,25,43,50,81–83,97,99] using 4×4 NoC as compared to other NoC sizes. This helped us to compare our work with the literature. Moreover, 2D mesh topology is selected because it is suitable for 2D silicon surface [80].

The clock frequency of the NoC is 500 MHz. The flit size is 4 bytes. The packet length is 10 flits. In every iteration, different sources are injecting various numbers of packets. A total of 100,000 packets or 1,000,000 flits are injected in the NoC to fully utilize the bandwidth of the NoC. The algorithm is tested using uniform and random traffic patterns. The simulation is run 50 ms per iteration. A wormhole switching mechanism is used in which the packet is divided into header, body, and tail flits.

We are calculating the QoS parameters of NoC. The QoS is mainly linked with the latency and throughput parameters [12]. We are also analyzing other parameters of bandwidth utilization and inter flit arrival time at sink. As our algorithm is detecting faults on per router basis, it is also working on a router to router basis, which makes the algorithm more efficient at high traffic and error (fault) rates [70].

Table 9.2 Saturation points of techniques (having different network size)

Techniques	Number of Faults							
	0	1	2	3	4	9	16	20
Typical Virtual Channel Management (5 × 5 NoC) [55]	0.18	—	0.16	—	—	—	—	—
PVS (5 × 5 NoC) [55]	0.22	—	0.20	—	—	—	—	—
FVS (5 × 5 NoC) [55]	0.23	—	—	—	—	—	—	—
A Reconfigurable Routing Algorithm (5 × 5 NoC) [110]	0.28	0.27	0.22	0.16	0.15	—	—	—
RAFT (5 × 5 NoC) [98]	—	—	—	—	—	0.28	—	—
Agent-based routing (5 × 5 NoC) [98]	—	—	—	—	—	0.3	—	—
XY (8 × 8 NoC) [109]	0.36	—	—	—	—	—	0.25	—
FFAR (8 × 8 NoC) [109]	0.35	—	—	—	—	—	0.28	—
OE (9 × 9 NoC) [69]	—	0.04	—	—	—	—	—	0.036
IOE (9 × 9 NoC) [69]	—	0.035	—	—	—	—	—	0.03
OE+IOE (9 × 9 NoC) [69]	—	0.06	—	—	—	—	—	0.065

The R-Software [95] is used to analyze the result statistically. With the help of "sqldf" package of R-Software multiple queries were run to analyze the various tables generated by HNOCS. The R software was run over a high performance computing server as it required processing of a large data set. The HPC server also solved the problem of RAM and large memory requirements.

The result shows that the algorithm is fault tolerant as the NoC quickly recovers from the faults. The bandwidth, throughput, and latency degrade gracefully during the network recovery from faults.

9.6.1 Bio-inspired NoC using GT services

The bio-inspired algorithm is also implemented using GT based communication services. In this architecture, the algorithm reserves the resources on the NoC. The slot table is used to allocate the interconnect bandwidth allocation time to a particular vc. The slot tables are present at the scheduler module of the port. There are five ports in the router that are connected with the four neighbor routers and one to the local PE device. The scheduler controls the output port while the inPort module is used to control the input port. There are parallel queues at the inPort to avoid the head of line blocking problem. Each queue is allocated to

Table 9.3 Saturation points of techniques (4 × 4 NoC)

Techniques	Number of faults							
	0	1	2	3	4	5	6	7
Bio-inspired algorithm using TDM [84,85,87]	0.52	0.32	0.28	0.25	0.22	0.2	0.18	0.17
Synaptogenesis algorithm [82,83,86]	0.39	0.31	0.29	0.28	0.27	0.26	0.25	0.24
Sprouting algorithm [81,83]	0.42	0.40	0.34	0.30	0.22	0.21	0.19	0.18
DBP network [50]	0.22	0.18	0.17	0.16	0.15	0.15	0.14	0.14
ODD-EVEN [18]	—	0.4	—	0.4	—	0.4	—	—
Inverted ODD-EVEN [18]	—	0.4	—	0.4	—	0.4	—	—
FTCAR [99]	0.31	—	—	0.25	—	0.24	—	—
HiPFaR [25]	0.37	0.35	—	—	—	—	—	—
Baseline [25]	0.38	0.32	—	—	—	—	—	—
RAFT2 algorithm using VOPD traffic [97]	0.26	—	0.23	—	—	—	—	—
RAFT2 algorithm using uniform traffic [97]	0.37	—	0.29	—	—	—	—	—
Fault Aware Dynamic Routing Algorithm [43]	0.2	0.2	0.2	—	—	—	—	—
Flooding Algorithm [43]	0.10	0.10	0.10	—	—	—	—	—

every virtual channel same as slot at the scheduler. The timer at the scheduler reads a particular slot, which in turn sends the request to a particular port of the router to send data flit from a particular virtual channel.

The results show that the algorithm is fault tolerant as the NoC quickly recovers from the faults and the bandwidth, throughput, and latency degrade gracefully during the network recovery from faults.

Figure 9.13 shows the saturation point of the NoC as the number of faults is increased. The saturation point is analyzed in Tables 9.1 and 9.2. The saturation point of the bio-inspired algorithm is better as comparable to other fault tolerant techniques. The GT based NoC algorithm has a saturation point of 0.52 at zero fault, which shows the efficiency and simplicity of the algorithm. This shows that the bio-inspired algorithm quickly finds the path from source PE to destination PE when there is no fault and data flits are sent over it. This saturation point is better as compared to all other techniques in Table 9.1 and Table 9.2. Although Table 9.1 shows the techniques that are modeling a different NoC network, the saturation point of the bio-inspired algorithm is still better as compared to them. The accepted traffic rate of the bio-inspired algorithm is better as compared to the technique mentioned in Ref. [50] for different numbers of faults. Technique [50] is the most thoroughly analyzed, tested, and presented technique

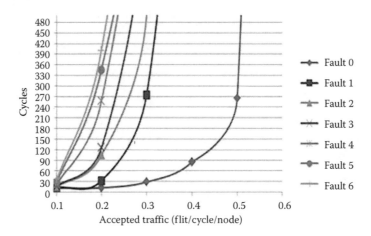

Figure 9.13 Accepted traffic for various numbers of faults.

in the literature. The accepted traffic rate of the TDM based bio-inspired NoC (GT communication) is less as compared to our previous work [81–83]. This is expected behavior as these techniques are based on BE scenario. In BE the bandwidth and throughput utilization are better as compared to TDM based techniques. The efficient utilization of the TDM based GT connections depends on the burst of traffic from source or from neighbor routers. Usually, TDM based virtual circuits are allocated to those PEs that need guaranteed and faster services, while BE techniques are packet switching techniques and utilize all available bandwidth without any reservation. The saturation point of the techniques mentioned in Ref. [18] is slightly higher in three faulty cases. The faulty cases are one, three, and five. The saturation point of 0.4 for three different faulty cases is not common behavior in the literature. Usually the saturation point decreases as the number of faults is increased. The network saturates early as the network has less paths to traverse through and they are occupied by all the PEs available on the NoC. Due to this, the NoC saturates early and it leads to overflow of buffers and queues. The bio-inspired NoC accepted traffic better than the techniques [25,43,97,99] for different faulty cases.

Figure 9.14 shows the throughput utilization graph of the bio-inspired algorithm. The overall percentage drop of the throughput is 14.05% for zero to seven faults. The flit/cycle/node behavior of the bio-inspired algorithm is overall better than the literature techniques mentioned in Table 9.2 and Table 9.3.

The bandwidth of the bio-inspired NoC is degrading gracefully as the numbers of faults is increased, as shown in Figure 9.15. The overall percentage drop of the bandwidth is 1.86% from zero to seven faults. Similarly, the overall latency of the bio-inspired algorithm is increased slightly by 16.63%, which shows the

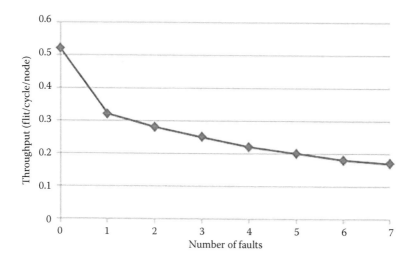

Figure 9.14 Throughput (flit/cycle/node) vs number of faults.

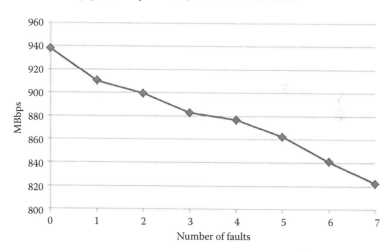

Figure 9.15 Bandwidth vs number of faults.

robustness of the algorithm. The algorithm quickly detects and recovers from the fault and tries to connect with the older synapse.

The inter-flit arrival time at the destination is increasing very slightly as the number of faults is increased, as shown in Figure 9.16. This shows that bio-inspired algorithms quickly recover from the faults and retain the available maximum throughput. The inter-flit arrival time is measured in nano second (ns). The overall inter-flit arrival time is only increased by 0.75% from zero to seven faults.

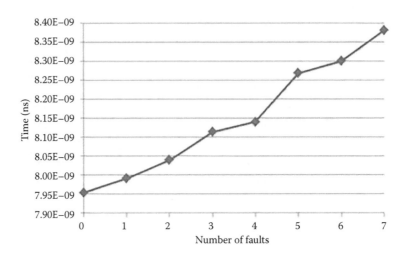

Figure 9.16 Inter-flit arrival time vs number of faults.

9.6.2 GT vs BE NoC performance

The BE based communication services complement the GT services. At this time, GT based communication channels do not efficiently utilize the bandwidth and throughput of the NoC. This leads to underutilization of the resources. In order to efficiently utilize the resources of the NoC, BE communication services are also provided along with GT services.

The bandwidth, throughput, interflit arrival time, and latency of BE and GT architectures depend upon the traffic pattern and injection rates. At higher injection rates GT connections are preferred, as can be seen in Figure 9.17. At 100% traffic injection rate, the bandwidth utilization is better as compared to the overall average TDM performance for 10% to 100% injection rates for multiple faults. The BE connections of sprouting and synaptogenesis are performing better as compared to TDM based GT connections.

The BE connections are efficiently utilizing 44.93% of the bandwith as compared to 100% injection rate for a TDM based connection. The bandwidth utilization is increased to 125.73% when BE bandwidth utilization is compared with the bandwidth utilization of average TDM connections. The 100% TDM connection is efficiently utilizing 55.75% of the bandwidth of the NoC as compared to average TDM base connections. The overall performance of BE based connections is better as compared to TDM based GT connections by 76.53%.

Figure 9.18 shows the magnified version of Figure 9.17, so that the performance of BE architecture is visible. It can be seen that the sprouting BE is utilizing bandwidth better as compared to synaptogenesis, BE. The average bandwidth

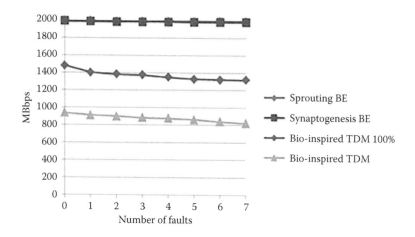

Figure 9.17 Bandwidth vs number of faults.

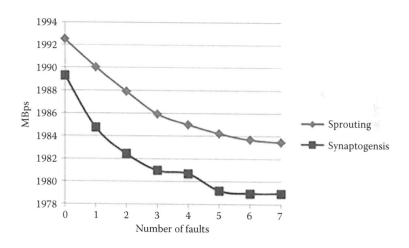

Figure 9.18 Bandwidth vs number of faults for BE connections.

utilization of the sprouting algorithm is better as compared to synaptogenesis algorithm by 0.24%.

In BE NoC, the synaptogenesis algorithm takes a greater number of cycles to detect the fault while the sprouting algorithm quickly discovers the fault, and a sprout emerges from the neighbor router. In synaptogenesis, as the number of faults increases from 5, the bandwidth almost becomes stable because of directed and shorter paths between source PE and destination PE while in sprouting the

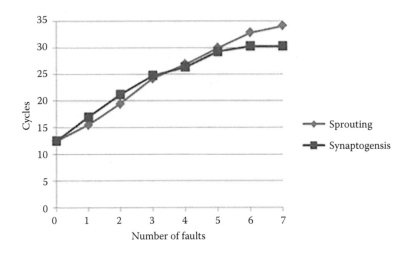

Figure 9.19 Latency vs number of faults in BE based NoC.

decrease is continuous but slow. Similarly, the latency of the bio-inspired techniques is increasing slightly as the numbers of faults are increasing as shown in Figure 9.19. As the numbers of faults increase from four, the latency of the sprouting increases as compared to synaptogenesis, but at the same time the bandwidth increase is higher for the sprouting algorithm as shown in Figure 9.18. The latency increase of the sprouting algorithm is 4.66% as compared to the synaptogenesis algorithm.

The 100% injection rate of the throughput utilization of the bio-inspired NoC is higher as compared to the overall average performance of TDM based GT connections as shown in Figure 9.20. This is normal behavior as there are no idle cycles between packets and flits, and they are received without any delay at destination. The 100% TDM connection is efficiently utilizing the throughput by 16.70% as compared to average TDM based connections. The BE service based synaptogenesis and sprouting algorithms are not utilizing the throughput efficiently as compared to GT based connections. The GT based connections are efficiently utilizing the throughput by 122.04% as compared to BE based connections. Moreover, the sprouting based BE connection is efficiently utilizing the throughput by 10.41% as compared to synaptogenesis based BE connections.

Figure 9.21 shows the interflit arrival time for GT and BE connections vs number of faults. At 100% traffic injection rate, the inter flit arrival time of TDM is shorter, as there are no idle cycles between packets and flits are reached at their destination more quickly as compared to average TDM connection, which has injection rates from 10% to 100% for multiple numbers of faults. The percentage increase of interflit arrival time is 128.99% for overall TDM based connections

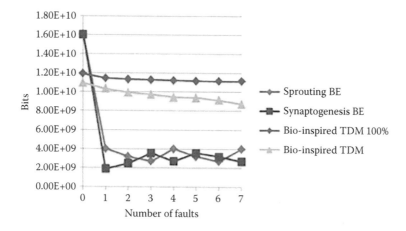

Figure 9.20 Throughput vs number of faults.

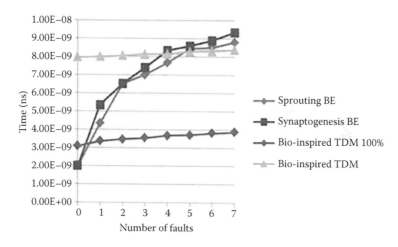

Figure 9.21 Interflit arrival time vs number of faults.

as compared to 100% TDM connection. The inter flit arrival time of BE connections grows gradually as the injection rate and number of faults increases, while the interflit arrival time of the BE based connections is 16.98% more as compared to TDM based GT connections. The interflit arrival time of the BE based synaptogenesis based algorithm is 5.88% greater as compared to the BE based sprouting algorithm.

Results show that the bio-inspired NoC algorithm using BE and GT communication services is performing better as compared to traditional fault tolerant techniques. The bandwidth, throughput, latency, and interflit arrival times are degrading gracefully during the recovery of faults. Future work includes combining the BE and GT base algorithms together to have completely bio-inspired NoC fault tolerant architecture.

9.7 Conclusion

Bio-inspired algorithms avoid deadlock and livelock situations as they have two hop information of the neighbours. The router avoids the path that is more congested or has a greater number of faulty routers. If the interconnect or router is busy for a particular time the various control signals and data flits are rescheduled to avoid the deadlock and congestion. If router or interconnect become faulty they bypass the complete row of routers, which helps to avoid congestion. This makes the bio-inspired algorithm reliable and efficient. As there are no routing tables in the bio-inspired algorithm, it takes less area as compared to the routing table techniques. But at the same time, the flow of control messages between neighbor routers makes bio-inspired algorithms a little bit complex. Bio-inspired NoC algorithms consume less power and energy as the "synap" signal moves in a directed path as controlled by the neighbors, information. Bio-inspired algorithms are implemented using BE and GT architectures. It was found that the BE architecture was efficiently utilizing the bandwidth as compared to GT services, while throughput utilization of GT services was better.

Bibliography

[1] Balal Ahmad, Ahmet T Erdogan, and Sami Khawam. Architecture of a dynamically reconfigurable NoC for adaptive reconfigurable MPSoC. In *Adaptive Hardware and Systems, 2006. AHS 2006. First NASA/ESA Conference on*, pp. 405–411. IEEE, 2006.

[2] John Bainbridge and Steve Furber. Chain: A delay-insensitive chip area interconnect. *IEEE Micro*, 22(5):16–23, 2002.

[3] TA Bartic, J-Y Mignolet, Vincent Nollet, Theodore Marescaux, Diederik Verkest, Serge Vernalde, and Rudy Lauwereins. Topology adaptive network-on-chip design and implementation. *IEE Proceedings-Computers and Digital Techniques*, 152(4):467–472, 2005.

[4] Edith Beigné, Fabien Clermidy, Hélène Lhermet, Sylvain Miermont, Yvain Thonnart, Xuan-Tu Tran, Alexandre Valentian, Didier Varreau, Pascal Vivet, Xavier Popon et al. An asynchronous power aware and adaptive NoC based circuit. *IEEE Journal of Solid-State Circuits*, 44(4):1167–1177, 2009.

[5] Edith Beigné, Fabien Clermidy, Pascal Vivet, Alain Clouard, and Marc Renaudin. An asynchronous NoC architecture providing low latency service and its multi-level design framework. In *11th IEEE International Symposium on Asynchronous Circuits and Systems, 2005*, pp. 54–63. IEEE, 2005.

[6] Shane Bell, Bruce Edwards, John Amann, Rich Conlin, Kevin Joyce, Vince Leung, John MacKay, Mike Reif, Liewei Bao, John Brown et al. Tile64-processor: A 64-core SoC with mesh interconnect. In *IEEE International Solid-State Circuits Conference, 2008. ISSCC 2008. Digest of Technical Papers.*, pp. 88–598. IEEE, 2008.

[7] Yaniv Ben-Itzhak, Eitan Zahavi, Israel Cidon, and Avinoam Kolodny. HNOCS: modular open-source simulator for heterogeneous NoCs. In *2012 International Conference on Embedded Computer Systems (SAMOS)*, pp. 51–57. IEEE, 2012.

[8] Davide Bertozzi and Luca Benini. Xpipes: A network-on-chip architecture for gigascale systems-on-chip. *IEEE Circuits and Systems Magazine* 4(2):18–31, 2004.

[9] Tobias Bjerregaard and Jens Sparso. Virtual channel designs for guaranteeing bandwidth in asynchronous network-on-chip. In *Proceedings Norchip Conference, 2004.* pp. 269–272. IEEE, 2004.

[10] Tobias Bjerregaard and Jens Sparso. A router architecture for connection-oriented service guarantees in the mango clockless network-on-chip. In *Proceedings Design, Automation and Test in Europe, 2005*, pp. 1226–1231. IEEE, 2005.

[11] Tobias Bjerregaard and Jens Sparsø. Scheduling discipline for latency and bandwidth guarantees in asynchronous network-on-chip. In *11th IEEE International Symposium on Asynchronous Circuits and Systems, 2005.* pp. 34–43. IEEE, 2005.

[12] Evgeny Bolotin, Israel Cidon, Ran Ginosar, and Avinoam Kolodny. Cost considerations in network on chip. *INTEGRATION, The VLSI Journal*, 38(1):19–42, 2004.

[13] Evgeny Bolotin, Israel Cidon, Ran Ginosar, and Avinoam Kolodny. QNoC: QoS architecture and design process for network on chip. *Journal of Systems Architecture*, 50(2):105–128, 2004.

[14] S Marc Breedlove, Neil Verne Watson, and Mark R Rosenzweig. *Biological Psychology*. Sinauer Associates, Incorporated Publishers, 2007.

[15] Michael D Byrne. How many times should a stochastic model be run? an approach based on confidence intervals. In *Proceedings of the 12th International Conference on Cognitive Modeling,* Ottawa, 2013.

[16] David Castells-Rufas, Jaume Joven, and Jordi Carrabina. A validation and performance evaluation tool for ProtoNoC. In *International Symposium on System-on-Chip, 2006,* pp. 1–4. IEEE, 2006.

[17] David Castells-Rufas, Jaume Joven, Sergi Risueño, Eduard Fernandez, and Jordi Carrabina. Nocmaker: a cross-platform open-source design space exploration tool for networks on chip. In *INA-OCMC Workshop,* Paphos, Cyprus, 2009.

[18] Helano S Castro and Otavio Alcantara de Lima. A fault tolerant NoC architecture based upon external router backup paths. In *2013 IEEE 11th International New Circuits and Systems Conference (NEWCAS),* pp. 1–4. IEEE, 2013.

[19] Jeremy Chan and Sri Parameswaran. Nocgen: A template based reuse methodology for networks on chip architecture. In *17th International Conference on VLSI Design, 2004. Proceedings,* pp. 717–720. IEEE, 2004.

[20] Kuei-Chung Chang, Jih-Sheng Shen, and Tien-Fu Chen. Evaluation and design trade-offs between circuit-switched and packet-switched NoCs for application-specific SoCs. In *Proceedings of the 43rd Annual Design Automation Conference,* pp. 143–148. ACM, 2006.

[21] Sudhanshu Choudhary and Shafi Qureshi. Performance evaluation of mesh-based NoCs: Implementation of a new architecture and routing algorithm. *International Journal of Automation and Computing,* 9(4):403–413, 2012.

[22] Marcello Coppola, Riccardo Locatelli, Giuseppe Maruccia, Lorem Pieralisi, and Alberto Scandurra. Spidergon: A novel on-chip communication network. In *2004 International Symposium on System-on-Chip, 2004.* Proceedings p. 15. IEEE, 2004.

[23] William J Dally and Brian Towles. Route packets, not wires: On-chip interconnection networks. In *Design Automation Conference, 2001. Proceedings,* pp. 684–689. IEEE, 2001.

[24] Falko Dressler and Ozgur B Akan. Bio-inspired networking: From theory to practice. *IEEE Communications Magazine,* 48(11):176–183, 2010.

[25] Mojtaba Ebrahimi, Masoud Daneshtalab, and Juha Plosila. High performance fault-tolerant routing algorithm for NoC-based many-core systems.

In *2013 21st Euromicro International Conference on Parallel, Distributed and Network-Based Processing (PDP)*, pp. 462–469. IEEE, 2013.

[26] Abd El Ghany, Magdy El-Moursy, Darek Korzec, Mohammed Ismail et al. High throughput architecture for octagon network on chip. In *16th IEEE International Conference on Electronics, Circuits, and Systems, 2009, ICECS 2009*, pp. 101–104. IEEE, 2009.

[27] Samuel Evain, J-P Diguet, and Dominique Houzet. μ spider: A CAD tool for efficient NoC design. In *Proceedings Norchip Conference, 2004*, pp. 218–221. IEEE, 2004.

[28] Al Faruque, Mohammad Abdullah, Thomas Ebi, and Jörg Henkel. AdNoC: Runtime adaptive network-on-chip architecture. *IEEE Transactions on Very Large Scale Integration (VLSI) Systems*, 20(2):257–269, 2012.

[29] Brett Stanley Feero and Partha Pratim Pande. Networks-on-chip in a three-dimensional environment: A performance evaluation. *IEEE Transactions on Computers*, 58(1):32–45, 2009.

[30] F Feliciian and Steve B Furber. An asynchronous on-chip network router with quality-of-service (QoS) support. In *IEEE International SOC Conference, 2004. Proceedings*, pp. 274–277. IEEE, 2004.

[31] Amlan Ganguly, Kevin Chang, Sujay Deb, Partha Pratim Pande, Benjamin Belzer, and Christof Teuscher. Scalable hybrid wireless network-on-chip architectures for multicore systems. *IEEE Transactions on Computers*, 60(10):1485–1502, 2011.

[32] Diana Göhringer, Michael Hübner, Laure Hugot-Derville, and Jürgen Becker. Message passing interface support for the runtime adaptive multiprocessor system-on-chip RAMPSoC. In *2010 International Conference on Embedded Computer Systems (SAMOS)*, pp. 357–364. IEEE, 2010.

[33] Kees Goossens, John Dielissen, and Andreea Radulescu. Æthereal network on chip: Concepts, architectures, and implementations. *IEEE Design & Test of Computers*, 22(5):414–421, 2005.

[34] Paul Gratz, Changkyu Kim, Karthikeyan Sankaralingam, Heather Hanson, Premkishore Shivakumar, Stephen W Keckler, and Danilo Burger. On-chip interconnection networks of the trips chip. *IEEE Micro*, 27(5):41–50, 2007.

[35] Boris Grot, Joel Hestness, Stephen W Keckler, and Onur Mutlu. Kilo-NoC: A heterogeneous network-on-chip architecture for scalability and service guarantees. *ACM SIGARCH Computer Architecture News*, 39(3):401–412, 2011.

[36] Pierre Guerrier and Alain Greiner. A generic architecture for on-chip packet-switched interconnections. In *Proceedings of the Conference on Design, Automation and Test in Europe*, pp. 250–256. ACM, 2000.

[37] Andreas Hansson, Mahesh Subburaman, and Kees Goossens. Aelite: A flit-synchronous network on chip with composable and predictable services. In *Proceedings of the Conference on Design, Automation and Test in Europe*, pp. 250–255. European Design and Automation Association, 2009.

[38] Atif Hashmi, Hugues Berry, Olivier Temam, and Mikko Lipasti. Automatic abstraction and fault tolerance in cortical microarchitectures. In *ACM SIGARCH Computer Architecture News*, Vol. 39, pp. 1–10. ACM, 2011.

[39] Tomas Henriksson, Daniel Wiklund, and Dake Liu. VLSI implementation of a switch for on-chip networks. In *Proceedings of the International Workshop on Design and Diagnostics of Electronic Circuits and Systems, Poznan, Poland* 2003.

[40] Clint Hilton and Brent Nelson. PNoC: A flexible circuit-switched NoC for FPGA-based systems. *IEE Proceedings-Computers and Digital Techniques*, 153(3):181–188, 2006.

[41] Jeff Hoffman, David Arditti Ilitzky, Anthony Chun, and Aliaksei Chapyzhenka. Architecture of the scalable communications core. In *Proceedings of the First International Symposium on Networks-on-Chip*, pp. 40–52. IEEE Computer Society, 2007.

[42] Mohammad Hosseinabady, Mohammad Reza Kakoee, Jimson Mathew, and Dhiraj K Pradhan. Reliable network-on-chip based on generalized de Bruijn graph. In *IEEE International High Level Design Validation and Test Workshop, 2007. HLVDT 2007*, pp. 3–10. IEEE, 2007.

[43] Amir Hosseini, Tamer Ragheb, and Yehia Massoud. A fault-aware dynamic routing algorithm for on-chip networks. In *IEEE International Symposium on Circuits and Systems, 2008. ISCAS 2008*, pp. 2653–2656. IEEE, 2008.

[44] Arun Janarthanan, Vijay Swaminathan, Karen Tomko et al. MoCReS: An area-efficient multi-clock on-chip network for reconfigurable systems. In *IEEE Computer Society Annual Symposium on VLSI, 2007. ISVLSI'07*, pp. 455–456. IEEE, 2007.

[45] Arun Janarthanan, Karen Tomko et al. MoCSYS: A multi-clock hybrid two-layer router architecture and integrated topology synthesis framework

for system-level design of FPGA based on-chip networks. In *21st International Conference on VLSI Design, 2008. VLSID 2008*, pp. 397–402. IEEE, 2008.

[46] Faraydon Karim, Anh Nguyen, and Sujit Dey. An interconnect architecture for networking systems on chips. *IEEE Micro*, 22(5):36–45, 2002.

[47] Heikki Kariniemi and Jari Nurmi. On-line reconfigurable XGFT network-on-chip designed for improving the fault-tolerance and manufacturability of the MPSoC chips. In *International Conference on Field Programmable Logic and Applications, 2006*. FPL'06, pp. 1–6. IEEE, 2006.

[48] Nikolay Kavaldjiev, Gerard JM Smit, Pierre G Jansen, and Pascal T Wolkotte. A virtual channel network-on-chip for GT and BE traffic. In *IEEE Computer Society Annual Symposium on Emerging VLSI Technologies and Architectures, 2006*, pp. 211–216. IEEE, 2006.

[49] Jongman Kim, Dongkook Park, Theo Theocharides, Narayanan Vijaykrishnan, and Chita R Das. A low latency router supporting adaptivity for on-chip interconnects. In *Proceedings of the 42nd Annual Design Automation Conference*, pp. 559–564. ACM, 2005.

[50] Michihiro Koibuchi, Hiroki Matsutani, Hideharu Amano, and Timothy Mark Pinkston. A lightweight fault-tolerant mechanism for network-on-chip. In *Proceedings of the Second ACM/IEEE International Symposium on Networks-on-Chip*, pp. 13–22. IEEE Computer Society, 2008.

[51] Yana E Krasteva, Eduardo De la Torre, and Teresa Riesgo. Reconfigurable networks on chip: DRNoC architecture. *Journal of Systems Architecture*, 56(7):293–302, 2010.

[52] Amit Kumar, Li-Shiuan Peh, Partha Kundu, and Niraj K Jha. Toward ideal on-chip communication using express virtual channels. *IEEE Micro*, 28(1):80–90, 2008.

[53] Shashi Kumar, Axel Jantsch, Juha-Pekka Soininen, Martti Forsell, Mikael Millberg, Johny Öberg, Kari Tiensyrjä, and Ahmed Hemani. A network on chip architecture and design methodology. In *IEEE Computer Society Annual Symposium on VLSI, 2002. Proceedings*, pp. 105–112. IEEE, 2002.

[54] Ying-Cherng Lan, Hsiao-An Lin, Shih-Hsin Lo, Yu Hen Hu, and Sao-Jie Chen. A bidirectional NoC (BiNoC) architecture with dynamic self-reconfigurable channel. *IEEE Transactions on Computer-Aided Design of Integrated Circuits and Systems*, 30(3):427–440, 2011.

[55] Khalid Latif, Amir-Mohammad Rahmani, Kameswar Rao Vaddina, Tiberiu Seceleanu, Pasi Liljeberg, and Hannu Tenhunen. Enhancing performance sustainability of fault tolerant routing algorithms in NoC-based architectures. In *2011 14th Euromicro Conference on Digital System Design (DSD)*, pp. 626–633. IEEE, 2011.

[56] Kangmin Lee, Se-Joong Lee, and Hoi-Jun Yoo. Low-power network-on-chip for high-performance SoC design. *IEEE Transactions on Very Large Scale Integration (VLSI) Systems*, 14(2):148–160, 2006.

[57] Anthony Leroy, Paul Marchal, Adelina Shickova, Francky Catthoor, Frédéric Robert, and Diederik Verkest. Spatial division multiplexing: A novel approach for guaranteed throughput on NoCs. In *Proceedings of the 3rd IEEE/ACM/IFIP International Conference on Hardware/Software Codesign and System Synthesis*, pp. 81–86. ACM, 2005.

[58] Jian Liang, Sriram Swaminathan, and Russell Tessier. aSOC: A scalable, single-chip communications architecture. In *International Conference on Parallel Architectures and Compilation Techniques, 2000. Proceedings*, pp. 37–46. IEEE, 2000.

[59] Andrew Lines. Asynchronous interconnect for synchronous SoC design. *IEEE Micro*, 24(1):32–41, 2004.

[60] Alexander Logvinenko, Carsten Gremzow, and Dietmar Tutsch. Recmin: A reconfiguration architecture for network on chip. In *2013 8th International Workshop on Reconfigurable and Communication-Centric Systems-on-Chip (ReCoSoC)*, pp. 1–6. IEEE, 2013.

[61] Mikael Millberg, Erland Nilsson, Rikard Thid, and Axel Jantsch. Guaranteed bandwidth using looped containers in temporally disjoint networks within the nostrum network on chip. In *Proceedings Design, Automation and Test in Europe Conference and Exhibition, 2004*, Vol. 2, pp. 890–895. IEEE, 2004.

[62] Prabhakar Mishra, A Nidhi, and JK Kishore. Custom network on chip architecture for map generation in autonomous navigating robots. In *2012 Annual IEEE India Conference (INDICON)*, pp. 086–091. IEEE, 2012.

[63] Fernando Moraes, Ney Calazans, Aline Mello, Leandro Möller, and Luciano Ost. Hermes: An infrastructure for low area overhead packet-switching networks on chip. *INTEGRATION, The VLSI Journal*, 38(1):69–93, 2004.

[64] Robert Mullins, Andrew West, and Simon Moore. The design and implementation of a low-latency on-chip network. In *Proceedings of the 2006*

Asia and South Pacific Design Automation Conference, pp. 164–169. IEEE Press, 2006.

[65] Christian Neeb and Norbert Wehn. Designing efficient irregular networks for heterogeneous systems-on-chip. *Journal of Systems Architecture*, 54(3):384–396, 2008.

[66] Chrysostomos Nicopoulos, Vijaykrishnan Narayanan, and Chita R Das. *Network-on-Chip Architectures: A Holistic Design Exploration*, Vol. 45. Springer Science & Business Media, 2009.

[67] I Miro Panades, Alain Greiner, Abbas Sheibanyrad, and G STMicroelcctronics. A low cost network-on-chip with guaranteed service well suited to the GALS approach. *Proceedings of NANONET*, p. 3, 2006.

[68] Partha Pratim Pande, Cristian Grecu, André Ivanov, and Res Saleh. High-throughput switch-based interconnect for future SoCs. In *The 3rd IEEE International Workshop on System-on-Chip for Real-Time Applications, 2003*, pp. 304–310. IEEE, 2003.

[69] Sudeep Pasricha, Yong Zou, Dan Connors, and Howard Jay Siegel. OE+IOE: A novel turn model based fault tolerant routing scheme for networks-on-chip. In *2010 IEEE/ACM/IFIP International Conference on Hardware/Software Codesign and System Synthesis (CODES+ ISSS)*, pp. 85–93. IEEE, 2010.

[70] Ahmad Patooghy, Seyed Ghassem Miremadi, and Mahdi Fazeli. A low-overhead and reliable switch architecture for network-on-chips. *INTEGRATION, The VLSI Journal*, 43(3):268–278, 2010.

[71] Christian Paukovits and Hermann Kopetz. Concepts of switching in the time-triggered network-on-chip. In *14th IEEE International Conference on Embedded and Real-Time Computing Systems and Applications, 2008. RTCSA'08*, pp. 120–129. IEEE, 2008.

[72] JC Peña-Ramos and Ramon Parra-Michel. Network on chip architectures for high performance digital signal processing using a configurable core. In *2011 International Conference on Reconfigurable Computing and FPGAs (ReConFig)*, p. 375–379. IEEE, 2011.

[73] Sandro Penolazzi and Axel Jantsch. A high level power model for the Nostrum NoC. In *9th EUROMICRO Conference on Digital System Design: Architectures, Methods and Tools, 2006. DSD 2006*, pp. 673–676. IEEE, 2006.

[74] Jacob Postman, Tushar Krishna, Christopher Edmonds, Li-Shiuan Peh, and Patrick Chiang. SWIFT: A low-power network-on-chip implementing

the token flow control router architecture with swing-reduced interconnects. *IEEE Transactions on Very Large Scale Integration (VLSI) Systems*, 21(8):1432–1446, 2013.

[75] Edwin Rijpkema, Kees Goossens, Andrei Rădulescu, John Dielissen, Jef van Meerbergen, Paul Wielage, and Erwin Waterlander. Trade-offs in the design of a router with both guaranteed and best-effort services for networks on chip. *IEE Proceedings-Computers and Digital Techniques*, 150(5):294–302, 2003.

[76] Erno Salminen, Tero Kangas, Timo D Hämäläinen, Jouni Riihimäki, Vesa Lahtinen, and Kimmo Kuusilinna. Hibi communication network for system-on-chip. *Journal of VLSI Signal Processing Systems for Signal, Image and Video Technology*, 43(2–3):185–205, 2006.

[77] Faizal A Samman, Thomas Hollstein, and Manfred Glesner. Networks-on-chip based on dynamic wormhole packet identity mapping management. *VLSI Design*, 2009:2, 2009.

[78] Henrik Samuelsson and Sudhakar Kumar. Ring road NoC architecture. In *Proceedings Norchip Conference, 2004*, pp. 16–19. IEEE, 2004.

[79] Christian Schuck, Stefan Lamparth, and Jurgen Becker. artNoC—A novel multi-functional router architecture for organic computing. In *International Conference on Field Programmable Logic and Applications, 2007*, pp. 371–376. IEEE, 2007.

[80] Mohammad Reza Seifi and Mohammad Eshghi. Clustered NoC, a suitable design for group communications in network on chip. *Computers & Electrical Engineering*, 38(1):82–95, 2012.

[81] Muhammad Athar Javed Sethi, Fawnizu Azmadi Hussin, and Nor Hisham Hamid. Implementation of biological sprouting algorithm for NoC fault tolerance. In *2013 IEEE International Conference on Circuits and Systems (ICCAS)*, pp. 39–44. IEEE, 2013.

[82] Muhammad Athar Javed Sethi, Fawnizu Azmadi Hussin, and Nor Hisham Hamid. Synaptogenesis based bio-inspired NoC fault tolerant interconnects. In *2013 IEEE International Conference on Control System, Computing and Engineering (ICCSCE)*, pp. 46–51. IEEE, 2013.

[83] Muhammad Athar Javed Sethi, Fawnizu Azmadi Hussin, and Nor Hisham Hamid. Bio-inspired NoC fault tolerant techniques. In *2014 5th International Conference on Intelligent and Advanced Systems (ICIAS)*, pp. 1–6. IEEE, 2014.

[84] Muhammad Athar Javed Sethi, Fawnizu Azmadi Hussin, and Nor Hisham Hamid. Bio-inspired NoC fault tolerant techniques using guaranteed throughput and best effort services. *Integration, The VLSI Journal*, 54:65–96, 2016.

[85] Muhammad Athar Javed Sethi, Fawnizu Azmadi Hussin, and Nor Hisham Hamid. Biologically inspired network on chip fault tolerant algorithm using time division multiplexing. In *2016 6th International Conference on Intelligent and Advanced Systems (ICIAS)*, pp. 1–6. IEEE, 2016.

[86] Muhammad Athar Javed Sethi, Fawnizu Azmadi Hussin, and Nor Hisham Hamid. Implementation and analysis of biological synaptogenesis technique on nodes and interconnects for NoC fault tolerance. *Maxwell Scientific Publication Corp.*, 2015(22):483–489, 2016.

[87] Muhammad Athar Javed Sethi, Fawnizu Azmadi Hussin, and Nor Hisham Hamid. Bio-inspired fault tolerant network on chip. *Integration, The VLSI Journal*, 58:155–166, 2017.

[88] Swati Sharma, Chandra Mukherjee, and Ashish Gambhir. A comparison of network-on-chip and buses. In *Proceedings of National Conference on Recent Advances in Electronics and Communication Engineering (RACE - 2014)*, 28–29 March 2014.

[89] Hao Shu, Jiang-Yi Shi, Yue Hao, Pei-Jun Ma, and Zhao Xu. DANoC: A dynamic adaptive network on chip architecture. In *2012 IEEE 11th International Conference on Solid-State and Integrated Circuit Technology (ICSICT)*, pp. 1–3. IEEE, 2012.

[90] D Siguenza-Tortosa and Jari Nurmi. Proteo: A new approach to network-on-chip. In *IASTED International Conference on Communication Systems and Networks (CSN'02)*. Citeseer, 2002.

[91] Vassos Soteriou, Noel Eisley, Hangsheng Wang, Bin Li, and Li-Shiuan Peh. Polaris: A system-level roadmap for on-chip interconnection networks. In *International Conference on Computer Design, 2006. ICCD 2006*, pp. 134–141. IEEE, 2007.

[92] Radu Andrei Stefan, Anca Molnos, and Kees Goossens. DAElite: A TDM NoC supporting QoS, multicast, and fast connection set-up. *IEEE Transactions on*, 63(3):583–594, 2014.

[93] Mikkel B Stensgaard and Jens Sparso. ReNoC: A network-on-chip architecture with reconfigurable topology. In *Second ACM/IEEE International Symposium on Networks-on-Chip, 2008. NoCS 2008*, pp. 55–64. IEEE, 2008.

[94] Michael Bedford Taylor, Jason Kim, Jason Miller, David Wentzlaff, Fae Ghodrat, Ben Greenwald, Henry Hoffman, Paul Johnson, Jae-Wook Lee, Walter Lee, et al. The raw microprocessor: A computational fabric for software circuits and general-purpose programs. *IEEE Micro*, 22(2):25–35, 2002.

[95] R Core Team. R: A language and environment for statistical computing. R: A language and environment for statistical computing. Vienna, Austria: R Foundation for Statistical Computing; 2014 Vienna, Austria, 2012, 2014.

[96] Leonel Tedesco, Aline Mello, Diego Garibotti, Ney Calazans, and Fernando Moraes. Traffic generation and performance evaluation for mesh-based NoCs. In *18th Symposium on Integrated Circuits and Systems Design*, pp. 184–189. IEEE, 2005.

[97] Mojtaba Valinataj. Evaluation of fault-tolerant routing methods for NoC architectures. In *2011 14th Euromicro Conference on Digital System Design (DSD)*, pp. 446–449. IEEE, 2011.

[98] Mojtaba Valinataj, Pasi Liljeberg, and Juha Plosila. Enhanced fault-tolerant network-on-chip architecture using hierarchical agents. In *2013 IEEE 16th International Symposium on Design and Diagnostics of Electronic Circuits & Systems (DDECS)*, pp. 141–146. IEEE, 2013.

[99] Mojtaba Valinataj, Siamak Mohammadi, Juha Plosila, and Pasi Liljeberg. A fault-tolerant and congestion-aware routing algorithm for networks-on-chip. In *2010 IEEE 13th International Symposium on Design and Diagnostics of Electronic Circuits and Systems (DDECS)*, pp. 139–144. IEEE, 2010.

[100] Sriram R Vangal, Jason Howard, Gregory Ruhl, Saurabh Dighe, Howard Wilson, James Tschanz, David Finan, Arvind Singh, Tiju Jacob, Shailendra Jain et al. An 80-tile sub-100-W teraflops processor in 65-nm CMOS. *IEEE Journal of Solid-State Circuits*, 43(1):29–41, 2008.

[101] Xin Wang, Tapani Ahonen, and Jari Nurmi. Applying CDMA technique to network-on-chip. *IEEE Transactions on Very Large Scale Integration (VLSI) Systems*, 15(10):1091–1100, 2007.

[102] Daniel Wiklund and Dake Liu. SoCBUS: Switched network on chip for hard real time embedded systems. In *International Parallel and Distributed Processing Symposium, 2003. Proceedings*, pp. 8–16. IEEE, 2003.

[103] Drew Wingard. Micronetwork-based integration for SOCs. In *Proceedings Design Automation Conference, 2001*, pp. 673–677. IEEE, 2001.

[104] Pascal T Wolkotte, Gerard JM Smit, Gerard K Rauwerda, and Lodewijk T Smit. An energy-efficient reconfigurable circuit-switched network-on-chip. In *19th IEEE International Parallel and Distributed Processing Symposium, 2005. Proceedings. 19th IEEE International*, p. 155a. IEEE, 2005.

[105] Li-Wei Wu, Wei-Xiang Tang, and Yarsun Hsu. A novel architecture and routing algorithm for dynamic reconfigurable network-on-chip. In *2011 IEEE 9th International Symposium on Parallel and Distributed Processing with Applications (ISPA)*, pp. 177–182. IEEE, 2011.

[106] Yoon Seok Yang, Reeshav Kumar, Gwan Choi, and Paul V Gratz. WaveSync: Low-latency source-synchronous bypass network-on-chip architecture. *ACM Transactions on Design Automation of Electronic Systems (TODAES)*, 19(4):34, 2014.

[107] Cesar Albenes Zeferino, Márcio Eduardo Kreutz, and Altamiro Amadeu Susin. RASoC: A router soft-core for networks-on-chip. In *Proceedings Design, Automation and Test in Europe Conference and Exhibition, 2004*, Vol. 3, pp. 198–203. IEEE, 2004.

[108] Cesar Albenes Zeferino and Altamiro Amadeu Susin. SoCIN: A: a parametric and scalable network-on-chip. In *16th Symposium on Integrated Circuits and Systems Design, 2003. SBCCI 2003. Proceedings*, pp. 169–174. IEEE, 2003.

[109] Shijian Zhang, Guodong Han, and Fan Zhang. Very fine-grained fault-tolerant routing algorithm of NoC based on buffer reuse. In *2013 4th IEEE International Conference on Software Engineering and Service Science (ICSESS)*, pp. 758–762. IEEE, 2013.

[110] Zhen Zhang, Alain Greiner, and Sami Taktak. A reconfigurable routing algorithm for a fault-tolerant 2D-mesh network-on-chip. In *45th ACM/IEEE Design Automation Conference, 2008. DAC 2008*. pp. 441–446. IEEE, 2008.

[111] Mounir Zid, Abdelkrim Zitouni, Adel Baganne, and Rached Tourki. New generic GALS NoC architectures with multiple QoS. In *International Conference on Design and Test of Integrated Systems in Nanoscale Technology, 2006. DTIS 2006*, pp. 345–349. IEEE, 2006.

Index

Page numbers followed by *f* indicate figures; those followed by *t* indicate tables.